OXFORD STUDIES IN NUCLEAR PHYSICS

GENERAL EDITOR
P. E. HODGSON

OXFORD STUDIES IN NUCLEAR PHYSICS
General Editor: P. E. Hodgson

1 J. McL. Emmerson: *Symmetry principles in particle physics* (1972)
2 J. M. Irvine: *Heavy nuclei, superheavy nuclei, and neutron stars* (1975)
3 I. S. Towner: *A shell-model description of light nuclei* (1977)
4 P. E. Hodgson: *Nuclear heavy-ion reactions* (1978)
5 R. D. Lawson: *Theory of the nuclear shell model* (1980)
6 W. E. Frahn: *Diffractive processes in nuclear physics* (1985)
7 S. S. M. Wong: *Nuclear statistical spectroscopy* (1986)
8 N. Ullah: *Matrix ensembles in the many-nucleon problem* (1987)
9 A. N. Antonov, P. E. Hodgson, and I. Zh. Petkov: *Nucleon momentum and density distributions in nuclei* (1988)
10 D. Bonatsos: *Interacting boson models of nuclear structure* (1988)
11 H. Ejiri and M. J. A. de Voigt: *Gamma-ray and electron spectroscopy in nuclear physics* (1989)
12 B. Castel and I. S. Towner: *Modern theories of nuclear moments* (1990)

Modern Theories of Nuclear Moments

B. CASTEL
Queen's University, Kingston, Canada

AND

I. S. TOWNER
Chalk River Nuclear Laboratories, Chalk River, Canada

CLARENDON PRESS · OXFORD

This book has been printed digitally and produced in a standard specification in order to ensure its continuing availability

OXFORD
UNIVERSITY PRESS

Great Clarendon Street, Oxford OX2 6DP
Oxford University Press is a department of the University of Oxford.
It furthers the University's objective of excellence in research, scholarship,
and education by publishing worldwide in
Oxford New York
Auckland Cape Town Dar es Salaam Hong Kong Karachi
Kuala Lumpur Madrid Melbourne Mexico City Nairobi
New Delhi Shanghai Taipei Toronto
With offices in
Argentina Austria Brazil Chile Czech Republic France Greece
Guatemala Hungary Italy Japan South Korea Poland Portugal
Singapore Switzerland Thailand Turkey Ukraine Vietnam

Oxford is a registered trade mark of Oxford University Press
in the UK and in certain other countries

Published in the United States
by Oxford University Press Inc., New York

© B. Castel and I. S. Towner, 1990

The moral rights of the author have been asserted

Database right Oxford University Press (maker)

Reprinted 2011

All rights reserved. No part of this publication may be reproduced,
stored in a retrieval system, or transmitted, in any form or by any means,
without the prior permission in writing of Oxford University Press,
or as expressly permitted by law, or under terms agreed with the appropriate
reprographics rights organization. Enquiries concerning reproduction
outside the scope of the above should be sent to the Rights Department,
Oxford University Press, at the address above

You must not circulate this book in any other binding or cover
And you must impose this same condition on any acquirer

ISBN 978-0-19-851728-3

'The nucleus is the only system which shows quantized collective surface motions and nuclear moments provide a unique source of information on these motions. So, let us enjoy the nuclear moments!'

Akito Arima, in *Proceedings of the International Conference on Nuclear Moments and Nuclear Structure,* Osaka, 1972

PREFACE

Nuclear moments have had a long history which parallels the development of nuclear physics as a whole. In 1937, Schmidt proposed a standard for magnetic moments based on the assumption that the free-nucleon magnetism persists in heavy nuclei. Then in the early 1950s attempts were made to explain deviations from the Schmidt estimates by invoking meson-exchange corrections or by modifying single-particle wavefunctions so as to include many-body effects. The early 1970s saw the emergence of fundamental problems in trying to reconcile quadrupole transitions and moments in light nuclei with shell-model predictions. Thus the study of nuclear moments could prove an excellent pedagogical tool to acquaint oneself with the complexities and elegance of some of the most current and powerful nuclear models. That is what we have attempted in this book. While refraining from presenting a compilation of theoretical calculations of nuclear moments, we have instead endeavoured to show *to what extent nuclear moments can be used as a stringent test of current nuclear models and of their predictive power*. We have thus set out to illustrate the role that nuclear moments have played in setting limits and sometimes inspiring the development of nuclear models and techniques.

The plan of the book is as follows. Chapters 1 and 2 set up the theoretical and experimental tools which have become necessary for the study of nuclear moments. More precisely Chapter 1 presents the necessary definitions relevant to the analyses of dipole and quadrupole moments and discusses their expectation values in simple systems like the deuteron and the $A = 3$ system. Chapter 2 then follows the historical evolution of experimental technique and presents an overview of the tools which have enabled the experimental physicist to collect the vast amount of data now available on nuclear moments. Chapters 3 to 5 are then broadly focused on *magnetic dipole moments*, with Chapter 3 devoted to *core-polarization effects*, Chapter 4 to *meson-exchange corrections* while Chapter 5 introduces the idea of *collective models* and concerns itself with elucidating the ways nuclear properties can be used as testing grounds for these models. Chapters 6 and 7 then analyse the various microscopic and collective model theories relevant to *quadrupole moments*. Chapter 6 focuses more on shell-model theories while Chapter 7 concerns itself with current aspects of macroscopic approaches. Finally we devote Chapter 8 to E0 and E4 transitions as tests of *bulk properties of nuclear matter* concluding with an overview of *current relativistic models* and their contribution to our understanding of nuclear excitations and moments in relativistic regimes.

In keeping with our desire to adopt a didactic approach to our study of nuclear models and moments, we have made every effort to make this book accessible to the graduate student interested in sub-atomic phenomena and who has already had some rudiments of nuclear and quantum physics at the graduate level. Of course we hope that more seasoned nuclear physics practitioners will also find something of value between the covers of this book and that some of the sense of excitement and wonder about the beauty of physics which we experienced in writing this book will be shared with them.

In writing this book we have benefited from the help of several colleagues. In particular we would like to thank Hiro Sagawa, Hiroshi Toki, and Larry Zamick for helpful comments and Gerald Dolling and Graham Lee-Whiting for a careful reading of the manuscript. Margaret Carey and Bernice Ison deserve the expression of our gratitude for their unfailing technical help.

Kingston and Chalk River   B. C.
                                                                I. S. T.

CONTENTS

1. Introduction   **1**
  1.1 Definitions   1
  1.2 Quark model of the nucleon   4
  1.3 Addition theorem   5
  1.4 Magnetic moment of the deuteron   8
  1.5 Magnetic moment of ^{3}He and ^{3}H   10
  1.6 Magnetic moment in odd-mass nuclei   12
  1.7 Electromagnetic transitions   15
  1.8 Effective operators   18
  1.9 Special case of ^{210}Po   22
  1.10 Isospin considerations   24
  1.11 Connection with nuclear β decay   26
  1.12 Buck–Perez plot   28
  1.13 Summary   32
  1.14 References   32

2. Measurement of nuclear moments   **35**
  2.1 Hyperfine structure in atomic spectra   35
  2.2 Nuclear magnetic resonance   38
  2.3 Nuclear orientation   39
  2.4 Perturbed angular distribution techniques   41
  2.5 Electric quadrupole hyperfine interactions   44
  2.6 Perturbed angular correlation techniques   44
  2.7 Reorientation in Coulomb excitation   46
  2.8 Nuclear magnetic resonance on β-emitting nuclei   47
  2.9 Transient fields   49
  2.10 Laser spectroscopy   50
  2.11 Summary   52
  2.12 References   53

3. Core polarization   **55**
  3.1 First-order core polarization   56
  3.2 Zero-range residual interaction   60
  3.3 Finite-range interactions   65
  3.4 Extension to RPA   67

CONTENTS

3.5	RPA in closed-shell nuclei	70
3.6	Core-polarization blocking	73
3.7	Second-order core polarization	79
3.8	Summary	84
3.9	References	84

4. Meson-exchange currents — 87
4.1	Minimal coupling	87
4.2	Magnetic moment operator	92
4.3	Relativistic corrections	96
4.4	Current conservation	98
4.5	Pion-exchange graphs	103
4.6	Construction of the MEC magnetic moment operator	120
4.7	Calculation in closed-shell-plus-one nuclei	126
4.8	MEC and core polarization	132
4.9	M1 and GT giant resonances in the Pb region	136
4.10	MEC in few-nucleon systems	139
4.11	Summary	143
4.12	References	143

5. Collective models — 146
5.1	Axially symmetric nuclei	147
5.2	Particle–rotor model	153
5.3	Collective states at high spin	158
5.4	Superdeformation at high spin	164
5.5	Vibrational spectra	167
5.6	Interacting boson model 1 (IBM-1)	170
5.7	Transition rates in IBM-1	173
5.8	Interacting boson model 2 (IBM-2)	175
5.9	Electric quadrupole transitions in IBM-2	179
5.10	g-factors of 2^+ states	181
5.11	Scissors mode	183
5.12	Summary	185
5.13	References	185

6. Single-particle and shell-model theories of quadrupole moments — 188
6.1	Single-particle quadrupole moments	188
6.2	Quadrupole moments of two-particle systems	190
6.3	Effective charge and the shell model	191
6.4	Mass dependence of the effective charge	194
6.5	Electric quadrupole moments in the shell model	198

CONTENTS

6.6	Shell-model calculations for the p shell	204
6.7	Expansion of the shell model and nuclear moments	205
6.8	The multiconfiguration Hartree–Fock model	207
6.9	Summary	209
6.10	References	210

7. Quadrupole moments and collective model theories — **212**

7.1	The harmonic vibration model	212
7.2	The anharmonic vibration model	214
7.3	Anharmonicity and the particle-vibration coupling model	215
7.4	Quadrupole moments of octupole vibrational states: the 3^- state in ^{208}Pb	216
7.5	Anharmonic vibrations and the time-dependent Hartree–Fock theory	220
7.6	The sum-rule method	225
7.7	The cranking model and quadrupole collectivity in strongly deformed nuclei	227
7.8	The Kumar–Baranger theory	230
7.9	Summary	232
7.10	References	233

8. Nuclear moments: future directions — **236**

8.1	Hexadecapole transitions and moments	236
8.2	Monopole transitions and moments	238
8.3	Relativistic theories of nuclear moments	248
8.4	References	254

Conclusion — **256**

Appendix A — **259**

A.1	Notation and definitions	259
A.2	Feynman rules	263

Index — **267**

1
INTRODUCTION

In 1956 Roger Blin-Stoyle wrote a very useful monograph[1] entitled *Theories of Nuclear Moments*. Since then the body of experimental data available on ground and excited states of nuclei has increased enormously. Not only are the spins, parities, and electromagnetic moments of ground states of most nuclei known, but today moments are routinely measured in short-lived isomers through advancements in such techniques as laser spectroscopy. Likewise, since 1956, there have been many theoretical advancements in the understanding of the properties of nuclei. Yet the aim of this text remains the same as that of Blin-Stoyle's: to discuss electromagnetic moments and their interpretation in terms of current nuclear theories.

1.1 Definitions

It has been found experimentally that a nucleus with a non-zero spin has a magnetic dipole moment given by

$$\mu = gI\mu_N, \qquad (1.1)$$

where g is known as the nuclear gyromagnetic ratio, I the total angular momentum of the nuclear state, and μ_N the nuclear magneton defined as $e\hbar/2Mc$ (M is the proton mass); μ_N has the value $0.126184 \, \text{MeV}^{1/2} \, \text{fm}^{3/2}$. In much of the following, we will assume nuclear magnetic moments are to be expressed in nuclear magneton units and will drop the explicit mention of μ_N in the equations and formulae.

Magnetization in a nucleus arises from two sources:

(1) orbiting charged particles (protons) generate a magnetic field (cf. a current flowing in a closed wire loop generates a magnetic field through the loop); and

(2) the intrinsic spin, $s = \tfrac{1}{2}$, of a nucleon generates its own intrinsic magnetic field.

Thus the magnetic dipole operator for a nucleus is a sum of two terms

$$\boldsymbol{\mu} = \sum_{k=1}^{A} g_L^{(k)} \boldsymbol{L}^{(k)} + \sum_{k=1}^{A} g_S^{(k)} \boldsymbol{S}^{(k)}, \qquad (1.2)$$

where $L^{(k)}$ and $S^{(k)}$ are the orbital and spin angular momentum operators for the kth nucleon, and the total operator is assumed to be the sum of all the individual nucleon operators, summed over all A nucleons in the nucleus. Here $g_L^{(k)}$ and $g_S^{(k)}$ are known as the orbital and spin gyromagnetic ratios. For a proton $g_L^{(\pi)} = 1$ and for a neutron $g_L^{(\nu)} = 0$ (in units of the nuclear magneton), i.e. orbital magnetization comes entirely from the protons. The spin gyromagnetic ratios are taken as parameters whose values are fixed by requiring that expression (1.2) leads to the correct magnetic moment for a free proton and a free neutron, namely $g_S^{(\pi)} = 5.587$ and $g_S^{(\nu)} = -3.826$. The magnetic moment is then obtained from eqn (1.2) by calculating the expectation value of the z-component of μ for the nuclear substate in which $M = I$. Thus, if the wavefunction for a nucleus of spin I in the magnetic substate M is ψ_{IM}, the magnetic moment μ is given by

$$\mu = gI = \int \psi_{II}^*(\mu)_z \psi_{II}$$

$$\equiv \langle II|(\mu)_z|II\rangle, \qquad (1.3)$$

where the integration is over the coordinates (position and spin) of all A nucleons. This expression can be simplified a little using the results of angular momentum algebra. In nuclear physics it is assumed that the nuclear Hamiltonian is rotationally invariant. Then the nuclear wavefunctions are eigenfunctions of I^2 and I_z. Furthermore the operators of all observables can be written as spherical tensors. For example the magnetic moment operator, being made up of the angular momentum operators $L^{(k)}$ and $S^{(k)}$, is a spherical tensor of rank one. A consequence of rotational invariance is that matrix elements of a tensor operator of rank k and magnetic projection q, T_{kq}, have a simple geometrical dependence on the magnetic quantum numbers, the Wigner–Eckhart theorem:

$$\langle IM|T_{kq}|I'M'\rangle = \langle I'M'kq|IM\rangle\langle I\|T_k\|I'\rangle. \qquad (1.4)$$

The directional properties are contained in the Clebsch–Gordan coefficients and the dynamics of the system appear only in the reduced matrix element, $\langle I\|T_k\|I'\rangle$. Note that in the literature there are several different notations for reduced matrix elements which differ from the version used here by factors of $(2I+1)^{1/2}$. Our notation will be that of Brink and Satchler.[2] Applying the Wigner–Eckhart theorem to the expression for the magnetic moment (eqn (1.3)) produces

$$\mu = \langle II10|II\rangle\langle I\|\mu\|I\rangle$$

$$= \left(\frac{I}{I+1}\right)^{1/2}\langle I\|\mu\|I\rangle. \qquad (1.5)$$

Note that from angular momentum selection rules, the Clebsch–Gordan coefficient is only non-vanishing if $2I \geq 1$. Even–even nuclei with ground-state spins 0^+ have no magnetic dipole moments.

Nuclei with spin greater than or equal to one may also have an electric quadrupole moment, which is a partial measure of the deviation of the nuclear charge distribution from spherical symmetry. The electric quadrupole moment operator Q is defined by

$$Q = e \sum_{k=1}^{A} g_L^{(k)}(3z_k^2 - r_k^2) \qquad (1.6)$$

with e the electric charge and z_k, r_k two of the position coordinates of nucleon k. The charge distribution arises entirely from the motion of protons, hence the formal introduction of $g_L^{(\pi)} = 1$ and $g_L^{(\nu)} = 0$ as before. The electric quadrupole moment is then conventionally defined as the expectation value of Q in the nuclear state with $M = I$,

$$Q = \int \psi_{II}^*(Q)_z \psi_{II} = \langle II | (Q)_z | II \rangle. \qquad (1.7)$$

The quadrupole moment operator can also be written as a spherical tensor of rank 2; thus with use of the Wigner–Eckhart theorem

$$Q = \langle II 20 | II \rangle \left(\frac{16\pi}{5}\right)^{1/2} \langle I \| \sum_k g_L^{(k)} r_k^2 Y_2(\theta_k, \phi_k) \| I \rangle$$
$$= \left(\frac{I(2I-1)}{(I+1)(2I+3)}\right)^{1/2} \left(\frac{16\pi}{5}\right)^{1/2} \langle I \| \sum_k g_L^{(k)} r_k^2 Y_2(\theta_k, \phi_k) \| I \rangle \qquad (1.8)$$

in units of the electric charge, e. Here $Y_{2,0}(\theta_k, \phi_k)$ is the spherical harmonic of rank 2 in the polar angles θ_k, ϕ_k of the position coordinate of the kth nucleon. Note that for a prolate (cigar shaped) charge distribution, symmetrical about the z-axis, the quadrupole moment is positive, whilst an oblate (disc shaped) charge distribution gives rise to a negative quadrupole moment.

In some circumstances, it is more convenient to introduce the idea of isospin rather than continually differentiating between protons and neutrons. Isospin symmetry is an exact symmetry in nuclear physics in the limit that nuclear forces are independent of the charge of the nucleon. (Coulomb forces naturally break the symmetry.) In an isospin formalism, the neutron and proton are regarded as two different states of a single particle. Thus the wavefunction for a nucleon will depend partly on the usual space and spin coordinates and partly on the isospin variable τ_z, which distinguishes between a proton ($\tau_z = +1$) and a neutron ($\tau_z = -1$). (Note that the reverse sign convention is often used.) The variable τ_z, or rather $t_z = \frac{1}{2}\tau_z$, are the two possible magnetic projections of a spin vector, t, of magnitude $t = \frac{1}{2}$. The proton, then, has total isospin $t = \frac{1}{2}$ and

z-component $t_z = +\frac{1}{2}$. As is the case with normal spin, the isospin matrices, $\boldsymbol{\tau}$, are represented as the three Pauli matrices.

With this notation, the magnetic dipole moment operator, eqn (1.2), can be rewritten as

$$\boldsymbol{\mu} = g_L^{(0)} \sum_{k=1}^{A} \boldsymbol{L}^{(k)} + g_L^{(1)} \sum_{k=1}^{A} \boldsymbol{L}^{(k)} \tau_z^{(k)}$$
$$+ g_S^{(0)} \sum_{k=1}^{A} \boldsymbol{S}^{(k)} + g_S^{(1)} \sum_{k=1}^{A} \boldsymbol{S}^{(k)} \tau_z^{(k)}, \tag{1.9}$$

where

$$\begin{aligned} g_L^{(0)} &= \tfrac{1}{2}(g_L^{(\pi)} + g_L^{(\nu)}) = 0.5, \\ g_L^{(1)} &= \tfrac{1}{2}(g_L^{(\pi)} - g_L^{(\nu)}) = 0.5, \\ g_S^{(0)} &= \tfrac{1}{2}(g_S^{(\pi)} + g_S^{(\nu)}) = 0.880, \\ g_S^{(1)} &= \tfrac{1}{2}(g_S^{(\pi)} - g_S^{(\nu)}) = 4.706. \end{aligned} \tag{1.10}$$

The gyromagnetic ratios with superscript zero are called the isoscalar g-factors (they are the coupling constants associated with operators that are scalars in isospin space), and those with superscript one are called isovector g-factors. Likewise the quadrupole moment operator, eqn (1.6), becomes in an isospin notation

$$Q = eg_L^{(0)} \sum_{k=1}^{A} (3z_k^2 - r_k^2) + eg_L^{(1)} \sum_{k=1}^{A} (3z_k^2 - r_k^2) \tau_z^{(k)}. \tag{1.11}$$

Again the first term is called the isoscalar and the second term the isovector quadrupole moment operators.

1.2 Quark model of the nucleon

We pause briefly to ask why the spin gyromagnetic ratio for a proton and a neutron are $g_S^{(\pi)} = 5.587$ and $g_S^{(\nu)} = -3.826$, respectively. Had it been assumed that the proton and neutron were point particles described by the Dirac equation, then quantum electrodynamics would have dictated that $g_S^{(\pi)} = 2.0$ and $g_S^{(\nu)} = 0.0$. For electrons and muons, which to a very good approximation are point Dirac particles, the departure of g_S from 2 is very small and is calculable in QED as radiative corrections. For protons and neutrons, the departure is very large, implying that they are not point particles but have some intrinsic structure.

One very naïve approach is to assume the nucleon comprises three quarks, each quark being a point Dirac particle. The nucleon magnetic moment operator is then

$$\boldsymbol{\mu} = \left(\tfrac{2}{3} \sum_{k=u} g_S^{(k)} \boldsymbol{S}^{(k)} - \tfrac{1}{3} \sum_{k=d} g_S^{(k)} \boldsymbol{S}^{(k)}\right) e\hbar/2m_q c, \tag{1.12}$$

where orbital contributions have been neglected, assuming the three quarks are arranged in a symmetric S-state. The first term is a sum over the 'up' quarks with charge $\frac{2}{3}e$, and the second term is a sum over the 'down' quarks with charge $-\frac{1}{3}e$. The spin gyromagnetic ratio for point quarks is $g_S^{(k)} = 2$ so the only unknown in eqn (1.12) is the quark mass, m_q. If a parameter ε, the ratio of nucleon to quark mass $\varepsilon = M/m_q$, is introduced then one might expect in such a naïve non-relativistic quark model that ε would be of the order of 3. (Relativistic quark models are quite different.) On taking the expectation value (eqn (1.3)) of the operator (eqn (1.12)) for a proton state (two up quarks and one down quark) the magnetic moment, after some angular momentum algebra, is calculated to be $\mu_p = \varepsilon$, whilst for a neutron (one up quark and two down quarks) it becomes $\mu_n = -2\varepsilon/3$. In particular, the ratio $\mu_p/\mu_n = -\frac{3}{2}$, independent of the value of ε, is a result in very close agreement to experiment, -1.46. This is one of the first early successes of the constituent quark model.

1.3 Addition theorem

Next we derive a general result for the magnetic moment of a composite nuclear state in terms of the magnetic moment of its constituents. The derivation is an example of the use of angular momentum algebra. Suppose a nuclear state of angular momentum I, magnetic substate M is constructed by vector coupling two components of angular momenta I_1 and I_2. Our notation for this is

$$|[I_1, I_2]IM\rangle = \sum_{M_1 M_2} \langle I_1 M_1 I_2 M_2 | IM \rangle |I_1 M_1\rangle |I_2 M_2\rangle. \quad (1.13)$$

Likewise suppose the tensor describing an observable operator is made up of a spherical tensor, $R_{k_1 q_1}(1)$, acting on one part and another spherical tensor, $S_{k_2 q_2}(2)$, acting on a second part. Then the composite tensor $T_{kq}(k_1, k_2)$, defined as

$$T_{kq}(k_1, k_2) = \sum_{q_1 q_2} \langle k_1 q_1 k_2 q_2 | kq \rangle R_{k_1 q_1}(1) S_{k_2 q_2}(2), \quad (1.14)$$

is a spherical tensor of rank k acting on the composite system. With these definitions, the general result from angular momentum algebra (see, for example, eqn (5.12) of Brink and Satchler[2]) is that the reduced matrix element of the composite system can be factorized in the following way:

$$\langle [I_1, I_2]I \| T_k(k_1, k_2) \| [I_1', I_2']I' \rangle$$
$$= \begin{bmatrix} I_1' & k_1 & I_1 \\ I_2' & k_2 & I_2 \\ I' & k & I \end{bmatrix} \langle I_1 \| R_{k_1} \| I_1' \rangle \langle I_2 \| S_{k_2} \| I_2' \rangle. \quad (1.15)$$

This result will be used many times in this text. In particular there are two limiting cases obtained on first setting $k_2 = 0$ and then $k_1 = 0$:

$$\langle [I_1, I_2]I \| R_{k_1}(1) \| [I'_1, I'_2]I' \rangle = (-)^{I'_1+I-I_1-I'} U(Ik_1 I'_2 I'_1; I' I_1) \langle I_1 \| R_{k_1}(1) \| I'_1 \rangle \delta_{I_2, I'_2} \quad (1.16a)$$

and

$$\langle [I_1, I_2]I \| S_{k_2}(2) \| [I'_1, I'_2]I' \rangle = U(I'_2 k_2 I'_1 I; I_2 I') \langle I_2 \| S_{k_2}(2) \| I'_2 \rangle \delta_{I_1, I'_1}. \quad (1.16b)$$

These limiting results are used in the situation when the tensor operator acts on only one part of a composite system. The notation for the angular momentum recoupling coefficients is as follows:[3]

$$U(abcd; ef) = (-)^{a+b+c+d}[(2e+1)(2f+1)]^{1/2} \begin{Bmatrix} a & b & e \\ d & c & f \end{Bmatrix} \quad (1.17)$$

and is the transformation coefficient for the recoupling of three angular momenta expressed here in terms of Wigner's 6j-symbol, whilst

$$\begin{bmatrix} a & b & c \\ d & e & f \\ g & h & i \end{bmatrix} = [(2c+1)(2f+1)(2g+1)(2h+1)]^{1/2} \begin{Bmatrix} a & b & c \\ d & e & f \\ g & h & i \end{Bmatrix} \quad (1.18)$$

is the transformation coefficient for the recoupling of four angular momenta expressed in terms of Wigner's 9j-symbol.

Returning now to the calculation of the magnetic moment of a composite state, $|[I_1, I_2]I\rangle$, the magnetic moment operator, being a one-body operator (eqn (1.2)), can be written as a sum of two terms $\mu = \mu_1 + \mu_2$, where μ_1 acts on one part of the composite system and μ_2 on the other part. Hence with the use of eqn (1.16) we obtain

$$\mu = \left(\frac{I}{I+1}\right)^{1/2} \langle [I_1, I_2]I \| \mu_1 + \mu_2 \| [I_1, I_2]I \rangle$$

$$= \left(\frac{I}{I+1}\right)^{1/2} \left[U(I1I_2I_1; II_1)\left(\frac{I_1+1}{I_1}\right)^{1/2} \mu_1 + U(I_21I_1I; I_2I)\left(\frac{I_2+1}{I_2}\right)^{1/2} \mu_2 \right], \quad (1.19)$$

where $\mu_1(\mu_2)$ is the magnetic moment of the first (second) constituent

$$\mu_1 = \left(\frac{I_1}{I_1+1}\right)^{1/2} \langle I_1 \| \mu_1 \| I_1 \rangle.$$

A closed algebraic form is available for the recoupling coefficients in eqn (1.19) (see Table 4 of Brink and Satchler[2]):

$$U(I1I_2I_1; II_1) = \frac{1}{2} \frac{I(I+1) + I_1(I_1+1) - I_2(I_2+1)}{[I(I+1)I_1(I_1+1)]^{1/2}}, \quad (1.20)$$

1.3 ADDITION THEOREM

leading to a simplification

$$\mu = I\left[\frac{1}{2}\left(\frac{\mu_1}{I_1} + \frac{\mu_2}{I_2}\right) + \frac{1}{2}\left(\frac{\mu_1}{I_1} - \frac{\mu_2}{I_2}\right)\frac{I_1(I_1+1) - I_2(I_2+1)}{I(I+1)}\right]. \quad (1.21)$$

The final and most elegant form is in terms of gyromagnetic ratios (g-factors), $g = \mu/I$, $g_1 = \mu_1/I_1$, $g_2 = \mu_2/I_2$:

$$g = \tfrac{1}{2}(g_1 + g_2) + \tfrac{1}{2}(g_1 - g_2)\frac{I_1(I_1+1) - I_2(I_2+1)}{I(I+1)}. \quad (1.22)$$

This result is known as the addition theorem and shows how the g-factor of a composite nuclear state can be built up knowing the g-factors of its constituents.

An alternative derivation (that is specific to rank-one tensors) can be obtained from what is known as the Landé formula. Recall from the Wigner–Eckhart theorem that the M-dependence of the matrix elements of any two tensor operators T_{kq} and U_{kq} of the same rank, k, is the same (and contained in the Clebsch–Gordan coefficient). Thus

$$\langle IM|\, T_{kq}\, |I'M'\rangle = A_{II'}\langle IM|\, U_{kq}\, |I'M'\rangle,$$

where the proportionality constant $A_{II'}$, which is the ratio of the two reduced matrix elements, will generally depend on I and I' and on the physical nature of T_k and U_k, but not on M and M'. In particular consider matrix elements diagonal in I, i.e. $I = I'$, and take for U_k the angular momentum vector $\boldsymbol{I}$ itself; then for any vector $\boldsymbol{T}$ we have

$$\langle IM|\, \boldsymbol{T}\, |IM'\rangle = A\langle IM|\, \boldsymbol{I}\, |IM'\rangle.$$

To find the coefficient A we calculate the diagonal matrix elements of the scalar product $\boldsymbol{T}\cdot\boldsymbol{I}$:

$$\langle IM|\, \boldsymbol{T}\cdot\boldsymbol{I}\, |IM\rangle = \sum_{M'} \langle IM|\, \boldsymbol{T}\, |IM'\rangle\langle IM'|\, \boldsymbol{I}\, |IM\rangle$$

$$= A\sum_{M'} \langle IM|\, \boldsymbol{I}\, |IM'\rangle\langle IM'|\, \boldsymbol{I}\, |IM\rangle$$

$$= A\langle IM|\, \boldsymbol{I}^2\, |IM\rangle$$

$$= AI(I+1).$$

Hence the value of A is determined and we obtain the Landé formula

$$\langle IM|\, \boldsymbol{T}\, |IM'\rangle = \frac{\langle IM|\, \boldsymbol{T}\cdot\boldsymbol{I}\, |IM\rangle}{I(I+1)}\langle IM|\, \boldsymbol{I}\, |IM'\rangle. \quad (1.23)$$

We use this formula now to calculate the magnetic moment of a composite state, $|[I_1, I_2]IM\rangle$, where the operator $\boldsymbol{T}$ is $\boldsymbol{\mu}_1 + \boldsymbol{\mu}_2 = g_1\boldsymbol{I}_1 + g_2\boldsymbol{I}_2$.

Evaluating the matrix element in the $M = I$ substate we have

$$\mu = gI = \langle [I_1, I_2]II| g_1 I_1 + g_2 I_2 |[I_1, I_2]II\rangle$$

$$= \frac{\langle [I_1, I_2]II| g_1 I_1 \cdot I + g_2 I_2 \cdot I |[I_1, I_2]II\rangle \langle II| I |II\rangle}{I(I+1)}$$

or that

$$g = \langle II| \tfrac{1}{2}g_1(I^2 + I_1^2 - I_2^2) + \tfrac{1}{2}g_2(I^2 + I_2^2 - I_1^2) |II\rangle / I(I+1)$$

$$= \tfrac{1}{2}g_1 \frac{I(I+1) + I_1(I_1+1) - I_2(I_2+1)}{I(I+1)} + \tfrac{1}{2}g_2 \frac{I(I+1) + I_2(I_2+1) - I_1(I_1+1)}{I(I+1)}$$

$$= \tfrac{1}{2}(g_1 + g_2) + \tfrac{1}{2}(g_1 - g_2) \frac{I_1(I_1+1) - I_2(I_2+1)}{I(I+1)} \quad (1.22)$$

and hence the addition theorem. We consider two examples of its use.

Example 1: A single nucleon in a shell-model orbital, j, is an example of a composite state, being the coupling of an orbital part and a spin part, $[l, \tfrac{1}{2}]j$. Likewise the magnetic moment operator is the sum of an orbital part, $g_L L$, and a spin part, $g_S S$. Then, from eqn (1.22), the g-factor for a single nucleon in orbital j is

$$g = \tfrac{1}{2}(g_L + g_S) + \frac{1}{2} \frac{l(l+1) - \tfrac{3}{4}}{j(j+1)} (g_L - g_S). \quad (1.24)$$

In particular, the cases when $j = l + \tfrac{1}{2}$ and $j = l - \tfrac{1}{2}$ simplify to

$$g(j = l \pm \tfrac{1}{2}) = g_L \pm \frac{g_S - g_L}{2l + 1}. \quad (1.25)$$

We will discuss this result shortly in connection with magnetic moments of odd-mass nuclei.

Example 2: Next, consider two nucleons in the identical shell-model orbital, j, coupled to a resultant angular momentum, I. The g-factor for this state, $g(I)$, in terms of the single-nucleon g-factor, $g(j)$, is immediately given, namely

$$g(I) = g(j). \quad (1.26)$$

The result can be generalized to n nucleons in a single shell-model orbital, $[j^n; I]$, to obtain $g(j^n; I) = g(j)$. That is, the g-factor for all configurations, j^n, for n identical nucleons are all equal, independent of n and I, and equal to the single-nucleon g-factor, $g(j)$. We will discuss this result again in Section 3.6.

1.4 Magnetic moment of the deuteron

The deuteron, the only bound state in the two-nucleon system, has spin 1, isospin 0 and accordingly it has both a magnetic dipole and electric

1.4 MAGNETIC MOMENT OF THE DEUTERON

quadrupole moment. Experimentally it is found that $\mu_d = 0.857438$ μ_N,[4] and $Q_d = 0.2860 \pm 0.0015$ $e\,\text{fm}^2$.[4,5] (The quadrupole moment is deduced from the hyperfine splitting in the HD and D_2 molecules. The theoretical analysis relies on the use of variational electron wavefunctions. The systematic uncertainty in this procedure is the present limit to the accuracy of Q_d. The earlier result of Reid and Vaida[6] has recently been improved by Bishop and Cheung[5] using more sophisticated wavefunctions; their result is quoted above.)

If only central forces were operative between nucleons the ground state configuration of the deuteron would be $^3S_1$, i.e. $L=0$, $S=1$. The magnetic moment comes solely from the spin magnetic moment of the two constituents. Thus from eqn (1.22) with $I_1 = I_2 = \frac{1}{2}$, we get $g_d = \frac{1}{2}(g_S^{(\pi)} + g_S^{(\nu)})$ or

$$\mu_d = \mu_p + \mu_n, \qquad (1.27)$$

where μ_p and μ_n are the proton and neutron magnetic moments, respectively. This gives $\mu_d = 0.879805\mu_N$,[4,7] and there is an obvious discrepancy well outside experimental error. Further for a $^3S_1$ state, the charge distribution is spherically symmetrical and $Q_d = 0$, again in conflict with experiment. These two discrepancies can be resolved by assuming that, in addition to the central force, there is a non-central force (e.g. a tensor force) between nucleons. This would lead to an admixture of the $^3D_1$ state in the deuteron ground-state wavefunction. Again we can use eqn (1.22) to find the contribution to the deuteron magnetic moment from the D-state admixture, taking g_1 as the contribution from the isoscalar orbital operator, $g_1 = g_L^{(0)} = 0.5$, and g_2 the contribution from the spin operator $g_2 = \frac{1}{2}(g_S^{(\pi)} + g_S^{(\nu)}) = g_S^{(0)} = \mu_p + \mu_n$. This gives

$$g_d(L=2) = \frac{1}{2}(\tfrac{1}{2} + \mu_p + \mu_n) + (\tfrac{1}{2} - \mu_p - \mu_n)$$
$$= \tfrac{1}{2}[\tfrac{3}{2} - (\mu_p + \mu_n)]. \qquad (1.28)$$

On multiplying eqn (1.27) by the probability the deuteron ground state is in an S-state, P_S, and eqn (1.28) by the D-state probability, P_D, and normalizing, $P_S + P_D = 1$, the final expression for the deuteron magnetic moment becomes

$$\mu_d = (\mu_p + \mu_n) - \tfrac{3}{2}(\mu_p + \mu_n - \tfrac{1}{2})P_D + \Delta\mu_d^{\text{MEC}} \qquad (1.29)$$

where $\Delta\mu_d^{\text{MEC}}$ represents an isoscalar meson-exchange current correction, which is small and positive. (We will discuss MEC corrections in some detail in later chapters.) There are many theoretical estimates for this correction in the literature[8] hovering around $\Delta\mu_d^{\text{MEC}} \simeq 0.02\mu_N$, but with a wide spread. Because of this difficulty, the deuteron magnetic moment does not provide a useful constraint on the D-state probability in the deuteron wavefunction. Taking $\Delta\mu_d^{MEC}$ as $(0.02\pm0.01)\mu_N$, the D-state probability is then determined from the magnetic moment to be $P_D = (7\pm2)\%$, slightly higher than currently accepted values, suggesting that the meson-exchange current correction is overestimated. (Note $\Delta\mu_d^{MEC} = 0$ corresponds to $P_D = 4\%$.) The D-state probability is not an experimentally measurable quantity; but it is trivially computed given deuteron wavefunctions obtained from the solution of coupled Schrödinger equations involving local nucleon–nucleon coordinate-space potentials. The D-state probability is a quantity often used in comparing one potential with another. The recent work of the Bonn group[9] for example, based on a model of mesons being exchanged by nucleons, leads to an energy-dependent momentum-space potential with $P_D = 4.25\%$. A slight readjustment of meson–nucleon coupling constants to produce an approximate energy-independent coordinate space potential with an equivalent acceptable fit to nucleon–nucleon phase shifts yields $P_D = 4.81\%$. Other semi-phenomenological potentials produce values in the range $P_D = 5.5$–6.5% (Paris:[10] 5.8%; Sprung:[11] 5.9%; Argonne:[12] 6.1%; Nijmegen:[13] 5.4% and 6.4%). The older purely phenomenological potentials all tend to have larger values of P_D (Reid soft-core:[14] 6.5%; Yale:[15] 7.0% Hamada–Johnston:[16] 7.0%).

The calculated quadrupole moment of the deuteron is quite sensitive to the D-state probability. If $u(r)$ and $w(r)$ are the S-wave and D-wave radial wavefunctions of the deuteron obtained in solving coupled Schrödinger equations with a local potential, then the quadrupole moment is

$$Q_d = \frac{1}{\sqrt{50}} \int dr\, r^2 \left(u(r)w(r) - \frac{1}{\sqrt{8}} w^2(r) \right) + \Delta Q_d^{MEC}. \qquad (1.30)$$

Again there is a small and positive meson-exchange current correction that has a fair amount of theoretical uncertainty associated with it. Present estimates[17] place $\Delta Q_d^{MEC} \simeq 0.010 \pm 0.005$ e fm². The quadrupole moment roughly scales as $(P_D)^{1/2}$, as evident in the first term in eqn (1.30). (The second term only contributes around -6% to the total.) A fit to the experimental Q_d, corrected for ΔQ_d^{MEC}, is obtained from local potentials with P_D in the range $5\pm1\%$. This is lower than, but consistent with, the value deduced from the magnetic moment. Note that the energy-dependent momentum-space Bonn potential[9] has the ability to combine a small P_D with a large Q_d; this is a characteristic feature of energy-dependent potentials.

1.5 Magnetic moment of ³He and ³H

Nuclei ³He and ³H are isospin mirrors (identical nuclear states except for the interchange of protons and neutrons) both having spin ½ and isospin

1.5 MAGNETIC MOMENT OF ³He AND ³H

$\frac{1}{2}$. They will therefore have a magnetic dipole moment but no higher moments. Experimentally[4] $\mu(^3\text{He}) = -2.1276\mu_N$ and $\mu(^3\text{H}) = 2.9790\mu_N$. The leading term in the ground-state wavefunction is the fully symmetric S state, $^2S_{1/2}$, for which the magnetic moment in ³He and ³H is just $\mu_n = -1.9130\mu_N$ and $\mu_p = 2.7928\mu_N$, respectively. The discrepancy is of order 10%. It is more convenient to discuss this question in terms of the isoscalar and isovector combination of magnetic moments, namely

$$\mu^{(0)}(A=3) = \tfrac{1}{2}(\mu(^3\text{H}) + \mu(^3\text{He})),$$
$$\mu^{(1)}(A=3) = \tfrac{1}{2}(\mu(^3\text{H}) - \mu(^3\text{He})). \quad (1.31)$$

Then the comparison between experiment and simple S-state estimates becomes: $\mu^{(0)}(\text{Expt}) = 0.4257\mu_N$, $\mu^{(0)}(L=0) = 0.4399\mu_N$, a discrepancy of 3%; $\mu^{(1)}(\text{Expt}) = 2.5533\mu_N$, $\mu^{(1)}(L=0) = 2.3529\mu_N$, a discrepancy of 9%. It is tempting therefore to attribute the discrepancy (as in the case of the deuteron) to the admixture of other states. As we will see, this alone will not cure the problem. For the three-nucleon system the dominant wavefunction configurations are the symmetric $^2S_{1/2}$, the mixed-symmetry $^2S'_{1/2}$, three P-states $^2P_{1/2}$, $^2P'_{1/2}$, $^4P_{1/2}$, and three $^4D_{1/2}$ states. If the small P-wave parts are dropped and only one of the three D-wave components retained, then the magnetic moments are given by the following simple expressions:[18,19]

$$\mu^{(0)}(A=3) = \tfrac{1}{2}(\mu_p + \mu_n)(P_S + P_{S'} - P_D) + \tfrac{1}{2}P_D + \Delta\mu_0^{\text{MEC}},$$
$$\mu^{(1)}(A=3) = \tfrac{1}{2}(\mu_p - \mu_n)(P_S - \tfrac{1}{3}P_{S'} + \tfrac{1}{3}P_D) - \tfrac{1}{6}P_D + \Delta\mu_1^{\text{MEC}}, \quad (1.32)$$

where P_S is the probability of the principal S-state, etc. Again meson-exchange corrections are indicated in the formulae. Typical values for these probabilities, as calculated for example with the Reid soft core potential by Harper et al.,[20] are $P_S = 89.7\%$, $P_{S'} = 1.7\%$, and $P_D = 8.6\%$ leading to magnetic moments (without the MEC correction) of $\mu^{(0)} = 0.408\mu_N$ and $\mu^{(1)} = 2.152\mu_N$. The discrepancy between this and experiment in the isoscalar case is $0.018\mu_N$, small and positive, and essentially the same size as the MEC correction calculated for the deuteron. In the isovector combination, the discrepancy, $0.402\mu_N$, is large (17%) and positive. At first sight this might seem a puzzle; in fact, it is not. Meson-exchange current theories predict large corrections to isovector magnetic moments and small corrections to isoscalar magnetic moments. This is because the leading terms come from the exchange of the lightest-mass meson (the lighter the meson mass the longer the range of the derived operator). Thus the leading terms, as we will discuss in Chapter 4, come from pion exchange and these are purely isovector. Isoscalar corrections come from the shorter-range heavy-meson exchange and their evaluation is much more model dependent and therefore much less certain. In the literature[19–21] there are several calculations of $\Delta\mu_1^{MEC}$, all quite close and well encompassed by an average value $\Delta\mu_1^{MEC} = (0.32 \pm 0.05)\mu_N$. This is just less than but consistent with the experimental discrepancy. Additional small short-range contributions tend to increase (in a negative sense) the $\Delta\mu_1^{MEC}$ and will improve the agreement between theory and experiment. Thus an understanding of the trinucleon magnetic moments is well in hand.

1.6 Magnetic moment in odd-mass nuclei

Apart from the very simple systems such as deuteron, ^{3}He, and ^{3}H it is not possible to obtain an exact wavefunction representing the structure of a nucleus even with full knowledge of the internucleon potential. Thus inevitably one is led to approximation schemes (nuclear models). Generally speaking there are two types of nuclear models:

(1) shell models, in which nuclear wavefunctions are built up from Slater determinants of single-nucleon functions; and

(2) collective models in which individual nucleon coordinates are replaced by a lesser number of collective coordinates.

It is not our purpose to review nuclear models here, but rather we will concentrate on the narrower goal of understanding the predictions that various models give for magnetic moments. Here we will say a few words about shell models, deferring collective models to Chapter 5.

Shell models are based on the idea that each nucleon within a nucleus moves in an average central potential generated by all the other nucleons. The eigenfunctions of the Schrödinger equation with this average potential provide the single-particle wavefunctions out of which Slater determinants are constructed. These single-particle wavefunctions are classified by a set of quantum numbers (n, l, j), where n is the principal quantum number (counting the number of nodes in the radial wavefunction), l the orbital angular momentum, which in turn is coupled to a spin angular momentum of 1/2 to give the resultant, j. In a given nucleus these (n, l, j) orbitals are filled up by protons and neutrons according to the Pauli exclusion principle, filling the lowest energy levels first. As in the atomic case, this leads to the concept of shells and by suitably adjusting the parameters of the average potential it is a simple matter to account for the so-called 'magic numbers' of nucleons 2, 8, 20, 28, 50, 82, and 126. Nuclei with these numbers of neutrons and protons are particularly stable and have distinctive properties which can be interpreted in terms of the complete filling of certain levels.

In light nuclei, the stable nuclei are those in which neutron and proton orbitals are filled to the same level. Furthermore the magic numbers occur with the orbitals $j = l + \frac{1}{2}$ and $j = l - \frac{1}{2}$, either both completely

1.6 MAGNETIC MOMENT IN ODD-MASS NUCLEI

occupied or both completely unoccupied. These shell fillings are referred to as *LS* closed shells. In heavier nuclei, the stable nuclei are those in which neutron and proton orbitals are filled to different levels and the shell closures occur such that for the higher spin orbital the $j = l + \frac{1}{2}$ orbital is occupied and the $j = l - 1/2$ orbital is not. These shell fillings are referred to as *jj* closed shells. As we will see the magnetic moment of a nucleus will depend quite sensitively on whether we are discussing light nuclei at *LS* closed shells or heavy nuclei at *jj* closed shells.

For a completely filled shell the Pauli principle guarantees that the total angular momentum of the nucleons in the shell is zero. On the other hand, if the shell is only partially filled then a considerable degeneracy exists. Naturally, the representation of all nucleon–nucleon interactions by an average central potential is an oversimplification. There will be a residual interaction left over, not the same as the free nucleon–nucleon interaction, but one that is much weaker such that perturbation theory in the residual interaction can be usefully used to split the degeneracy just mentioned. One extreme version of this uses a pairing force as the residual interaction. Then all like nucleons (i.e. all the protons and separately all the neutrons) are grouped in pairs and the residual interaction will lower the energy of the angular momentum zero pairs relative to all the others. In this model the ground-state spin of an even–even nucleus is 0^+, in an odd–even or even–odd nucleus the ground-state spin is that of the unpaired nucleon, while in an odd–odd nucleus the ground-state spin is formed from coupling the angular momentum of the odd proton to that of the odd neutron according to Nordheim's rules.[22] Likewise in this extreme model, the magnetic moment of an odd-mass nucleus is given by that of the unpaired nucleon.

In eqn (1.25) we gave the expression for the magnetic moment (or rather the *g*-factor) for a single nucleon in orbital *nlj*. Writing $g_L^{(\pi)} = 1$, $g_L^{(\nu)} = 0$, $g_S^{(\pi)} = 2\mu_p$, and $g_S^{(\nu)} = 2\mu_n$, where μ_p and μ_n are the free proton and free neutron magnetic moments, the equation can be rewritten as follows:

Odd proton
$$\begin{cases} \mu = (j - \tfrac{1}{2}) + \mu_p & j = l + \tfrac{1}{2}, \\ \mu = \dfrac{j}{(j+1)}[(j + \tfrac{3}{2}) - \mu_p] & j = l - \tfrac{1}{2}, \end{cases}$$

Odd neutron
$$\begin{cases} \mu = \mu_n & j = l + \tfrac{1}{2}, \\ \mu = -\dfrac{j}{j+1}\mu_n & j = l - \tfrac{1}{2}. \end{cases}$$

(1.33)

It is common to plot curves of μ against j for neutrons, and for protons, such curves being referred to as Schmidt diagrams and the resulting lines

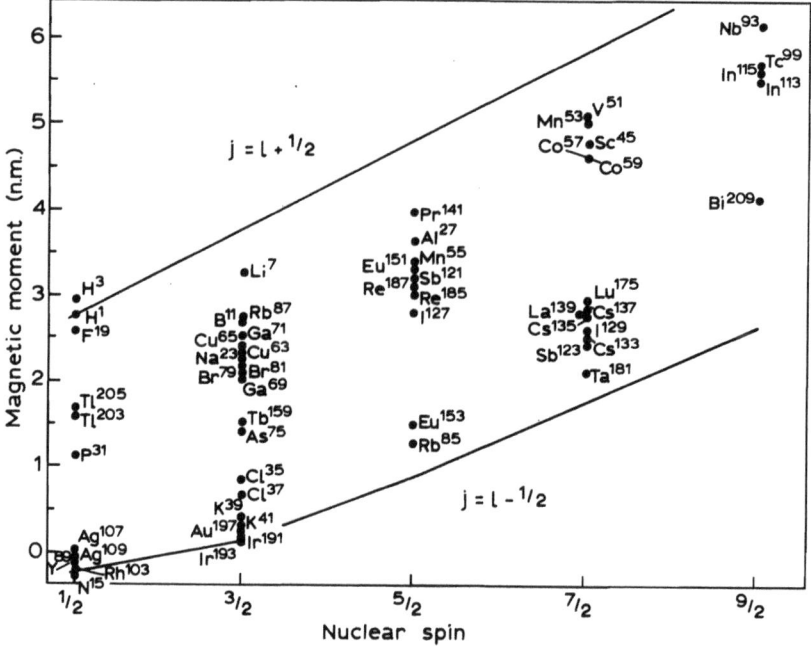

Fig. 1.1 Schmidt diagram for odd-proton nuclei. (From Blin-Stoyle.[1])

as the Schmidt[23] lines. The Schmidt diagrams are given in Figs. 1.1 and 1.2 on which are also plotted the measured magnetic moments of odd-mass nuclei. Note the following points from the diagrams:

1. The moments of most nuclei deviate from the Schmidt line by amounts varying from 0.5 to $1.5\mu_N$.
2. Apart from a few cases in light nuclei, the deviations of the magnetic moments are all inwards from the Schmidt lines.
3. The average deviation of the odd-proton nuclei is about 20% larger than the average deviation of odd-neutron nuclei.
4. The only nuclei (apart from cases in light nuclei) that deviate from the Schmidt line by less than $0.2\mu_N$ are nearly all $p_{1/2}$ nuclei. (A reason for this will become evident in Chapter 3.)

The fact that in general there are large deviations between experimental magnetic moments in odd-mass nuclei and the Schmidt values may be broadly attributed to two possible causes. Firstly the single-particle model wavefunction is certainly not the correct nuclear wavefunction and therefore there should be no surprise that there is poor agreement. In Chapter 3 we will discuss in detail a perturbation theory approach to correcting single-particle wavefunctions. Secondly the magnetic moment

1.7 ELECTROMAGNETIC TRANSITIONS

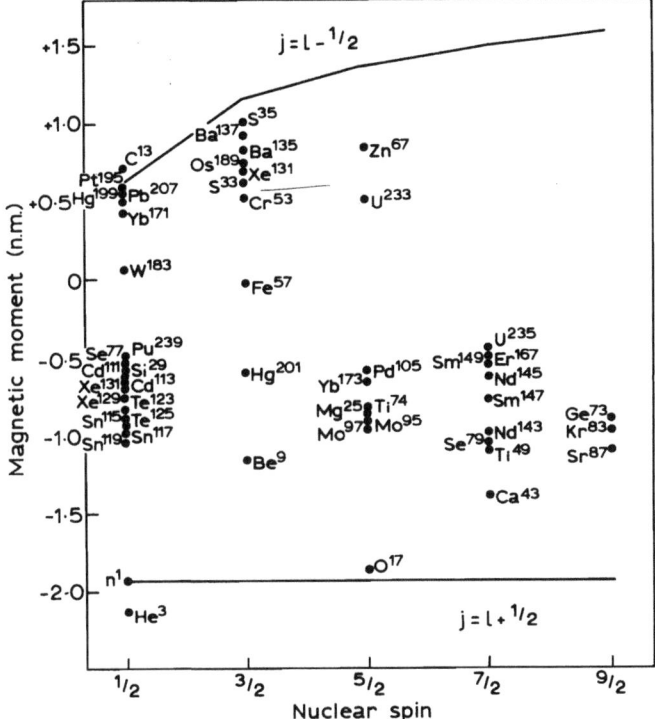

Fig. 1.2 Schmidt diagram for odd-neutron nuclei. (From Blin-Stoyle.[1])

operator, eqn (1.2), is in itself an approximation. Modifications arise because nucleons in nuclei interact through the exchange of mesons and this exchange can be disturbed by the presence of an electromagnetic field. Such modifications lead to corrections to the magnetic moment operator, known as meson-exchange current (MEC) corrections. Since the disturbance requires at least two nucleons to be involved, the corrections lead to two-body magnetic moment operators (in contrast to one-body operators evident in eqn (1.2)). We will discuss MEC further in Chapter 4.

1.7 Electromagnetic transitions

The body of available magnetic moment data can be extended on considering electromagnetic transitions of multipolarity $L = 1$, and no parity change. These are the M1 transitions. In the long-wavelength approximation (in which the spherical Bessel function $j_L(kr)$ in the multipole decomposition of the photon plane-wave field is replaced by its leading term in a series expansion), the transition operator is closely related to the magnetic moment operator. The approximation is a good one whenever the photon momentum $k = E/\hbar c$ is small. In nuclei, typical photon energies, E, are a few MeV, while $\hbar c \simeq 200$ MeV fm.

In general, electromagnetic transitions of multipolarity L are of two types, electric and magnetic, satisfying selection rules:

$$|I_i - I_f| \leq L; \quad \Delta(\text{parity}) = (-1)^L \quad \text{for } EL \text{ radiation,}$$
$$|I_i - I_f| \leq L; \quad \Delta(\text{parity}) = (-1)^{L-1} \quad \text{for } ML \text{ radiation.} \quad (1.34)$$

In a gamma-ray transition from an initial nuclear state of spin I_i, magnetic substate M_i, to a final state, $I_f M_f$, the transition probability is proportional to the square of the matrix element of a multipole operator, $T_{LM}(\xi)$, summed over final and averaged over initial magnetic substates. (The symbol ξ differentiates between electric, $\xi = E$, and magnetic, $\xi = M$, transitions.) Since this transition probability is dependent on the energy of the emitted gamma ray it is convenient to define a reduced transition probability (first introduced by Bohr and Mottelson):[24]

$$B(\xi L; I_i \to I_f) = \frac{2L+1}{4\pi} \frac{1}{2I_i + 1} \sum_{M_i M_f M} |\langle I_i M_i | T_{LM}(\xi) | I_f M_f \rangle|^2$$
$$= \frac{2L+1}{4\pi} |\langle I_i \| T_L(\xi) \| I_f \rangle|^2. \quad (1.35)$$

Note our choice of using the Brink and Satchler[2] definition of reduced matrix element and the convention of placing the initial state on the left in the matrix element. For detailed derivations and a discussion of the different conventions in use in the literature, the reader is referred to a review article by Rose and Brink.[25] The multipole operators for electric and magnetic transitions (denoted by Q_{LM} and $(M_{LM} + M'_{LM})$ respectively by Rose and Brink) are

$$T_{LM}(E) = eg_L \left(\frac{4\pi}{2L+1}\right)^{1/2} r^L Y_{LM}(\hat{r}),$$

$$T_{LM}(M) = \mu_N \left\{ -2ig_L \left(\frac{L}{L+1}\right)^{1/2} \left(\frac{4\pi}{2L+1}\right)^{1/2} r^L [Y_L(\hat{r}), \mathbf{p}]_{LM} \right. \quad (1.36)$$
$$\left. + g_s L^{1/2} (4\pi)^{1/2} r^{L-1} [Y_{L-1}(\hat{r}), \mathbf{S}]_{LM} \right\}.$$

In particular, the E2 multipole operator reduces to

$$T_{20}(E) = \tfrac{1}{2} eg_L (3z^2 - r^2) \quad (1.37)$$

and is half the quadrupole moment operator, eqn (1.6); while the M1 multipole operator reduces to

$$T_{10}(M) = \mu_N \{g_L \mathbf{L} + g_s \mathbf{S}\}, \quad (1.38)$$

1.7 ELECTROMAGNETIC TRANSITIONS

where $L = r \times p$, and is precisely the magnetic moment operator, eqn 1.2).

The reduced transition probability satisfies a hermiticity relationship that connects photon emission with photon absorption processes:

$$(2I_1 + 1)B(\xi L; I_1 \to I_2) = (2I_2 + 1)B(\xi L; I_2 \to I_1). \quad (1.39)$$

To make contact with the literature, we need to introduce a few more definitions. The transition probability per unit time, ω, for electromagnetic decay is proportional to the reduced transition probability. The factors involved in the proportionality depend on the density of photon states and the normalization of the photon wavefunction. In general, for a transition of multipolarity L the relation is

$$\omega(\xi L) = \frac{8\pi(L+1)}{L\{(2L+1)!!\}^2} \frac{1}{\hbar} \left(\frac{E}{\hbar c}\right)^{2L+1} B(\xi L; I_i \to I_f) \quad (1.40)$$

and specifically for E2 and M1 transitions becomes

$$\omega(\text{E2}) = 1.23 \times 10^9 E^5 B(\text{E2}; I_i \to I_f) \text{ s}^{-1},$$
$$\omega(\text{M1}) = 1.76 \times 10^{13} E^3 B(\text{M1}; I_i \to I_f) \text{ s}^{-1}, \quad (1.41)$$

where E, the energy of the emitted photon in MeV, is the difference in energy between the initial and final states, $E = E_i - E_f$. The mean lifetime of the state is given by the inverse

$$\tau_m = 1/\omega. \quad (1.42)$$

Note the half-life is related to the mean lifetime $\tau_{1/2} = 0.693 \tau_m$.) Another related quantity frequently used is the gamma width, Γ_γ, defined as $\Gamma_\gamma = \hbar/\tau_m = \hbar\omega$. In particular,

$$\Gamma_\gamma(\text{E2}) = 8.06 \times 10^{-7} E^5 B(\text{E2}; I_i \to I_f) \text{ eV},$$
$$\Gamma_\gamma(\text{M1}) = 1.16 \times 10^{-2} E^3 B(\text{M1}; I_i \to I_f) \text{ eV}, \quad (1.43)$$

where, as before, E is in MeV. A convenient measure of whether a particular transition is fast or slow compared to a norm of the same energy and type is afforded by the Weisskopf unit (Weisskopf;[26] Wilkinson[27]). This unit is based on an extreme one-particle model, in which a single proton of orbital angular momentum L and total spin $= L + \frac{1}{2}$ makes a transition to a final state of zero orbital angular momentum. The reduced matrix element is trivially evaluated

$$(2L + 1) |\langle j = L + \tfrac{1}{2} \| T_L(E) \| j = \tfrac{1}{2}\rangle|^2 = e^2 \langle r^L \rangle^2, \quad (1.44)$$

and the radial integral $\langle r^L \rangle$ estimated by taking the radial wavefunctions as simple rectangles of radial extension, R, which is identified with the radius of the nucleus. Then the Weisskopf unit for electric transitions becomes

$$\Gamma_{W\gamma}(EL) = e^2 \left(\frac{E}{\hbar c}\right)^{2L+1} \frac{2(L+1)}{L[(2L+1)!!]^2} \left(\frac{3}{L+3}\right)^2 R^{2L}, \quad (1.45)$$

with $R = r_0 A^{1/3}$ and $r_0 = 1.2$ fm. The Weisskopf unit for magnetic transitions of multipolarity L is obtained by applying a correction factor $10(\hbar/mcR)^2$ to the electric estimate, namely

$$\Gamma_{W\gamma}(ML) = 10\left(\frac{e\hbar}{mcR}\right)^2 \left(\frac{E}{\hbar c}\right)^{2L+1} \frac{2(L+1)}{L[(2L+1)!!]^2} \left(\frac{3}{L+3}\right)^2 R^{2L}, \quad (1.46)$$

where m is the proton mass. In particular we have

$$\begin{aligned} \Gamma_{W\gamma}(E2) &= 4.79 \times 10^{-8} A^{4/3} E^5 \text{ eV}, \\ \Gamma_{W\gamma}(M1) &= 2.07 \times 10^{-2} E^3 \text{ eV}. \end{aligned} \quad (1.47)$$

Measured E2 transitions of the order of 1 Weisskopf unit (W.u.) are said to be of single-particle strength. In deformed nuclei, many E2 transitions have been measured to be tens of W.u.; these are said to be collective transitions. For M1, most transitions are less than 1 W.u. The only exceptions occur in very light nuclei where there are a few cases of transitions between 1 and 10 W.u. The strongest known M1 transition is in ^{18}F, $1.04 \to 0$ MeV, $0^+; T=1 \to 1^+; T=0$, of 10.3 ± 1.5 W.u. (Ajzenberg-Selove[28]).

1.8 Effective operators

The failure of the extreme single-particle model to explain quantitatively the magnetic moments of odd-A nuclei does not mean we should give up completely on the single-particle model. Rather we use the model as a starting point and build up corrections to it using perturbation theory. In a certain limiting case (in which configuration mixing corrections arise solely from central, spin-independent residual interactions) we can expect the form of the one-body operator to remain the same, only its g-factors are no longer given by the free-nucleon values but rather are renormalized because of the presence of the other nucleons in the nucleus. The same is true for quadrupole moments. In particular, the free-nucleon values of the coefficients of the quadrupole operator, eqn (1.6), of $eg_L = e$ for a proton and $eg_L = 0$ for a neutron imply that quadrupole moments of all odd-neutron nuclei in an extreme independent particle model should be zero, which is far from true. For example, the quadrupole moment of ^{17}O is $Q = -2.578$ e fm^2 (Schaefer et al.[29]), not far short of a single-proton value. Thus we are led to the idea that the odd neutron in ^{17}O has an 'effective charge'. This effective charge arises from configuration interactions between the odd neutron and the protons in the ^{16}O closed-shell core. We postpone to Chapter 6 any further discussion of effective charge.

The magnetic moment and the electromagnetic M1 transition operators, eqns (1.2) and (1.38), satisfy certain selection rules. For example, a matrix element evaluated between single-particle states, (n_1, l_1, j_1) and (n_2, l_2, j_2), is non-zero only when $n_1 = n_2$ and $l_1 = l_2$, i.e. the operator is diagonal in orbital space. Thus single-particle magnetic moment matrix elements only involve a specific shell-model orbit $(n, l, j = l + \frac{1}{2})$, and its spin–orbit partner $(n, l, j = l - \frac{1}{2})$. For a given n, l value there are three non-zero linearly independent single-particle matrix elements: (a) $\langle j = l + \frac{1}{2} \| \mu \| j = l + \frac{1}{2} \rangle$, (b) $\langle j = l - \frac{1}{2} \| \mu \| j = l - \frac{1}{2} \rangle$, and (c) $\langle j = l + \frac{1}{2} \| \mu \| j = l - \frac{1}{2} \rangle$. A fourth matrix element with $j = l - \frac{1}{2}$ on the left and $j = l + \frac{1}{2}$ on the right is not independent of (c), but is related to it by hermiticity, eqn (1.39). Thus from completely general arguments it can be reasoned that an effective magnetic moment operator can have at most three terms, each operator having multipolarity $L = 1$, and parity selection rule $\Delta \pi =$ no. Arima and Huang-Lin[30] were the first to exploit this idea. They write the effective M1 operator as

$$\mu_{\text{eff}} = g_{L,\text{eff}} L + g_{S,\text{eff}} S + g_{P,\text{eff}} [Y_2, S], \qquad (1.48)$$

where $g_{L,\text{eff}} = g_L + \delta g_L$, etc., i.e. a free-nucleon value and a correction to it. Note the presence of a new one-body operator $[Y_2, S]$, which is a spherical harmonic of multipolarity 2 coupled to the spin operator to form a spherical tensor of rank 1. Such a term is absent in the free-nucleon magnetic moment operator. Suppose that in some model a calculation were to be done for the effect of configuration admixtures or the effect of MEC currents on the magnetic moment matrix elements of an odd-mass nucleus. Then it would be convenient to express the result of such a calculation in terms of δg_L, δg_S, δg_P. This can be done by simply performing the calculation three times for the three matrix elements (a), (b), and (c) and solving a set of three linear equations for the three unknowns. Defining

$$\mu_> = \left(\frac{2l+1}{2l+3}\right)^{1/2} \langle l + \tfrac{1}{2} \| \mu \| l + \tfrac{1}{2} \rangle,$$

$$\mu_< = \left(\frac{2l-1}{2l+1}\right)^{1/2} \langle l + \tfrac{1}{2} \| \mu \| l - \tfrac{1}{2} \rangle, \qquad (1.49)$$

$$M = \left(\frac{2l+1}{l}\right)^{1/2} \langle l + \tfrac{1}{2} \| \mu \| l - \tfrac{1}{2} \rangle,$$

where the first two represent corrections to magnetic moments and the third is related to the correction to the off-diagonal matrix element, the solution to the linear equations is

$$\delta g_L = \frac{1}{(2l+1)^2}[(2l+3)\mu_> + (2l+1)\mu_< + 2M]$$

$$\delta g_S = \frac{2}{3}\frac{1}{(2l+1)^2}[(l+1)(2l+3)\mu_>$$
$$- l(2l+1)\mu_< - 4l(l+1)M]$$

$$(8\pi)^{-1/2}\delta g_P = \frac{2}{3}\frac{(2l+3)}{(2l+1)^2}[(2l-1)\mu_> - (2l+1)\mu_< + (2l-1)M].$$
(1.50)

In the literature there are two very complete calculations of configuration admixtures and MEC corrections to magnetic moments evaluated to second-order in perturbation theory for nuclei predominantly described as closed-shell-plus (or minus)-one nucleon. They are by Towner[31] and by Arima et al.[32] (to be denoted by ASBH). In both cases the result of the calculation is expressed in terms of δg_L, δg_S, and δg_P using eqn (1.50). In Table 1.1 we quote some of their results for the jj closed-shell core of ^{208}Pb. Note there is a very good agreement between the two calculations, particularly for δg_L. Although not strictly justified, we have averaged the tabulated values to derive a state-independent effective

Table 1.1 Corrections to the magnetic moment operator eqn (1.48) for single-particle states in the Pb region as calculated by Towner[31] and ASBH.[32]

	Towner			ASBH	
	δg_L	δg_S	δg_P	δg_L	δg_S
Proton					
0h	0.11	−2.35	0.30	0.13	−2.31
0i	0.07	−2.48	0.11	0.13	−2.07
1d	0.17	−1.86		0.15	−2.10
1f	0.16	−1.66	1.10	0.12	−1.92
2s		−1.83	−0.11		−1.83
Average	0.13(2)	−2.04(16)	0.4(3)	0.13(1)	−2.05(8)
Neutron					
0i	−0.07	1.74	0.45	−0.07	2.11
1f	−0.10	1.36	−0.08	−0.09	1.74
1g	−0.06	1.48	−0.59	−0.07	1.80
2p	−0.05	1.50	−0.30	−0.09	1.55
Average	−0.07(1)	1.52(9)	−0.1(2)	−0.08(1)	1.80(12)

operator. For the proton, a 13% increase in $g_L^{(\pi)}$, namely $\delta g_L^{(\pi)} = 0.13 \pm 0.02$ is calculated while a reduction is obtained in the neutron value, namely $\delta g_L^{(\nu)} = -0.07 \pm 0.01$. The first experimental indication that the proton g_L value is enhanced in nuclei over its free-nucleon value came in the measurement of Yamazaki et al.[33] of the magnetic moment of the 11^- isomer in ^{210}Po. We discuss this result in more detail in the next section. Nagamiya and Yamazaki[34] later showed the enhancement to be a general phenomenon present throughout the whole mass region. A systematic analysis[35] of all the experimental magnetic moment data in the Pb region with an effective magnetic moment operator produces as best-fit values: $\delta g_L^{(\pi)} = 0.16 \pm 0.02$, $\delta g_L^{(\nu)} = -0.06 \pm 0.02$, in very good agreement with calculations. For the spin terms δg_S and δg_P in the effective operator, it is more difficult to derive unambiguous average values. This is because the calculations show a hint that δg_S increases with the orbital angular momentum of the state. (Yamazaki[35] gets a fit to the experimental data using $\delta g_S = G_S \langle F(r) \rangle$ with G_s state-independent and $F(r)$ a state-dependent radial function.) The calculated δg_P values are very variable but generally small compared with the δg_S value. We will ignore δg_P in the Pb region. The average δg_S value expressed as a percentage of the free-nucleon value is around $\delta g_S/g_S \approx -40\%$, a significant quenching in the spin matrix element for both protons and neutrons.

In light nuclei at LS closed shells it is more convenient to discuss the effective magnetic moment operator in its isoscalar and isovector combination. In Table 1.2 we again quote calculated results from Towner[31] and Arima et al.,[32] but the agreement between the two is less satisfactory than it was in Table 1.1. For the orbital part of the operator Towner obtains a larger enhancement to the g_L value in isovector than in isoscalar operators, $\delta g_L^{(1)}/g_L^{(1)} = 14\%$ and $\delta g_L^{(0)}/g_L^{(0)} = 2.6\%$ on the average. This is a consequence of the isovector nature of the dominant pion MEC correction. For the spin part of the operator there is a fairly obvious state dependence in the tabulated results, nevertheless an average value shows a quenching of about $\delta g_S/g_S = -12\%$ for both the isoscalar and isovector operators. Note there is a very significant δg_P value for the isovector operator.

Another way of obtaining the average effective magnetic moment operator is to treat the g-factors as parameters and determine them in a fit of calculation with experiment over a large number of magnetic moment and electromagnetic transition probability data. This approach has been carried out most efficiently in the sd shell by Brown and Wildenthal,[36] where untruncated shell-model calculations in the complete $0d_{5/2}$–$1s_{1/2}$–$0d_{3/2}$ space have been done for 49 magnetic moment and 114 transition data. The authors assume the effective magnetic moment operator to have a smooth mass dependence in its parameters

Table 1.2 Corrections to the magnetic moment operator eqn (1.48) for single-particle states at LS closed shells as calculated by Towner[31] and ASBH[32]

	Towner			ASBH		
	δg_L	δg_S	δg_P	δg_L	δg_S	δg_P
Isovector						
$A = 3$ 0s		0.34				
$A = 5$ 0p	0.069	−0.38	1.47			
$A = 15$ 0p	0.104	−0.30	0.94	0.061	−0.13	1.03
$A = 17$ 0d	0.068	−0.54	1.10	0.040	−0.36	1.10
$A = 17$ 1s		−0.61				
$A = 39$ 0d	0.080	−0.63	0.90	0.017	−0.37	0.90
$A = 39$ 1s		−0.68				
$A = 41$ 0f	0.066	−0.62	0.96	0.017	−0.44	0.68
$A = 41$ 1p	0.033	−0.78	0.63			
Average	0.070(9)	−0.57(6)	0.98(8)	0.034(11)	−0.33(7)	0.93(9)
Isoscalar						
$A = 3$ 0s		−0.06				
$A = 5$ 0p	0.020	−0.06	−0.07			
$A = 15$ 0p	0.022	−0.12	−0.03	0.033	−0.15	0.00
$A = 17$ 0d	0.010	−0.10	−0.03	0.024	−0.11	0.02
$A = 17$ 1s		−0.10				
$A = 39$ 0d	0.014	−0.14	−0.02	0.028	−0.16	0.01
$A = 39$ 1s		−0.14				
$A = 41$ 0f	0.005	−0.12	−0.02	0.021	−0.12	0.03
$A = 41$ 1p	0.007	−0.12	−0.02			
Average	0.013(4)	−0.11(1)	−0.03(1)	0.027(5)	−0.13(1)	0.01(1)

and scale the δg values by a factor $A^{0.35}$. (In principle, away from the closed LS shells the renormalized operator would have not just one-body terms but many-body terms as well. Empirically there appears to be little need for many-body terms beyond that incorporated implicitly into the mass dependence.) The results of the Brown and Wildenthal fit for a 0d particle at $A = 17$ and $A = 39$ is given in Table 1.3. It is in reasonable accord with the calculated values of Towner and ASBH given in Table 1.2, although a number of subtle differences are analysed in detail by Towner.[31]

1.9 Special case of ^{210}Po

Is there any experimental evidence which irrefutably confirms the enhancement in nuclei of the proton g_L value over its free-nucleon value? The answer is yes. It follows from the beautiful idea of Yamazaki[35] to examine a two-proton configuration $(j_1, j_2)I$, for which $j_1 = l_1 + \frac{1}{2}$ and

1.9 SPECIAL CASE OF ^{210}Po

Table 1.3 Empirical values of the effective magnetic moment operator in the sd shell deduced by Brown and Wildenthal[36]

	δg_L	δg_S	δg_P
Isovector			
$A = 17$ 0d	0.089(14)	−0.59(5)	1.6(2)
$A = 17$ 1s		−0.39(8)	
$A = 39$ 0d	0.122(14)	−0.79(7)	2.2(2)
$A = 39$ 1s		−0.52(10)	
Isoscalar			
$A = 17$ 0d	0.016(3)	−0.10(2)	0.13(10)
$A = 17$ 1s		−0.04(3)	
$A = 39$ 0d	0.021(4)	−0.14(2)	0.17(13)
$A = 39$ 1s		−0.06(4)	

$j_2 = l_2 - \frac{1}{2}$. The magnetic moment of such configurations is dominated by the orbital moments, because the spin moments from the two particles are opposite in sign and cancelling.

For strong singlet forces, the lowest member of the multiplet $(j_1, j_2)I$ is the stretched configuration $I = j_1 + j_2 = l_1 + l_2$. Furthermore, this state is isomeric because of a fortunate spin gap between this level and any of the low-lying levels. Thus its magnetic moment is amenable to experiment by measuring the time differential perturbed angular distribution of the delayed gamma rays. Our case in point is the $\pi(h_{9/2}, i_{13/2})11^-$ isomeric state in ^{210}Po which feeds the $\pi(h_{9/2}^2)8^+$ state. This 24-ns isomer was found, with the 8^+ isomer, in ^{208}Pb$(\alpha, 2n)^{210}$Po reaction by Yamazaki and Ewan,[37] and its magnetic moment measured by Yamazaki et al.[33] and confirmed by Häusser et al.[38]

Theoretically, the g-factor for the two-particle state $(j_1, j_2)I$ is given by the addition theorem, eqn (1.22):

$$g = \tfrac{1}{2}(g_1 + g_2) + \frac{j_1(j_1 + 1) - j_2(j_2 + 1)}{I(I + 1)} \tfrac{1}{2}(g_1 - g_2), \quad (1.51)$$

where $g_1(g_2)$ is the g-factor for the single-particle state $j_1(j_2)$, eqn (1.25):

$$g_1 = g_{L,\text{eff}} \pm (g_{S,\text{eff}} - g_{L,\text{eff}})/(2l_1 + 1) \quad (1.52)$$

for $j_1 = l_1 \pm \frac{1}{2}$. Here $g_{L,\text{eff}} = g_L + \delta g_L$ and $g_{S,\text{eff}} = g_S + \delta g_S$, and a $g_{P,\text{eff}}$ term is neglected. Furthermore if it is assumed these effective g-factors are state independent, then the g-factor for the isomer with $j_1 = l_1 + \frac{1}{2}$, $j_2 = l_2 - \frac{1}{2}$ and $I = l_1 + l_2$ reduces to

$$g = g_{L,\text{eff}} + \frac{1}{I(2l_2 + 1)}(g_{S,\text{eff}} - g_{L,\text{eff}}). \quad (1.53)$$

The key here is that the second term is a small correction, so a measurement of g leads directly to $g_{L,\text{eff}}$.

The experimentally observed 11^- state in ^{210}Po will not be just a pure two-proton state, but can conceivably have components $\pi(h_{9/2}^2)8^+$ and $\pi(h_{9/2}f_{7/2})8^+$ each coupled to a 3^- vibration in ^{208}Pb. These configurations are suggested by the strong $B(E3; 11^- \to 8^+)$ transitions to the two known 8^+ states in ^{210}Po, each of which is strongly enhanced over the single-particle Weisskopf estimate, and each of which is characteristic of the strong $B(E3, 3^- \to 0^+)$ known in ^{208}Pb. Yamazaki[35] estimated the isomeric state wavefunction to be

$$|11^-\rangle = 76\% \, |h_{9/2}, i_{13/2}\rangle + 20\% \, |h_{9/2}, f_{7/2} \otimes 3^-\rangle + 4\% \, |h_{9/2}^2 \otimes 3^-\rangle. \quad (1.54)$$

Knowing or estimating the g-factors of the 8^+ two-proton configurations and the 3^- vibration, the experimental g-factor for the 11^- state can be corrected for these admixtures to yield a net g-factor for the two-nucleon configuration:

$$g[(h_{9/2}, i_{13/2})11^-]_{\text{expt}} = 1.18 \pm 0.01. \quad (1.55)$$

This value is then equated with eqn (1.53), where an estimate has to be made for the second term on the right-hand side. We will use the average values given in Table 1.1, namely $g_{S,\text{eff}} - g_{L,\text{eff}} = 2.42 \pm 0.16$, which when divided by $I(2l_2 + 1)$ gives a correction to the g-factor of 0.02. Therefore, we obtain

$$g_{L,\text{eff}} = 1.16 \pm 0.01, \quad (1.56)$$

or $\delta g_L = 0.16 \pm 0.01$, which is a clear indication that the free-nucleon proton g_L value is enhanced in a heavy nucleus.

1.10 Isospin considerations

It is useful to clarify the formalism for the reduced transition probability when isospin is being used as a good quantum number. In eqn (1.35) the reduced transition probability for an M1 transition between an initial state of spin I_i and isospin T_i and a final state of spin I_f and isospin T_f is recorded. With isospin included the equation is interpreted as follows:

$$B(M1; I_i T_i \to I_f T_f) = \frac{3}{4\pi} |\langle I_i T_i \| \mu \| I_f T_f \rangle|^2, \quad (1.57)$$

where now

$$\langle I_i T_i \| \mu \| I_f T_f \rangle = \sum_{T=0,1} \langle T_f N_f T 0 | T_i N_i \rangle \langle I_i T_i \| \mu^{(T)} \| I_f T_f \rangle.$$

Here $\langle T_f N_f T 0 | T_i N_i \rangle$ is an isospin Clebsch–Gordan coefficient with N_i and N_f representing the z-component of isospin, and the triply reduced matrix element on the right is a reduced matrix element in both spin and isospin spaces. Further

$$\mu^{(0)} = g_L^{(0)}L + g_S^{(0)}S,$$
$$\mu^{(1)} = g_L^{(1)}L\tau + g_S^{(1)}S\tau, \qquad (1.58)$$

are the isoscalar and isovector magnetic moment operators, eqn (1.9). If we replace L by $I - S$, then these operators can be rewritten as

$$\mu^{(0)} = g_L^{(0)}I + (g_S^{(0)} - g_L^{(0)})S,$$
$$\mu^{(1)} = g_L^{(1)}I\tau + (g_S^{(1)} - g_L^{(1)})S\tau, \qquad (1.59)$$

where I is the total angular momentum operator, and $I\tau$ is a shorthand notation for $\sum I^{(k)}\tau_z^{(k)}$ summed over $k=1$ to A, the single-nucleon coordinates of the A nucleons in the nucleus. For an M1 transition in a nucleus for which $I_i \neq I_f$, the term I in the isoscalar operator will not contribute. Thus the isoscalar part of the M1 transition operator is proportional to $g_S^{(0)} - g_L^{(0)} = 0.38$ using free-nucleon values. In the isovector operator, the term in $I\tau$ has a much smaller coefficient than the term in $S\tau$. Thus the isovector part of the M1 operator is roughly proportional to $g_S^{(1)} - g_L^{(1)} = 4.2$. In a transition in which $T_i = T_f \neq 0$ with both isoscalar and isovector parts of the operator contributing it is evident that the isovector part will dominate over the isoscalar part of the M1 operator. This has been expressed as Morpurgo's rule:[39]

An isoscalar M1 transition, for example between $T = 0$ states in self-conjugate nuclei (nuclei with an equal number of neutrons and protons) should be much slower by a factor of 100 (approximately $(0.38/4.2)^2$) than the average M1 transition strength in the same or neighbouring nucleus in which the isovector part of the operator contributes.

An example of this rule (Lawson[40]) is provided by the decay of the 3.95 MeV $I = 1^+ T = 0$ state in ^{14}N. The decay to the $I = 1^+ T = 0$ ground state has a radiative width of $(3.6 \pm 0.7) \times 10^{-4}$ eV and that to the $I = 0^+ T = 1$ 2.31 MeV level has $\Gamma_\gamma = 0.079 \pm 0.010$ eV.[41] Thus the ratio of widths is

$$R = \frac{\Gamma_\gamma(3.95(1^+, 0) \to g.s.(1^+, 0))}{\Gamma_\gamma(3.95(1^+, 0) \to 2.31(0^+, 1))} = (4.6 \pm 1.0) \times 10^{-3}.$$

Naïvely one would have anticipated the ratio being approximately the ratio of the cube of the transition energies

$$R = \left(\frac{3.95}{3.95 - 2.31}\right)^3 = 14.$$

Many examples of this inhibition, although not as marked as this case, have been observed; see the surveys of Endt.[42]

1.11 Connection with nuclear β decay

An allowed β-decay transition between an initial nucleus of charge Z_i and a final nucleus of charge Z_f,

$$(Z_i, A) \rightarrow (Z_f = Z_i \pm 1, A) + e^{\mp} + \nu_e, \quad (1.60)$$

is one in which there is no orbital angular momentum carried away by the leptons. (The upper sign represents electron emission, the lower sign positron emission.) This leads to the following selection rules for a nucleus undergoing an allowed β transition: for a Fermi transition (only between analogue states, defined as states in neighbouring nuclei with identical configurations in all respects except for the z-component of the total isospin)

$$\Delta L = 0, \quad \Delta S = 0, \quad \Delta I = 0, \quad \Delta T = 0,$$

and for the Gamow–Teller (GT) transitions

$$\Delta L = 0, \quad \Delta S = 1, \quad \Delta I = 1, \quad \Delta T = 0, 1,$$

and in both cases there is no parity change. See textbooks[43] on nuclear β-decay for further details. The β-transition strength is characterized by its ft value defined as

$$ft = \frac{K}{B(\text{F}) + B(\text{GT})} \quad (1.61)$$

where f is a phase space factor (statistical rate function), t is the half-life of the transition, and K is an overall universal normalization. Its value is determined from the 'corrected' ft values measured in superallowed $0^+ \rightarrow 0^+$ Fermi decays, $K = 2(\mathcal{F}t)$, where $\mathcal{F}t = ft(1 + \delta_R)(1 - \delta_C)$ and δ_R and δ_C are small calculable electromagnetic corrections.[44] The most recent determination of K (following a revision[45] of the radiative correction δ_R) is given by Hardy et al:[44] $K = 6146.6 \pm 7.0$ s. Lastly $B(\text{F})$ and $B(\text{GT})$ are squares of nuclear matrix elements for Fermi and Gamow–Teller operators defined as:

$$B(\text{F}; I_i T_i \rightarrow I_f T_f) = |\langle T_f N_i 1 \pm 1 | T_i N_i \rangle \langle I_f T_f ||| (1/\sqrt{2})\tau ||| I_f T_f \rangle|^2$$
$$= (T_i \pm N_i)(T_i \mp N_i + 1)\delta_{I_i, I_f}\delta_{T_i, T_f}, \quad (1.62)$$

$$B(\text{GT}; I_i T_i \rightarrow I_f T_f) = |\langle T_f N_i 1 \pm 1 | T_i N_i \rangle \langle I_f T_f ||| \text{GT} ||| I_f T_f \rangle|^2,$$

where the Gamow–Teller operator is

$$\text{GT} = \mp \frac{1}{\sqrt{2}} g_A \sigma \tau_{\pm}. \quad (1.63)$$

Here g_A is the axial-vector coupling constant (relative to the vector coupling constant, $g_A = G_A/G_V$) determined from free-neutron decay to be $g_A = 1.26$[46] and σ is the Pauli spin operator, $\sigma = 2S$. Again, as was the case with magnetic moments, we envisage the need for an effective operator in finite nuclei, which we parameterize in the same way, namely

$$(GT)_{eff} = \mp \frac{1}{\sqrt{2}} \{g_{LA,eff} L + g_{A,eff} \sigma + g_{PA,eff}[Y_2, \sigma]\} \tau_{\pm} \quad (1.64)$$

where $g_{A,eff} = g_A + \delta g_A$ etc. Calculations of the effective operator from configuration admixtures and MEC corrections are to be found in Towner[31] and Arima et al.[32] Average values for light nuclei, assuming unjustifiably that the effective operator is state independent, are:[31]

$$\delta g_{LA} = 0.011 \pm 0.004; \quad \delta g_A = -0.19 \pm 0.02; \quad \delta g_{PA} = 0.21 \pm 0.02. \quad (1.65)$$

Note a significant quenching in g_A: $\delta g_A/g_A = -15\%$. Fitted values to sd-shell nuclei by Brown and Wildenthal[36] obtain for ^{17}O: $\delta g_{LA} = 0.01 \pm 0.01$, $\delta g_A = -0.26 \pm 0.01$, $\delta g_{PA} = 0.09 \pm 0.04$, and produce an even larger quenching in g_A: $\delta g_A/g_A = -20\%$.

It is evident from the structure of the operators that there is an intimate connection between a Gamow–Teller β transition and the corresponding isovector M1 γ transition. For example consider the case where the initial state for β-decay is the isospin analogue of the initial state for γ-decay, the final states being the same for both decays. Then the reduced transition probabilities are related in the following way

$$B(M1; I_i T_i \to I_f T_f) = \frac{3}{8\pi} \left|\frac{g_S^{(1)}}{g_A}\right|^2 \left|\frac{\langle CG \rangle_\gamma}{\langle CG \rangle_\beta}\right|^2 B(GT; I_i T_i \to I_f T_f)$$

$$\times \left(1 + \frac{g_L^{(1)}}{g_S^{(1)}} \frac{\langle I_i T_i \|\| L\tau \|\| I_f T_f \rangle}{\langle I_i T_i \|\| S\tau \|\| I_f T_f \rangle}\right)^2, \quad (1.66)$$

where $\langle CG \rangle_\gamma = \langle T_f N_f 1 0 | T_i N_i \rangle$ and $\langle CG \rangle_\beta = \langle T_f N_f 1 \pm 1 | T_i N_i \pm 1 \rangle$ are the appropriate isospin Clebsch–Gordan coefficients. If further we neglect the ratio of matrix elements $\langle L\tau \rangle/\langle S\tau \rangle$ because its coefficient $g_L^{(1)}/g_S^{(1)} \simeq 0.11$ is small then we get an approximate relation[47] that is useful as a test of nuclear structure:

$$B(M1; I_i T_i \to I_f T_f) \simeq 1.67 \left|\frac{\langle CG \rangle_\gamma}{\langle CG \rangle_\beta}\right|^2 B(GT; I_i T_i \to I_f T_f). \quad (1.67)$$

We used free-nucleon values of the coupling constants. This result can be expressed in terms of experimental measurable quantities, using eqns (1.43) and (1.61):

$$\Gamma_\gamma \simeq 1.18 \times 10^2 (E_\gamma)^3 (ft)^{-1} \left|\frac{\langle CG \rangle_\gamma}{\langle CG \rangle_\beta}\right|^2 \text{ eV}, \quad (1.68)$$

Table 1.4 Gamma ray M1 widths (in Weisskopf units) deduced from the analogue β-transition and compared with experimental widths

Nucleus	γ-decay $E_{xi} \to E_{xf}$	$I_i T_i \to I_f T_f$	β-decay $\log ft$	Γ_γ(W.u.) From 'β'	Expt
^{12}C	15.11→0.00	$1^+1 \to 0^+0$	4.09±0.03	0.45±0.03	0.53±0.01
	15.11→4.44	$1^+1 \to 2^+0$	5.13±0.02	0.042±0.002	0.038±0.005
^{20}Ne	10.27→1.63	$2^+1 \to 2^+0$	4.99±0.01	0.06±0.01	0.30±0.03
	10.27→7.42	$2^+1 \to 2^+0$	4.19±0.05	0.36±0.04	0.64±0.08
^{6}Li	3.56→0.00	$0^+1 \to 1^+0$	2.91±0.01	6.93±0.03	8.62±0.18
^{14}N	2.31→0.00	$0^+1 \to 1^+0$	7.27±0.01	0.00±0.00	0.028±0.003
	3.95→2.31	$1^+0 \to 0^+1$	3.15±0.02	1.33±0.06	0.90±0.10
^{18}F	1.04→0.00	$0^+1 \to 1^+0$	3.09±0.01	4.52±0.04	10.9±2.1
	1.70→1.04	$1^+0 \to 0^+1$	4.43±0.06	0.06±0.01	0.08±0.01
^{22}Na	0.66→0.58	$0^+1 \to 1^+0$	3.65±0.01	1.26±0.04	3.9±0.6
	1.94→0.66	$1^+0 \to 0^+1$	3.43±0.01	0.69±0.01	
^{26}Al	1.06→0.23	$1^+0 \to 0^+1$	3.53±0.02	0.55±0.03	1.4±0.3
	1.85→0.23	$1^+0 \to 0^+1$	3.81±0.03	0.29±0.02	0.18±0.05
^{30}P	0.67→0.00	$0^+1 \to 1^+0$	4.38±0.02	0.24±0.01	0.67±0.11
^{34}Cl	0.46→0.00	$1^+0 \to 0^+1$	5.31±0.05	0.01±0.00	0.04±0.00
^{42}Sc	0.61→0.00	$1^+0 \to 0^+1$	3.17±0.12	1.3±0.4	2.0±0.9
^{7}Li	0.48→0.00	$\frac{1}{2}^-\frac{1}{2} \to \frac{3}{2}^-\frac{1}{2}$	3.55±0.01	1.59±0.04	2.75±0.14
^{13}C	15.11→0.00	$\frac{3}{2}^-\frac{3}{2} \to \frac{1}{2}^-\frac{1}{2}$	4.01±0.01	0.37±0.01	0.31±0.02
	15.11→3.69	$\frac{3}{2}^-\frac{3}{2} \to \frac{1}{2}^-\frac{1}{2}$	4.45±0.04	0.13±0.01	0.58±0.08

where the photon energy, E_γ, is expressed in MeV. In Table 1.4 are a number of data from light nuclei that test this relation. Using the experimental β-decay ft-value an estimate is obtained for the corresponding M1 γ width (expressed in Weisskopf units) and compared with the experimental width. In some cases the comparison is very good, in other cases very poor. Cases where eqn (1.68) is badly broken give indications of where the orbital contribution, from the term $L\tau$, is important. This information is something that can be tested in shell-model calculations.

1.12 Buck–Perez plot

Another interrelation between β decay and in this case magnetic moments that goes back more than three decades is the one revived recently by Buck and Perez.[48] The comparison concerns the magnetic moments of odd-mass nuclei that are isospin mirrors of each other and the nuclear β-decay matrix element between the two. For example the ground states of ^{17}F and ^{17}O are isospin mirrors (the wavefunction of one is given by the wavefunction of the other by simply interchanging the role of neutrons and protons). Their magnetic moments have been measured and the ground-state to ground-state β decay, $^{17}F(\beta^+)^{17}O$, observed. Although we will be discussing light nuclei it is convenient to discuss this phenomenon in the proton–neutron formalism. We introduce the following notation

$$L_e = \left(\frac{I}{I+1}\right)^{1/2} \langle I \| \sum_k^{\text{even}} L^{(k)} \| I \rangle,$$
$$L_o = \left(\frac{I}{I+1}\right)^{1/2} \langle I \| \sum_k^{\text{odd}} L^{(k)} \| I \rangle, \qquad (1.69)$$

where L_e is the expectation value of the orbital angular momentum operator for the type of nucleon (proton or neutron) of which there is an even number in the nucleus. For example in ^{17}O, L_e is the expectation value of the proton's orbital angular momentum. Likewise L_o is the expectation value for the type of nucleon for which there is an odd number in the nucleus. Similar definitions follow for the spin, S_e and S_o, and total angular momentum, I_e and I_o, expectation values. Writing μ_p and μ_n for the magnetic moments of the odd-proton and odd-neutron members of a mirror pair and γ_p and γ_n as their corresponding g-factors, $\gamma_p = \mu_p/I$ and $\gamma_n = \mu_n/I$, then there are four interrelated equations:

$$\gamma_p = (g_L^{(\pi)}L_o + g_S^{(\pi)}S_o + g_L^{(\nu)}L_e + g_S^{(\nu)}S_e)/I,$$
$$\gamma_n = (g_L^{(\nu)}L_o + g_S^{(\nu)}S_o + g_L^{(\pi)}L_e + g_S^{(\pi)}S_e)/I, \qquad (1.70)$$
$$\gamma_\beta = g_A(S_o - S_e)/I,$$
$$I = L_e + S_e + L_o + S_o = I_e + I_o.$$

Here γ_β relates to the Gamow–Teller matrix element and is obtained from the ft value (see eqn (1.61))

$$\gamma_\beta = \frac{1}{2}\left(\frac{K}{ft} - B(F)\right)^{1/2} \frac{1}{[I(I+1)]^{1/2}}. \qquad (1.71)$$

If further it is assumed that in the ground states of odd-mass nuclei the expectation values of odd-rank multipole operators in the variables of the nucleons with even number are vanishingly small, namely $L_e = S_e = I_e = 0$, then eqns (1.70) are simplified

$$\gamma_p = g_L^{(\pi)} + (g_S^{(\pi)} - g_L^{(\pi)})S_o/I,$$
$$\gamma_n = g_L^{(\nu)} + (g_S^{(\nu)} - g_L^{(\nu)})S_o/I, \qquad (1.72)$$
$$\gamma_\beta = g_A S_o/I,$$

where $I = L_o + S_o$ is used to eliminate L_o. These equations are very familiar in the extreme single-particle model in which it is assumed further that only the last odd nucleon is active. With $S_o = \frac{1}{2}$ for $I = L_o + \frac{1}{2}$

and $S_o = -I/(2I+2)$ for $I = L_o - \frac{1}{2}$ they lead to the Schmidt estimates for magnetic moments. The strong deviations of actual nuclear moments from these estimates can be understood, suggest Buck and Perez, if it is assumed instead that all the odd nucleons are potentially active so that S_o does not necessarily take on the single-particle values. This odd-group model (see de Shalit[49]) can be tested independently of the values of S_o by eliminating S_o through combining eqns (1.72) in pairs

$$\gamma_p = A_p \gamma_\beta + B_p,$$
$$\gamma_n = A_n \gamma_\beta + B_n, \quad (1.73)$$
$$\gamma_p = A \gamma_n + B,$$

where $A_p = (g_S^{(\pi)} - g_L^{(\pi)})/g_A$, $B_p = g_L^{(\pi)}$, $A_n = (g_S^{(\nu)} - g_L^{(\nu)})/g_A$, $B_n = g_L^{(\nu)}$, $A = (g_S^{(\pi)} - g_L^{(\pi)})/(g_S^{(\nu)} - g_L^{(\nu)})$ and $B = g_L^{(\pi)} - Ag_L^{(\nu)}$. Plots of γ_p versus γ_β, γ_n versus γ_β, and γ_p versus γ_n are shown in Fig. 1.3 and each shows a remarkable straight-line correlation. These we will call the Buck–Perez plots. It is significant that the determined values of the slopes and intercepts differ from the free-nucleon values. For example in the γ_p versus γ_n plot, the deduced $A = -1.145 \pm 0.012$ and $B = 1.056 \pm 0.021$ differ from the Schmidt values $A = -1.199$ and $B = 1$. The differences of more than two standard deviations reflect the contributions from MEC two-body operators or the possible small contributions from even-nucleon expectation values. If we assume that L_e/I, S_e/I and I_e/I are small and nucleus-independent, but not zero, then eqns (1.73) still apply with the same expressions for the slopes but the intercepts are now

$$B_p = g_L^{(\pi)} + (g_L^{(\nu)} - g_L^{(\pi)})I_e/I + (g_S^{(\nu)} - g_L^{(\nu)} + g_S^{(\pi)} - g_L^{(\pi)})S_e/I,$$
$$B_n = g_L^{(\nu)} - (g_L^{(\nu)} - g_L^{(\pi)})I_e/I + (g_S^{(\nu)} - g_L^{(\nu)} + g_S^{(\pi)} - g_L^{(\pi)})S_e/I. \quad (1.74)$$

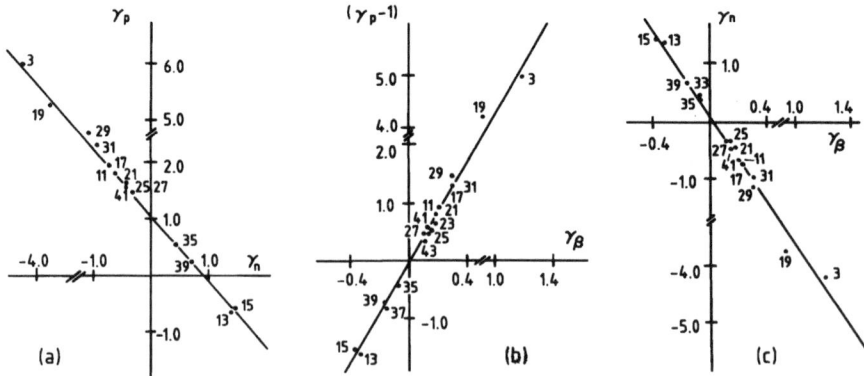

Fig. 1.3 (a) Plot of γ_p versus γ_n, the fitted straight line is $\gamma_p = -(1.145 \pm 0.012)\gamma_n + (1.056 \pm 0.021)$. (b) Plot of $\gamma_p - 1$ versus γ_β, the fitted line is $\gamma_p - 1 = (4.38 \pm 0.10)\gamma_\beta - (0.061 \pm 0.046)$. (c) Plot of γ_n versus γ_β, the fitted line is $\gamma_n = -(3.82 \pm 0.10)\gamma_\beta + (0.101 \pm 0.043)$.

1.12 BUCK–PEREZ PLOT

It is convenient to recast the equations into an isoscalar and isovector form with $\gamma_0 = 1/2(\gamma_p + \gamma_n)$ and $\gamma_1 = 1/2(\gamma_p - \gamma_n)$, namely

$$\gamma_1 = A_1 \gamma_\beta + B_1,$$
$$\gamma_0 = A_0 \gamma_\beta + B_0, \quad (1.75)$$

with

$$A_1 = (g_S^{(1)} - g_L^{(1)})/g_A, \qquad A_0 = (g_S^{(0)} - g_L^{(0)})/g_A,$$
$$B_1 = g_L^{(1)} - 2g_L^{(1)} I_e/I, \qquad B_0 = g_L^{(0)} + 2(g_S^{(0)} - g_L^{(0)})S_e/I, \quad (1.76)$$

and interpret the coupling constants as being the effective coupling constants. Let us concentrate on the isoscalar equation. We do this because in the calculations of effective operators discussed in Section 1.8, the isoscalar coupling constant $g_{P,\text{eff}}^{(0)}$ is small compared to $\delta g_S^{(0)}$ and can be neglected. Likewise in the β-decay effective operator, $g_{LA,\text{eff}}$ and $g_{PA,\text{eff}}$ are small. Thus the form of the isoscalar equation is preserved with the coupling constants replaced by their effective values. This is not true of the isovector equation because $g_{P,\text{eff}}^{(1)}$ is not negligible. The best fit to mirror magnetic moments and ground-state β decay is (Perez[50]):

$$\gamma_0 = (0.280 \pm 0.022)\gamma_\beta + (0.520 \pm 0.010). \quad (1.77)$$

Let us concentrate on the slope. First with free-nucleon values $A_0 = (g_S^{(0)} - g_L^{(0)})/g_A = 0.30$ and is in agreement with the fitted value. On the other hand, both $g_S^{(0)}$ and g_A are quenched from the free-nucleon value and so not much can be learned from the slope other than that the quenching must be comparable in both. However, there is less uncertainty in the calculation of isoscalar effective operators, so if we accept the calculated value of $g_{S,\text{eff}}^{(0)}$ and $g_{L,\text{eff}}^{(0)}$ from Table 1.2, then the slope can be used to determine $g_{A,\text{eff}}$. Thus with $\delta g_L^{(0)} = 0.013 \pm 0.004$ and $\delta g_S^{(0)} = -0.11 \pm 0.01$ we obtain

$$g_{A,\text{eff}} = 0.92 \pm 0.08 \quad (1.78)$$

or $\delta g_A = -0.34 \pm 0.08$. It is evident that the axial-vector coupling constant is strongly quenched in nuclei. This analysis leads to significantly more quenching than that calculated by Towner[31] or Arima et al.[32] and is only just consistent with the empirically determined value in the sd shell of $\delta g_A = -0.26 \pm 0.01$ of Brown and Wildenthal.[36]

A similar analysis to deduce $g_{A,\text{eff}}$ from magnetic moments and β decay in mirror nuclei has been given in a series of papers by Wilkinson.[51] Rather than treat the expectation values L_e and S_e as small and nucleus-independent, Wilkinson explicitly calculates them case by case using the best shell-model wavefunctions available. His result is $g_{A,\text{eff}} = 1.13 \pm 0.06$ or $\delta g_A = -0.13 \pm 0.06$ and much closer to the theoretical calculations.

32   INTRODUCTION

An alternative use of eqn (1.77) is to predict a magnetic moment for a case when it has not been measured using as input the magnetic moment of its mirror and the ft value of the β-transition. In this way Perez[50] predicts the magnetic moments of ^{23}Mg, ^{33}Cl, ^{37}Ar, ^{43}Ti, ^{51}Fe, and ^{55}Ni to be -0.548 ± 0.021, 0.851 ± 0.022, 1.221 ± 0.022, -0.784 ± 0.052, -0.780 ± 0.035, and -0.994 ± 0.045 nuclear magnetons respectively.

1.13 Summary

In this chapter we have set up the necessary definitions for the study of dipole and quadrupole moments and looked at their expectation values in simple systems like the deuteron and the $A = 3$ system.

In finite nuclei, we are faced with the problem of nuclear model dependence. The most obvious example being the departure of nuclear moments in odd-A nuclei from their extreme single-particle values. This has led us to introduce the concept of effective operators.

We have then discussed the origin of calculations involving effective operators but have postponed to Chapter 3 the details of how these calculations are carried out. The end results of these calculations show very good agreement with experiment and with the results deduced from systematic studies of the sd shell by Brown and Wildenthal.

We have concluded this chapter with a discussion of the connection between magnetic and Gamow–Teller transitions in nuclear β decay as seen for instance in the Buck and Perez plots.

Before developing further the models of nuclear structure we should also give some consideration to the methods of experimental measurements. This is the aim of our next chapter.

1.14 References

1. Blin-Stoyle, R. J. (1957). *Theories of nuclear moments*. Oxford University Press.
2. Brink, D. M. and Satchler, G. R. (1968). *Angular momentum*. Clarendon, Oxford.
3. Towner, I. S. (1977). *A shell model description of light nuclei*. (Clarendon, Oxford).
4. Raghavan, P. (1989). *At. Data and Nucl. Data Tables*, **42**, 189.
5. Bishop, D. M. and Cheung, L. (1979). *Phys. Rev.*, **A20**, 381; Ericson, T. E. O. and Rosa-Clot, M. (1983). *Nucl. Phys.*, **A405**, 497.
6. Reid, V. R. and Vaida, M. L. (1973). *Phys. Rev.*, **A7**, 1841; Reid, V. R. and Vaida, M. L. (1975). *Phys. Rev. Lett.*, **34**, 1064.
7. Greene, G. L., *et al.* (1977). *Phys. Lett.*, **71B**, 297; Phillips, W. D., *et al.* (1975). *Phys. Rev. Lett.*, **35**, 1619.
8. Fabian, W., Arenhövel, H., and Miller, H. G. (1974). *Z. Phys.*, **271**, 93;

Gari, M., Hyuga, H., and Sommer, B. (1976). *Phys. Rev.*, **C14,** 2196; Hadjimichael, E. (1978). *Nucl. Phys.*, **A312,** 341; Jaus, W. (1979). *Nucl. Phys.*, **A314,** 287; Vassanji, M. G., Khanna, F. C., and Towner, I. S. (1981). *J. Phys. G: Nucl. Phys.*, **7,** 1029; Sato, T., Kobayashi, M., and Ohtsubo, H. (1982). *Prog. Theor. Phys.*, **68,** 840; Sitarski, W. P., Blunden, P. G., and Lomon, E. L. (1987). *Phys. Rev.*, **C36,** 2479.
9. Machleidt, R., Holinde, K., and Elster, Ch. (1987). *Phys. Rep.*, **149,** 1.
10. Lacombe, M., et al. (1975). *Phys. Rev.*, **D12,** 1495.
11. de Tourreil, R., Rouben, B., and Sprung, D. W. L. (1975). *Nucl. Phys.*, **A242,** 445.
12. Wiringa, R. B., Smith, R. A., and Ainsworth, T. L. (1984). *Phys. Rev.*, **C29,** 1207.
13. Nagels, M. M., Rijken, T. A., and de Swart, J. D. (1979). *Phys. Rev.*, **D17,** 768; *Phys. Rev.*, **D20,** 1633.
14. Reid, R. V. (1968). *Ann. Phys.* (N.Y.), **50,** 411.
15. Lassila, K. E., et al. (1962). *Phys. Rev.*, **126,** 881.
16. Hamada, T. and Johnston, I. D. (1962). *Nucl. Phys.*, **34,** 382.
17. Hadjimichael, E. (1978). *Nucl. Phys.*, **A312,** 341; Vassanji, M. G., Khanna, F. C., and Towner, I. S. (1981). *J. Phys. G: Nucl. Phys.*, **7,** 1029; Kohno, M. (1983). *J. Phys. G: Nucl. Phys.*, **9,** L85.
18. Friar, J. L., et al. (1988). *Phys. Rev.*, **C37,** 2852.
19. Tomusiak, E. L., et al. (1985). *Phys. Rev.*, **C32,** 2075.
20. Harper, E. P., Kim, Y. E., and Tubis, A. (1972). *Phys. Rev. Lett.*, **28,** 1533; Harper, E. P., et al. (1972). *Phys. Lett.*, **40B,** 533.
21. Maize, M. A. and Kim, Y. E. (1984). *Nucl. Phys.*, **A420,** 365; Strueve, W., Hajduk, Ch., and Sauer, P. U. (1983). *Nucl. Phys.*, **A405,** 620; Torre, J. and Goulard, B. (1983). *Phys. Rev.*, **C28,** 529; Riska, D. O. (1980). *Nucl. Phys.*, **A350,** 227; Hadjimichael, E., Goulard, B., and Bornais, R. (1983). *Phys. Rev.*, **C27,** 831.
22. Nordheim, L. W. (1951). *Rev. Mod. Phys.*, **23,** 322.
23. Schmidt, T. (1937). *Z. Phys.*, **106,** 358.
24. Bohr, A. and Mottelson, B. R. (1953). *K. Danske Vidensk. Selsk. Mat.-Fys. Medd.*, **27,** 1.
25. Rose, H. J. and Brink, D. M. (1967). *Rev. Mod. Phys.*, **39,** 306.
26. Weisskopf, V. F. (1951). *Phys. Rev.*, **83,** 1073.
27. Wilkinson, D. H. (1960). In *Nuclear spectroscopy* Part B (ed. F. Ajzenberg-Selove), p. 852. Academic Press, New York.
28. Ajzenberg-Selove, F. (1983). *Nucl. Phys.*, **A392,** 1.
29. Schaefer, H. F., Klemm, R. A., and Harris, F. E. (1969). *Phys. Rev.*, **181,** 137.
30. Arima, A. and Huang-Lin, L. J. (1972). *Phys. Lett.*, **B41,** 429; (1972). **B41,** 435.
31. Towner, I. S. (1987). *Phys. Rep.*, **155,** 263.
32. Arima, A., Shimizu, K., Bentz, W., and Hyuga, H. (1987). *Adv. Nucl. Phys.*, **18,** 1.
33. Yamazaki, T., Nomura, T., Nagamiya, S., and Katou, T. (1970). *Phys. Rev. Lett.*, **25,** 547.

34. Nagamiya, S. and Yamazaki, T. (1971). *Phys. Rev.*, **C4**, 1961.
35. Yamazaki, T. (1979). In *Mesons in nuclei* (ed. D. H. Wilkinson and M. Rho), p. 651. North-Holland, Amsterdam.
36. Brown, B. A. and Wildenthal, B. H. (1987). *Nucl. Phys.*, **A474**, 290.
37. Yamazaki, T. and Ewan, G. T. (1967). *Phys. Lett.*, **B24**, 278.
38. Häusser, O., et al. (1976). *Nucl. Phys.*, **A273**, 253.
39. Morpurgo, G. (1958). *Phys. Rev.*, **110**, 721.
40. Lawson, R. D. (1980). *Theory of the nuclear shell model*. Clarendon, Oxford.
41. Ajzenberg-Selove, F. (1988). *Nucl. Phys.*, **A490**, 1; Ajzenberg-Selove, F. (1985). *Nucl. Phys.*, **A433**, 1; Ajzenberg-Selove, F. (1986). *Nucl. Phys.*, **A449**, 1; Ajzenberg-Selove, F. (1986). *Nucl. Phys.*, **A460**, 1; Ajzenberg-Selove, F. (1987). *Nucl. Phys.*, **A475**, 1.
42. Endt, P. M. and van der Leun, C. (1978). *Nucl. Phys.*, **A310**, 1; Endt, P. M. (1979). *At. Data Nucl. Data Tables*, **23**, 3.
43. Konopinski, E. J. (1966). *The theory of beta radioactivity*. Clarendon, Oxford; Schopper, H. (1966). *Weak interactions and nuclear beta decay*. North-Holland, Amsterdam.
44. Hardy, J. C., Towner, I. S., Koslowsky, V. T., Hagberg, E., and Schmeing, H. (1990). *Nucl. Phys.* (To be published.); Towner, I. S., Hardy, J. C., and Harvey, M. (1977). *Nucl. Phys.*, **A284**, 269.
45. Sirlin, A. (1987). *Phys. Rev.*, **D35**, 3423; Sirlin, A. and Zucchini, R. (1986). *Phys. Rev. Lett.*, **57**, 1994; Jaus, W. and Rasche, G. (1987). *Phys. Rev.* **D35**, 3420.
46. Bopp, P., et al. (1986). *Phys. Rev. Lett.*, **56**, 919.
47. Hanna, S. S. (1969). In *Isospin in nuclear physics* (ed. D. H. Wilkinson), p. 591. North-Holland, Amsterdam.
48. Buck, B. and Perez, S. M. (1983). *Phys. Rev. Lett.*, **50**, 1975.
49. de-Shalit, A. (1953). *Phys. Rev.* **90**, 83.
50. Perez, S. M. (1987). *Phys. Rev.*, **C36**, 1202.
51. Wilkinson, D. H. (1973). *Phys. Rev.*, **C7**, 930; Wilkinson, D. H. (1973). *Nucl. Phys.*, **A209**, 470; Wilkinson, D. H. (1974). *Nucl. Phys.*, **A225**, 365.

2
MEASUREMENT OF NUCLEAR MOMENTS

Methods of measuring nuclear moments in most cases depend on the interaction of the nuclear moment with either an applied external magnetic field or with internal atomic or molecular fields. As a consequence the experimental investigation determines a product of the moment and the magnetic field strength at the nucleus. Thus the extraction of one part, say the nuclear moment, requires detailed knowledge of the counterpart in the product, the field strength.

When the field at the nucleus is an externally applied field then there is little uncertainty in knowing its strength and the moment is determined with good accuracy. The nuclear magnetic resonance technique, to be discussed in Section 2.2, is an example of such a case. However, only in the case of a magnetic dipole interaction are external fields of the required strength available to allow direct determination of nuclear moments. Furthermore the classical methods such as nuclear magnetic resonance are restricted to long-lived nuclear states of at least several minutes. For this reason we will not say much on techniques based on nuclear magnetic resonance. The reader is referred to standard texts.[1-3] Most of the development in the past 30 years has concentrated instead on measuring moments in short-lived excited states or β-unstable ground states by observing a perturbation caused by the externally applied magnetic field on the angular distribution of the de-excitation γ-rays or β particles. These techniques are discussed in Sections 2.4 to 2.7. But even these methods are limited to states whose lifetimes are longer than nanoseconds. For short-lived states of picosecond lifetimes a new method based on the very large internal magnetic field acting on a nucleus recoiling in a ferromagnetic material has been developed. These transient field techniques are outlined in Section 2.8. Finally, when only a few atoms of the species under study are available laser spectroscopy methods, introduced in Section 2.9, prove to be both powerful and versatile.

In this chapter we will only sketch the ideas; no attempt will be made to provide detailed experimental information. References will direct the reader to source material.

2.1 Hyperfine structure in atomic spectra

A nucleus with spin greater than or equal to $\frac{1}{2}$ has a magnetic dipole and possibly higher multipole moments. Thus it is expected there will be an interaction between the nuclear moment and the magnetic field associated with the spin and orbital motion of the extra-nuclear electrons. This is the hyperfine interaction. It is convenient to express this interaction Hamiltonian in terms of a multipole expansion:

$$H_{hf} = H(E0) + H(M1) + H(E2) + \ldots, \quad (2.1)$$

where $H(E0)$ denotes the electric monopole, $H(M1)$ the magnetic dipole, $H(E2)$ the electric quadrupole term, and so on. Note there are no E1, M2, ... terms. Symmetry arguments (principally time-reversal invariance) requires nuclear moments of these multipolarities to vanish. Each term in eqn (2.1) is a product of a nuclear moment and the corresponding term of the electromagnetic field. Thus, the investigation of this hyperfine interaction yields information on both the nuclear moments and the solid-state surroundings.

Consider first the magnetic dipole interaction. Denote the nuclear spin operator by I, the total angular momentum of the electrons, J, then an operator $F = I + J$ can be constructed for the total angular momentum of the nuclear–electronic system. In the absence of any hyperfine interaction between the nucleus and the electrons both I and J are constants of motion. The magnetic dipole hyperfine interaction is taken to have the form

$$H(M1) = AI \cdot J, \quad (2.2)$$

which is well known to have eigenvalues

$$E_F = \tfrac{1}{2} A[F(F+1) - I(I+1) - J(J+1)], \quad (2.3)$$

where A depends on the nuclear magnetic moment, μ, and the magnetic field at the nucleus due to the surrounding environment, $B(0)$:

$$A = \mu B(0)/IJ = g\mu_N B(0)/J. \quad (2.4)$$

Here g is the g-factor for the nuclear state, $g\mu_N = \mu/I$, and μ_N the nuclear magneton unit. This means that a given electronic level with total angular momentum J will be split into $2I + 1$ (if $I < J$) or $2J + 1$ (if $J < I$) hyperfine structure component and that the splitting between these components is dependent on the sign and magnitude of the nuclear magnetic moment. In a hydrogen-like atom, the magnetic field at a nucleus with atomic number Z produced by a single electron with quantum numbers nlj can be calculated,[2] and in a non-relativistic approximation is

$$B(0) = 12.5Z^3/[n^3(l+1/2)(j+1)] \text{ T} \quad (2.5)$$

and is typically of order of tens of tesla. Note 1 T is 10^4 G.

If in addition to the electronic magnetic field, there is an applied external magnetic field B_0 then the magnetic interaction of a free

2.1 HYPERFINE STRUCTURE IN ATOMIC SPECTRA

atom–nucleus system becomes

$$H(M1) = A\bm{I} \cdot \bm{J} + g_J \mu_B \bm{B}_0 \cdot \bm{J} - g\mu_N \bm{B}_0 \cdot \bm{I} \tag{2.6}$$

where g_J is the g-factor for electrons and μ_B the atomic Bohr magneton unit. In the limit of very strong magnetic fields, m_J and m_I, the magnetic substate quantum numbers, tend to become good quantum numbers and the eigenvalues of eqn (2.6) are

$$E(m_J, m_I) = A m_I m_J + g_J \mu_B B_0 m_J - g\mu_N B_0 m_I. \tag{2.7}$$

In the common case of $|g_J \mu_B J| \gg |g\mu_N I|$, the eigenstates of electronic spin J separate into $(2J + 1)$ components, each component further separating into $(2I + 1)$ components. The energy spacing between two adjacent sublevels ($\Delta m_J = 0$, $\Delta m_I = 1$) is

$$\Delta E = A m_J - g\mu_N B_0 \equiv g\mu_N B, \tag{2.8}$$

and magnetic dipole transitions can take place between them. Note the equal spacing between levels. Here B is a combination of $B(0)$ and B_0.

Next consider the electric quadrupole interaction. The electric interaction between a nuclear charge distribution and an electrostatic potential V produced by external charges can be decomposed with a multipole expansion. The quadrupole terms will depend on the product of the nuclear quadrupole moment, Q, and the second derivative of the electrostatic potential, V, and is known as the electronic field gradient, q. Measurements of quadrupole energy splittings essentially determine the product e^2Qq. Macroscopic electric field gradients that can be produced in the laboratory are too weak by orders of magnitude to cause measurable quadrupole interactions in nuclei. Sufficiently large field gradients may, however, be experienced by probe nuclei substituted in solids with regular lattice structure. In regular crystals principal axes can be chosen so that all mixed derivatives of the potential vanish, e.g. $V_{xy} = \partial^2 V / \partial x\, \partial y = 0$. Furthermore with the Laplace equation $V_{xx} + V_{yy} + V_{zz} = 0$, the quadrupole interaction depends on only two quantities, V_{zz} and $\eta = (V_{xx} - V_{yy})/V_{zz}$. The quadrupole interaction is written:[2]

$$H(E2) = e^2 q Q [3 I_z^2 - I(I+1) + (\eta/2)(I_+^2 + I_-^2)]/[4I(2I-1)], \tag{2.9}$$

with $V_{zz} = eq$ and I_x, I_y, I_z the components of the nuclear spin along the principal axes of the crystal. In addition $I_+ = I_x + iI_y$ and $I_- = I_x - iI_y$. For non-axially symmetric electric field gradients ($\eta \neq 0$), the Hamiltonian must be diagonalized numerically. In the axially symmetric case ($\eta = 0$) the Hamiltonian is diagonal in the I, m_I representation with eigenvalues

$$E(m_I) = e^2 q Q [3 m_I^2 - I(I+1)]/[4I(2I-1)]. \tag{2.10}$$

It is seen that, in the absence of other perturbations, magnetic substates with $\pm m_I$ are degenerate and that the energy splittings depend on the value of the nuclear spin, I. For adjacent levels ($\Delta m_I = 1$) the energy splitting is

$$\Delta E = 3e^2qQ(2m_I + 1)/[4I(2I - 1)]$$
$$\equiv 3(2m_I + 1)\hbar\omega_Q \tag{2.11}$$

and depends linearly on m_I. In contrast to magnetic hyperfine splitting, the separation between the levels is not equidistant.

2.2 Nuclear magnetic resonance

One of the earliest methods of measuring magnetic moments is the technique of nuclear magnetic resonance initiated in 1946 by Purcell, Torrey and Pound[4] and pursued extensively since. In this method, the nuclei under investigation are subjected to a magnetic field of a few tesla. If the nuclei are in statistical equilibrium at a temperature T then the population of the different nuclear magnetic substates, m_I, will follow a Boltzmann distribution with probabilities proportional to $\exp(g\mu_N m_I B/kT)$, where B is the magnetic field strength at the nucleus and k the Boltzmann constant. Note that at high temperatures all magnetic states are more or less equally populated, while the lower the temperature the more uneven becomes the population among the substates.

The energy separation between adjacent magnetic substates, eqn (2.8), is $g\mu_N B$ and magnetic dipole transitions can take place between them. These transitions can be induced by subjecting the nuclei to an oscillating electromagnetic field in the plane normal to the steady field. A maximum effect is obtained when the frequency of the oscillating field exactly matches the energy separation: $\hbar\omega_L = g\mu_N B$. This is the resonance condition and ω_L is called the Larmor precessional frequency. The electromagnetic field can produce either stimulated emission or absorption with equal probability. However, since the nuclei are distributed among the various magnetic substates according to Boltzmann's law, the lower levels are slightly more populated than the upper ones and a small net absorption of electromagnetic power is to be expected. Either the frequency of the applied electromagnetic field or the strength of the static magnetic field can be varied. Thus for example for a fixed frequency ω_L, measurement of the static field strength B for which maximum power is absorbed enables the nuclear g-factor to be determined from the Larmor frequency condition.

As mentioned, the classical methods such as this are restricted to stable or long-lived nuclear states. For the rest of this chapter we will concentrate on the techniques to measure moments in excited states and short-lived isomers.

2.3 Nuclear orientation

Just about all methods of using hyperfine fields to measure nuclear moments require that the state of interest be produced with an unequal population of its magnetic substates and that this anisotropy be detected by some measurement on its subsequent radioactive decay. Thus in this section we will say a few words about methods of producing anisotropy and methods of detection.

First a few definitions. Let $W(m_I)$ be the probability that a certain magnetic substate is populated. The sum of probabilities is normalized: $\sum W(m_I) = 1$. If each magnetic substate is equally populated, i.e. $W(m_I) = (2I+1)^{-1}$ independent on m_I, the nucleus is said to be unoriented. If $W(m_I)$ depends on m_I and the substate populations are unequal the nucleus is oriented. In particular, if $W(+m_I) \neq W(-m_I)$, the nucleus is said to be polarized, while if $W(+m_I) = W(-m_I)$ but is still different for different m_I, e.g. $W(m_I)$ is proportional to m_I^2, the nucleus is said to be aligned. For example, for a state of spin $I = 1$, the polarization of the state is defined as $W(+1) - W(-1)$, and the alignment by $W(+1) + W(-1) - 2W(0)$. In general polarization is defined as $P = I^{-1} \sum m_I W(m_I)$ which is simply the excess of spins pointing in one direction.

There are three common ways to produce nuclear orientation:

1. *Low temperature methods.* This is the method used in the nuclear magnetic resonance technique, Section 2.2. In a strong magnetic field the nuclear magnetic substates are split and the population of the states follows the Boltzmann distribution law. For a measurable effect the splitting of the states should be of order kT, namely $g\mu_N B = kT$. For a state with $g\mu_N$ equal to one nuclear magneton, the condition is that $B/T = 2.8 \times 10^3$ T/K. The largest practical external fields are about 5 T, so temperatures as low as 0.002 K are required. Special tricks can raise this temperature some, but by and large very low temperatures are required with external fields.

Alternatively use can be made of the very large internal magnetic fields at the nucleus due to the surrounding electrons. A modest external field will orient the electrons and because of the hyperfine coupling the nuclear spins will likewise be oriented. Typically internal fields are of the order of 10–100 T.

2. *Optical pumping.* Another method again uses the fact that if the electronic spins can be oriented the hyperfine coupling will cause the nuclear spin to be oriented too. Consider free atoms (not bound in a solid); they can be oriented by absorption of resonance radiation. This leads to a change in the population of levels in the atom as can be demonstrated in the following example. Consider an atom with a ground

state spin $F (= I + J) = \frac{1}{2}$ and an excited state with $F = \frac{3}{2}$. If the ground state is illuminated with right circularly polarized light of the appropriate frequency the excited state is populated in transitions in which the change in m_F is +1, which is the property of the circularly polarized light. Thus the induced transitions are

$$(F = \tfrac{1}{2}, m_F = \tfrac{1}{2}) \to (F = \tfrac{3}{2}, m_F = \tfrac{3}{2}),$$
$$(F = \tfrac{1}{2}, m_F = -\tfrac{1}{2}) \to (F = \tfrac{3}{2}, m_F = \tfrac{1}{2}),$$

and only the $+\frac{3}{2}$ and $+\frac{1}{2}$ magnetic substates are populated in the upper level. These states decay back to the ground state by dipole transitions but there is no restriction on the magnetic state population other than $\Delta m_F = \pm 1, 0$. Thus the decaying transitions are

$$(F = \tfrac{3}{2}, m_F = \tfrac{3}{2}) \to (F = \tfrac{1}{2}, m_F = \tfrac{1}{2}),$$
$$(F = \tfrac{3}{2}, m_F = \tfrac{1}{2}) \to (F = \tfrac{1}{2}, m_F = \tfrac{1}{2})$$
$$\searrow (F = \tfrac{1}{2}, m_F = -\tfrac{1}{2}),$$

and we see that by roughly a factor of two to one the $m_F = \frac{1}{2}$ substate is preferred to the $m_F = -\frac{1}{2}$ substate. So by this optical pumping the population of the $+\frac{1}{2}$ substate in the ground state is increased with respect to the $-\frac{1}{2}$ substate. This can go on at room temperature. The nuclear spin will follow the orientation of the electron spins and a nuclear orientation is produced. This is the principle used in laser spectroscopy.

3. *Nuclear reactions.* In a reaction such as

$$a + X \to Y + b$$

where a projectile nucleus, a, impinges on a target nucleus, X, the dynamics of the reaction are quite complicated. However, the reaction process is governed by the strong interactions and these are known to be spin dependent. Thus even though the projectile and target are unoriented nuclei, the outgoing particle, b, and residual nucleus, Y, will be polarized to some extent. It is not easy to calculate the degree of polarization, but for the purpose of measuring nuclear moments it usually does not matter. It is sufficient that the residual nucleus have some polarization for the method to be applicable.

When the prepared nucleus is in an excited state one way to detect whether nuclear orientation has taken place is to observe the emission of γ radiation as a function of angle. The probability of γ-ray emission in general depends on the angle θ between the expectation value $\langle I \rangle$ of the angular momentum vector I of the radiating system and the direction k of the observed photon. For an unoriented ensemble of nuclei their angular momentum vectors are oriented at random, i.e. $\langle I \rangle = 0$, and the emission of radiation is isotropic in space. Accordingly the observation of an anisotropic angular distribution is evidence for nuclear orientation of the decaying state.

If the polarization of the photon is not observed, then the directional angular distribution of the photon is given by [5]

$$W(\theta) = \sum_{\lambda \text{ even}} B_\lambda(I) A_\lambda(\gamma) P_\lambda(\cos \theta), \quad (2.12)$$

where $B_\lambda(I)$ are known as the orientation coefficients and are given in terms of the population probabilities $W(m)$. The first few coefficients are

$$B_0(I) = \sum_m W(m) = 1,$$

$$B_1(I) = -\sum_m mW(m)[\tfrac{1}{3}I(I+1)]^{-1/2}, \quad (2.13)$$

$$B_2(I) = 3\sum_m [m^2 W(m) - \tfrac{1}{3}I(I+1)][\tfrac{1}{5}I(2I-1)(I+1)(2I+3)]^{-1/2},$$

and in general in terms of Clebsch–Gordan coefficients

$$B_\lambda(I) = \sum_m (2I+1)\langle I-mIm \mid 00\rangle\langle I-mIm \mid \lambda 0\rangle W(m).$$

The coefficients $A_\lambda(\gamma)$ are not written out in detail here but depend on the matrix element of the γ-ray transition operator connecting the initial and final nuclear state, the multipolarity of the γ ray, L, and certain angular momentum recoupling coefficients. Note that only even powers of λ occur in the sum. This is because the polarization of the photon has not been observed. As a consequence the angular distribution is not sensitive to $B_1(I)$, which is a measure of the polarization, but is sensitive to $B_2(I)$ which is a measure of the alignment. The sum in eqn (2.12) terminates at $\lambda = 2L$. Since only even powers of $\cos \theta$ appear, the distribution $W(\theta)$ is symmetric about 90°. This is a key feature as will be seen as we discuss a few of the various techniques.

2.4 Perturbed angular distribution techniques

Consider a case where an isomeric nuclear state is populated by a nuclear reaction. The projectile beam from an accelerator is a pulsed beam and the time of each pulse sets the $t = 0$ scale. The nuclear reaction excites the isomeric levels and simultaneously orients the nuclear spins. The degree of orientation depends on the details of the reaction process. Assuming that the particle beam is unpolarized, an alignment is generally produced where the beam direction is selected as the quantization axis. The angular distributions of the γ radiations emitted from these aligned excited states are anisotropic for a nuclear spin $I \geq 1$. The details of this angular distribution, eqn (2.12), depend on the nuclear alignment, the spins of the nuclear levels involved and the multipolarity of the γ radiation.

Now in addition suppose there is an applied external magnetic field at some angle to the beam direction. This external field will alter the m-state population from that produced in the nuclear reaction. This re-population will reflect in a change in the angular distribution of the γ radiation. In a semiclassical picture the nuclear moments and hence the angular distribution precess around the magnetic field axis with the Larmor frequency $\hbar\omega_L = g\mu_N B$. Thus the number of γ rays detected at some angle θ with respect to the beam axis will fluctuate with time according to

$$I(t, \theta, B) = I_0 \exp(-t/\tau) W(t, \theta, B), \qquad (2.14)$$

where I_0 is the intensity at time $t=0$, τ is the mean life characterizing the exponential decay of the excited nuclei, and $W(t, \theta, B)$ the oscillating intensity modulation due to the rotating angular distribution:

$$W(t, \theta, B) = \sum_{\lambda, \text{even}} B_\lambda(I) A_\lambda(\gamma) P_\lambda[\cos(\theta - \omega_L t)]. \qquad (2.15)$$

Here $B_\lambda(I)$ are the orientation parameters eqn (2.13), depending on the degree of alignment produced by the nuclear reaction, and $A_\lambda(\gamma)$ are parameters depending on the details of the γ decay. In most cases, the coefficients $B_\lambda(I)$ are negligible for $\lambda \geq 4$, so the angular distribution is characterized by the parameters $B_2(I)$ and $A_2(\gamma)$. The nuclear g-factor can be extracted conveniently from the intensity ratio of two measurements at angles θ and $\theta + \pi/2$:

$$\begin{aligned} R(t, \theta, B) &= \frac{I(t, \theta, B) - I(t, \theta + \pi/2, B)}{I(t, \theta, B) + I(t, \theta + \pi/2, B)} \\ &= \frac{3A_{22}}{4 + A_{22}} \cos[2(\theta - \omega_L t)], \end{aligned} \qquad (2.16)$$

where A_{22} stands for the product $B_2(I)$ and $A_2(\gamma)$. Fitting the measured intensity ratio as a function of the delay time, t, to an oscillatory pattern will yield the Larmor frequency ω_L. Assuming the magnetic field acting at the nucleus is known, the corresponding g-factor is determined. The first experiments along these lines were performed by Hrynkiewicz,[6] Matthias and Lundquist,[7] and Freeman.[8]

The application of the method requires the mean lifetime of the excited state be short compared with the pulse repetition time T_0 of the particle accelerator, otherwise the incoherent superposition of successive decay intensities will smear out the pattern. Furthermore the width of the beam pulses, ΔT_0, must be much smaller than the Larmor period, $\Delta T_0 \ll 1/\omega_L$, to ensure the anisotropy is fully observed. This leads to the conditions $\Delta T_0 \ll 1/\omega_L < \tau \ll T_0$ and restricts the range of application to nuclear

2.4 PERTURBED ANGULAR DISTRIBUTION TECHNIQUES

levels with mean lifetimes in the range $10^{-9}\,\text{s} < \tau < 10^{-5}\,\text{s}$. Within this range the perturbed angular distribution method has been applied extremely successfully in the determination of moments of excited nuclear states.

In the case that the mean lifetime τ of the isomeric state is short compared with the Larmor frequency, $\tau < 1/\omega_L$, one may obtain an average precession angle from the time-integrated perturbed angular correlation. Rewriting eqn (2.15) as an expansion in $\cos(\theta - \omega_L t)$,

$$W(t, \theta, B) = \sum_{\lambda \text{ even}} b_\lambda \cos[\lambda(\theta - \omega_L t)], \qquad (2.17)$$

and integrating the intensity over the individual lifetimes yields

$$I(\theta, B) = \frac{I_0}{\tau} \int_0^\infty \exp(-t/\tau)[1 + b_2 \cos 2(\theta - \omega_L t) + \ldots]\, dt$$

$$= I_0\left(1 + \frac{b_2}{[1 + (2\omega_L\tau)^2]^{1/2}} \cos 2(\theta - \Delta\theta_2)\right.$$

$$\left. + \frac{b_4}{[1 + (4\omega_L\tau)^2]^{1/2}} \cos 4(\theta - \Delta\theta_4) + \ldots\right), \qquad (2.18)$$

with $\Delta\theta_N$ being given by

$$\tan(N\Delta\theta_N) = N\omega_L\tau.$$

This angle $\Delta\theta$ is known as the precession angle. For $\omega_L\tau \ll 1$ one obtains $\Delta\theta = \omega_L\tau$ and the yield takes the form

$$I(\theta, B) = I_0[1 + b_2 \cos 2(\theta - \omega_L\tau) + \ldots], \qquad (2.19)$$

which is the unperturbed angular distribution rotated through an angle $\Delta\theta = \omega_L\tau$. For larger values of $\omega_L\tau$ this distribution is attenuated by factors $[1 + (2\omega_L\tau)^2]^{-1/2}$ etc. To extract the g-factor from a measurement of $I(\theta, B)$, the mean lifetime τ has to be known. Small precession angles $\Delta\theta$ are usually measured from the change in yield when the external field B_0 is reversed. It is convenient to define a yield ratio

$$\varepsilon = \frac{I(\theta, B) - I(\theta, -B)}{I(\theta, B) + I(\theta, -B)}. \qquad (2.20)$$

Normalization uncertainties can be avoided when pairs of detectors are mounted at angles symmetric to the beam direction. Then the double ratio

$$r = \left(\frac{I(\theta_1, B)/I(\theta_1, -B)}{I(\theta_2, B)/I(\theta_2 - B)}\right)^{1/2} \qquad (2.21)$$

is related to ε by $(r-1)/(r+1)$.

2.5 Electric quadrupole hyperfine interactions

An electric field gradient acting on a nucleus at a lattice position in a non-cubic lattice results in a splitting of the nuclear levels due to the electric quadrupole hyperfine interaction, eqn (2.11), between the nuclear quadrupole moment, Q, and the field gradient, q. This splitting can be observed in a perturbed angular distribution apparatus, but the required field strength of the electric field gradient cannot be applied externally. Therefore one must use internal field gradients in non-cubic crystalline solids and rely for the extraction of Q on theoretical calculations of the gradients.

The influence of this electric quadrupole hyperfine interaction on the angular distribution of the γ radiation is more complex than in the case of a magnetic interaction. There is still a precession of the spin ensemble around the field axis. In the case of a unique field gradient with axial symmetry, the angular distribution of γ rays is given by

$$W(t, \theta) = \sum_{\lambda \text{ even}} B_\lambda(I) A_\lambda(\gamma) G_{\lambda\lambda}(t) P_\lambda(\cos \theta) \qquad (2.22)$$

where $B_\lambda(I)$ and $A_\lambda(\gamma)$, as before, specify the nuclear orientation and the characteristics of the γ ray. Information on the hyperfine interaction is included in the perturbation factor, $G_{\lambda\lambda}(t)$.[9] In simple cases, this factor can be expanded in a finite Fourier series

$$G_{\lambda\lambda}(t) = \sum_n s_{\lambda n} \cos(n\omega_0 t), \qquad (2.23)$$

where the amplitudes $s_{\lambda n}$ depending on the nuclear spin and radiation parameter, λ, are tabulated,[9] and are only combinations of angular momentum vector coupling coefficients. The basic frequency ω_0 is related to the quadrupole frequency ω_Q, eqn (2.11), by $\omega_0 = 6\omega_Q$ for I half integer, and $\omega_0 = 3\omega_Q$ for I integer.

The angular distribution of γ rays will then have a sinusoidal behaviour with t and a measurement of a ratio of rates, such as eqn (2.16), will yield a value for ω_Q. With a knowledge of the nuclear spin and a calculation of the electric field gradient, q, the quadrupole moment can be determined. A first measurement of this kind was performed by Sugimoto et al.[10] Reviews of electric quadrupole interactions in non-cubic metals has been given by Witthuhn and Engel,[11] and by Kaufmann and Vianden.[12] Note only the magnitude of $|e^2qQ|$ is determined. To obtain the sign the initial state must be polarized, rather than just aligned.

2.6 Perturbed angular correlation techniques

In the preceding two sections a nuclear reaction is used to produce the nuclear alignment needed for the observation of hyperfine interactions. A

2.6 PERTURBED ANGULAR CORRELATION TECHNIQUES

second possiblity is to use the γ ray from the decay of a higher-lying state to prepare the alignment in the state under investigation. That is, the state of interest is an intermediate state in a γ–γ cascade. Then the subsequent radiation emitted by the nucleus shows a characteristic angular distribution with respect to the first radiation. The populating γ transition selects a substate population with a certain alignment while the angular distribution of the second γ ray then reflects this substate population.

The angular correlation function is given by an expression similar to eqn (2.15) for the perturbed angular distribution experiments and is

$$W(t, \theta, B) = \sum_{\lambda \text{ even}} A_\lambda(\gamma_1)A_\lambda(\gamma_2)P_\lambda[\cos(\theta - \omega_L t)], \quad (2.24)$$

where θ is the angle between the directions of the two γ rays. Furthermore it is understood that the time of emission of the first γ-ray signals the formation of the oriented state and defines the origin, $t = 0$, of the time scale. The coefficient $A_\lambda(\gamma_1)$ depends on the initial and intermediate nuclear spins and the radiation parameters of the preceding γ ray, γ_1, while the coefficients $A_\lambda(\gamma_2)$ are given by the corresponding characteristics of the decay radiation, γ_2. The hyperfine interaction between the nuclear moments of the intermediate state and the electromagnetic interaction fields again results in a characteristic perturbation of the angular correlation. The intensity ratio, R, eqn (2.16), is constructed from the experimental data and the Larmor precession frequency, ω_L, determined.

In both angular correlation and angular distribution techniques it is required to know the magnetic field at the nucleus. Only the externally applied magnetic field, B_0, in general is known. Thus the g-factors deduced from the measured Larmor frequency have to be corrected for internal magnetic fields, $B(0)$, namely

$$g_{\text{corrected}} = g(1 + B(0)/B_0)^{-1}. \quad (2.25)$$

The most common internal field is that resulting from the diamagnetic current density induced by the external field in the atomic electron cloud. This is calculable and correction factors have been tabulated.[13] Much larger shifts may arise in paramagnetic atoms having unfilled inner electronic shells as found, for example, in the transition elements, rare earths, and actinides. Again the effect is calculable and paramagnetic enhancement factors have been given by Günther and Lindgren[14] for the rare earths and by Kalish et al.[15] for the actinides.

While on the subject of corrections, in nuclear magnetic resonance experiments one finds a large shift in the resonance frequency when comparing insulators and metals. The shift in metals is generally known as the Knight shift,[16] and arises from the polarization of conduction-electron spins at the nucleus. The Knight shift is always positive, and may amount to several per cent in heavy nuclei. It cannot be calculated reliably and is best determined experimentally. An example can be found in the work of Beene et al.[17]

2.7 Reorientation in Coulomb excitation

A quite different way to measure quadrupole moments of excited states in stable nuclei is based on the reorientation process in Coulomb excitation. The topic is really an aside to our general development in this chapter since the method does not make use of hyperfine interactions.

Briefly, when a moving charged nuclear particle comes close to a nucleus it is deflected in the long-range Coulomb field arising from the nuclear electric charge. If the energy of the particle is somewhat less than that of the Coulomb barrier, the short-range nuclear forces do not interact significantly with the particle even at its distance of closest approach. Under these circumstances the particles may undergo elastic scattering, their angular distributions being given by the Rutherford formula. However, the relative motion of the two charged bodies produces time-varying electromagnetic fields at both particle and nucleus which may induce transitions to excited states in them. Thus some of the particles may undergo an inelastic scattering process, known as Coulomb excitation. The cross-section for the process can be accurately calculated in terms of the nuclear electromagnetic matrix elements because only the well known properties of the electromagnetic field are involved in this interaction. Reviews on Coulomb excitation can be found in Refs 18–20.

The reorientation process is a second-order process that interferes with the leading Coulomb excitation process. It is best described by way of an example. Suppose in an even–even nucleus the Coulomb excitation process excites the nucleus from its ground state to the 2^+ state. However, the electromagnetic field generated by the relative motion of projectile and nucleus will lead to a splitting of the magnetic substates in the 2^+ state and an electromagnetic transition could take place from one substate to another. This would be indistinguishable from the first-order process. Thus there is a correction to the Coulomb excitation probability which can be written[21]

$$P_{if}^{(2)} = P_{if}^{(1)}[1 + \rho(\theta, \xi)Q],$$

where $P_{if}^{(1)}$ is the first-order probability and $P_{if}^{(2)}$ the corrected probability depending on the quadrupole moment of the 2^+ state, Q. In addition, ρ is a calculable function depending on the charges and masses involved, the scattering angle θ, and a parameter ξ that depends on the relative velocity and distance of closest approach. Hence a measurement of this interference term gives information on the magnitude and sign of the quadrupole moment of the excited state. Furthermore to deduce Q from a measured $P_{if}^{(2)}$, the transition probability $B(E2; 0^+ \to 2^+)$ needs to be known since it is proportional to $P_{if}^{(1)}$. Thus at least two independent measurements are required to determine Q. The measurements have to be made with high accuracy since the reorientation effect is usually small. For example, suppose ρQ represents a 15% correction to P_{if}, then $P_{if}^{(1)}$ and $P_{if}^{(2)}$ have to be determined to 1% accuracy to get a 10% accuracy in Q. Such precision is not easy to achieve. The principal uncertainty in the analysis arises from the fact that the reorientation term is only one of many similar interference terms. An estimate of other possible two-step routes, such as Coulomb excitation to the second 2^+ state followed by γ-emission to the first 2^+ state, may be required. Despite these caveats many quadrupole moments have been measured by the reorientation process; including the 3^- state in ^{208}Pb the physics of which we discuss in Chapter 7.

2.8 Nuclear magnetic resonance on β-emitting nuclei

The perturbed angular distribution and correlation experiments depend on the alignment in the state under investigation because the polarization of the resulting γ ray is not detected and its angular distribution given in terms of even powers of $\cos \theta$, eqn (2.12). Thus only magnetic moments of states of spin $I \geq 1$ can be determined in this way; spin $I = \frac{1}{2}$ nuclei are not amenable to these techniques. If, on the other hand, the decaying radiation were β rays then because of parity non-conservation in the weak interaction, the resulting angular distribution would contain even and odd powers of $\cos \theta$ and the experiment would be sensitive to both polarization and alignment in the prepared nucleus. Furthermore, in γ-ray techniques the isomeric lifetimes range from 10^{-9} to 10^{-3} s, while lifetimes of β-emitting nuclei are considerably longer with those suitable for these experiments being in the range 10^{-2} to 10^3 s.

Again a nuclear polarization has to be induced in the β-unstable nucleus under investigation. Nuclear reactions are mainly used. However, with unpolarized beams and the beam direction taken as the quantization axis only alignment is achieved in the prepared nucleus. To get polarization there are two techniques in use. In the first, the β-unstable nuclei are produced in a nuclear reaction using unpolarized particle beams, but the target foils are thin and the recoiling nuclei are not stopped in the target. Just a fraction of the recoiling nuclei occurring at a preselected angle in the reaction plane are retained and implanted into a host lattice and investigated there. The polarization of the recoiled nuclei is normal to the reaction plane and depends on beam energy and recoil angle. This technique was first used by Sugimoto et al.[22] The second possibility is the use of polarized particle beams to initiate the nuclear reaction. Here a selected recoil angle is unnecessary, but the degree of polarization is generally smaller. Polarized light-ion beams were first applied in these experiments by Minaminoso et al.[23]

As one example of these types of techniques we will briefly describe how the magnetic moment of ^{39}Ca was measured by Minaminoso et al.[24] The principle is as follows. The polarization of the β-emitting nucleus is measured from the angular distribution of the electrons. A small oscillating electromagnetic field is applied and at a certain critical frequency the induced hyperfine transitions will destroy the polarization. This critical frequency matches the hyperfine splitting and hence leads to a measure of the g-factor.

The β-emitting ^{39}Ca nuclei are polarized in the reaction ^{39}K($\vec{p}$, n)^{39}Ca initiated with polarized protons on thick polycrystalline targets such as KBr, which also served as host for the recoiling ^{39}Ca nuclei. A strong magnetic field B_0 is applied parallel to the beam polarization axis. This plays the role of holding the polarization in the recoiling nucleus and defining the quantization axis. The polarization of the ^{39}Ca nuclei is detected by means of two particle detectors placed at 0° and 180° to B_0. The incident polarized beam is pulsed and β particles only counted during the beam-off period. The angular distribution of the β-particles is

$$W(\theta) = 1 + P \cos \theta,$$

where P is the polarization and θ the angle between the polarization direction (parallel to B_0) and the β detector. The sign of P could also be reversed by reversing the beam polarization direction. The polarization is determined from the observed anisotropy in the detectors, namely

$$P = (r-1)/(r+1),$$
$$r = \left(\frac{W_u(0)W_d(\pi)}{W_u(\pi)W_d(0)}\right)^{1/2} \quad (2.26)$$

where the subscripts u and d refer to the beam polarization direction up or down.

A small r.f. magnetic field B_1 is applied perpendicular to the field B_0 and its frequency varied. The anisotropy in the β detectors is measured as a function of the frequency and at a certain critical value the polarization goes to zero.

Ideally the interaction of a strong magnetic field B_0 along the z-direction with the magnetic moment of the β-emitting nucleus is governed by the Hamiltonian $H = -g\mu_N B_0 I_z$, where g is the nuclear g-factor. The r.f. field of amplitude B_1 applied perpendicular to the static field, say in the x-direction, introduces a perturbing term in the Hamiltonian of $H_{\text{pert}} = -g\mu_N B_1 I_x \cos \omega t$. The operator I_x has matrix elements between states m_I and m_I' such that $m_I' = m_I \pm 1$. Resonance occurs when the frequency of the alternating magnetic field matches the energy spacing, $\hbar\omega = g\mu_N B_0$, and induces transitions. At saturation the magnetic-state populations have all become equalized, thus destroying any polarization in the recoil nucleus produced in the nuclear reaction. As all energy spacings are degenerate, there is a single resonance, the Zeeman resonance, in the measured anisotropy in the β detectors. Off-resonance there is no redistribution of magnetic-state populations. Knowing the applied magnetic field B_0 and measuring the resonance frequency ω leads to a determination of the g-factor. A correction for diamagnetism is applied.

2.9 Transient fields

The determination of nuclear g-factors of short-lived excited states by the methods mentioned so far is limited by the strength of the highest available external magnetic fields. The lower limit for the accessible mean life is about 0.1 ns. In order to study states of shorter lifetimes attempts have been made to use internal fields produced at the nucleus in ferromagnetic materials where field strengths of order 100 T have been observed. Unfortunately only a few metals are ferromagnetic. There are two problems that have to be overcome:

1. It is necessary to implant the isotope of study as an impurity atom into the environment of a ferromagnetic host lattice in a reproducible way. A check is required that a successful implantation has been achieved and the lattice position of the implanted isotopes are unique. These problems are discussed by Bodenstedt.[25]

2. The effective hyperfine field has to be calibrated at the nuclear site.

The nature of the static hyperfine field acting on impurities embedded in ferromagnetics is fairly well understood and has been used in the methods discussed in the previous sections to measure magnetic moments of nuclear states with lifetimes longer than nanoseconds. It was not, however, until the discovery that very much larger effective magnetic fields can be produced by the interaction between very fast ions and magnetic media that the measurements of nuclear magnetic moments of very short-lived nuclei took on a new prominence. The existence of these strong magnetic fields acting on moving ions inside a ferromagnetic was discovered by Borchers et al.[26] They are known as transient magnetic fields, and they only operate while the recoil ions are in motion.

A typical experimental set-up is as follows. A particle beam is allowed to impinge on a target deposited on a ferromagnetic foil. Ions excited to the nuclear level of interest recoil into the ferromagnet and either stop in it or emerge into a non-magnetic backing. The angular distribution of the de-excitation γ rays, $W(\theta)$, is measured, either in coincidence with scattered particles at a particular angle or with respect to a chosen quantization axis. A weak external magnetic field B_0 polarizes the ferromagnetic foil and establishes a direction along which the hyperfine field acts. The integrated angular distribution, eqn (2.19), yields a new angular correlation $W(\theta - \Delta\theta)$ shifted by the precessional angle, $\Delta\theta$. A measurement of the angle $\Delta\theta$ yields information on the magnetic interaction between the nuclear moment and the transient field.

The measured $\Delta\theta$ is compared with the expression

$$\Delta\theta = g\mu_N \int_0^T B[v(t)]\exp(-t/\tau)\,dt, \quad (2.27)$$

where g is the g-factor of the nuclear level, τ its mean lifetime, $B(v)$ the strength of the transient field at ion velocity v, and T the flight time of the ion through the ferromagnetic material. The duality of hyperfine studies is again illustrated; if g is known from other experiments one learns about the function $B(v)$. Conversely knowing the magnetic field $B(v)$ one can measure nuclear magnetic moments.

The transient field is generally assumed to be linear with velocity

$$B(v) = c(Z)v \quad (2.28)$$

where $c(Z)$ describes the atomic number dependence of the transient field. Until about 1975 the experiments were analysed by the theory of Lindhard and Winther[27] (LW) based on the consequences of Coulomb scattering of unbound, polarized electrons by the moving recoils. The LW field is proportional to $1/v$ (in contrast to eqn (2.28)) and becomes largest at the end of the recoil path. Precession angles, $\Delta\theta$, much larger than predicted by the LW theory were observed in experiments in 1975 with light high-velocity ions.[28] Since then, considerable experimental effort has been directed toward mapping out and exploiting the large transient fields at high recoil velocities. Detailed discussions of these developments can be found in articles by van Middelkoop[29] and by Benczer-Koller et al.[30] The gross dependence of the transient field on velocity and atomic number of the recoil is now roughly known and varies smoothly with v and Z. Many g-factors on states with sub-picosecond lifetimes have been measured. One example is the work of Häusser et al.[31] where the transient field was calibrated against a known g-factor and then used to measure the magnetic moments of the two excited states, 3/2+ and 5/2+, in ^{205}Tl and ^{203}Tl.

2.10 Laser spectroscopy

One of the major developments of the past decade has been the application of laser spectroscopy to the measurement of nuclear moments and nuclear radii. Because these techniques require only a few atoms of the nucleus under investigation, they are particularly important in the study of rare nuclei well removed (on a chart of the nuclides) from the band of stable nuclei.

With single-mode tuneable lasers it is possible to scan the hyperfine structure of atomic transitions by a very sharp, tuneable optical excitation. Because of the excellent energy resolution and high power of laser light, and because of the large cross-sections for atomic excitations, extremely low atomic densities and even single atoms may be detected and polarized. The basic polarization technique is that of optical pumping discussed in Section 2.3.

In optical pumping experiments, the atoms are placed in a magnetic field which is sufficiently weak, 10^{-3} T, that the Zeeman splitting is smaller than the natural linewidth of the optical transition. The atoms are then irradiated along the direction of the magnetic field (quantization axis) by circularly polarized laser light. If the circular polarization is left handed (σ^- light), then the selection rule $\Delta m_F = -1$ applies to the absorption, whereas the spontaneous radiative decay from the excited state occurs according to statistical probabilities with the selection rule $\Delta m_F = 0, \pm 1$. Consequently after the first pumping cycle there will be a greater population of hyperfine states with negative m_F values than with positive m_F values. If the process of absorption and re-emission can be repeated many times without loss of polarization the atoms will eventually be pumped into their lowest substate, $m_F = -I - J$, with complete polarization of both atomic and nuclear spins.

One of the first examples of these techniques was the measurement[31] of the spins, isotope shifts, and magnetic moments in a long chain of sodium isotopes ranging from ^{21}Na to ^{31}Na. Only ^{23}Na is stable and its magnetic moment well known to six-digit precision. Thus $\mu(^{23}$Na) provides the reference value for the experiment. The sodium isotopes are produced as reaction products in the interaction of high-energy protons with a heavy target of uranium oxide; the sodium atoms then diffuse through the target backing. Behind the target is a region of a weak magnetic field, B_0, of known field strength. The collimated beam of sodium atoms is intersected at right angles by the σ^- light beam from a tuneable dye laser. The laser frequency scans across the region of the sodium D_1 line. The atomic beam then passes into a region of a strong magnetic field that has been designed to focus atoms with $m_J > 0$, but defocus atoms with $m_J < 0$. A signal proportional to the number of focused atoms of the sodium isotope studied is obtained by counting them with an electron multiplier after they have been ionized and passed through a mass spectrometer.

The sodium D_1 line ($^2P_{1/2} \rightarrow {}^2S_{1/2}$) has four hyperfine components. As the frequency of the σ^- laser light matches one of these hyperfine transitions, optical pumping sets in, the number of $m_I < 0$ atoms increases, and the number of $m_I > 0$ detected decreases. From the resonance frequencies, the energy splittings of the hyperfine field are determined. From the centroids of the four hyperfine components, the strength of the hyperfine interaction, A, eqns (2.2)–(2.4), for the $^2S_{1/2}$ and $^2P_{1/2}$ atomic states are obtained. These values are proportional to $\mu B_0/I$. By taking ratios with the reference nucleus ^{23}Na, the dependence on the magnetic field B_0 cancels out. Thus the magnetic moment of the other sodium isotopes is given by

$$\frac{\mu}{I} = \frac{\mu(^{23}\text{Na})}{I(^{23}\text{Na})} \frac{A}{A(^{23}\text{Na})},$$

where I is the nuclear spin of the sodium isotope. The spin, I, can also be determined by a separate measurement[32] with the same apparatus using an atomic beam resonance experiment.

Since these first experiments the advances in laser spectroscopy have been spectacular, especially for the spectroscopy of nuclei far from stability. Two recent conferences[33,34] demonstrate the proliferation of novel production and detection techniques. Long isotope chains of alkali elements, rare-earth elements, and elements near closed proton shells have been studied. A review has been given by Otten.[35] As a sample of recent achievements[36] we show in Fig. 2.1 the measurement of quadrupole and magnetic moments of the odd-mass indium isotopes from mass 105 to 127 for the $I = 9/2^+$ ground states and for seven cases of the $I = 1/2^-$ isomeric states. Data such as these help a great deal in understanding the systematics in nuclear spectroscopy.

2.11 Summary

Most methods of measuring nuclear moments depend on the interaction of the nuclear moment with magnetic fields. Thus we give a brief summary of the physics of hyperfine fields to obtain the basic formulae for magnetic and electric interactions. When the magnetic field is externally applied there is little uncertainty in its strength and the magnetic moment is determined with good accuracy by such classic techniques as nuclear magnetic resonance. The technique, however, is limited to long-lived nuclear states. The chapter proceeds by outlining the newer developments of the last 30 years of obtaining moments of short-lived excited states. These include the perturbed angular correlation techniques in which the angular distribution of de-excitation γ rays or β particles are detected following a polarization of the decaying nuclear state. For states with lifetimes less than nanoseconds, externally applied magnetic fields of sufficient strength cannot be obtained in the laboratory, but are available on recoiling nuclei moving in ferromagnetic

2.12 REFERENCES

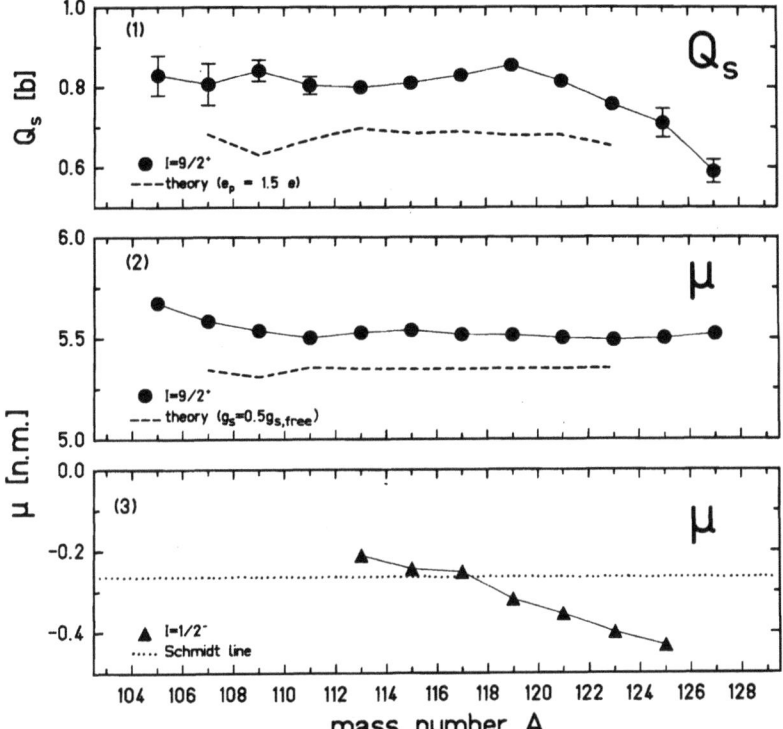

Fig. 2.1 Nuclear moments in the odd indium isotopes measured with laser spectroscopy techniques: (1) quadrupole moments of $I = 9/2^+$ ground states; (2) magnetic moments of $I = 9/2^+$ ground states; and (3) magnetic moments of $I = 1/2^-$ isomeric states. The broken lines are theoretical calculations based on core-particle coupling models (see Chapters 5 and 7) using effective coupling constants given in the figure. The Schmidt limit is shown as a dotted line. (From Eberz et al.[36])

hosts. These are the transient field methods. But the growth area of the past decade has been in the area of laser spectroscopy, where measurements can be made on just a few atoms of sample. This has led to measurements in rare nuclear species and in particular on long chains of isotopes. Systematic studies such as these help enormously in the understanding of nuclear spectroscopy as will be evident in the ensuing chapters.

2.12 References

1. Ramsey, N. F. (1955). *Molecular beams*. Oxford University Press.
2. Kopfermann, H. (1958). *Nuclear moments*. Academic Press, New York.
3. Abragam, A. (1961). *Nuclear magnetism*. Oxford University Press.
4. Purcell, E. M., Torrey, H., and Pound, R. V. (1946). *Phys. Rev.*, **69**, 37.
5. Steffen, R. M. and Alder, K. (1975). In *The electromagnetic interaction in*

nuclear spectroscopy (ed. W. D. Hamilton), pp. 583–644. North-Holland, Amsterdam.
6. Hrynkiewcz, A. Z. (1960). *Postepy Fiz.*, **11**, 521.
7. Matthias, E. and Lundquist, T. (1961). *Nucl. Instrum. Meth.*, **13**, 356.
8. Freeman, R. (1961). *Nucl. Phys.*, **25**, 446.
9. Frauenfelder, H. and Steffen, R. M. (1965). In *Alpha-, beta- and gamma-ray spectroscopy*, Vol. 2, (ed. K. Siegbahn). North-Holland, Amsterdam.
10. Sugimoto, K., Mizobuchi, A., and Nakai, K. (1964). *Phys. Rev.*, **B134**, 539.
11. Witthuhn, W. and Engel, W. (1983). In *Hyperfine interactions of radioactive nuclei* (ed. J. Christiansen), pp. 205–89. Springer, Heidelberg.
12. Kaufmann, E. N. and Vianden, R. J. (1979). *Rev. Mod. Phys.*, **51**, 161.
13. Fuller, G. H. (1976). *J. Phys. Chem. Ref. Data*, **5**, 835.
14. Günther, C. and Lindgren, I. (1964). In *Perturbed angular correlations* (ed. E. Karlsson, E. Matthias, and K. Siegbahn), pp. 357–75. North-Holland, Amsterdam.
15. Kalish, R., Shreter, U., and Grunzweig-Genossar, J. (1975). *Hyperfine Interact.*, **1**, 65.
16. Knight, W. D. (1947). *Phys. Rev.*, **76**, 1259; (1956). *Solid State Phys.*, **2**, 97.
17. Beene, J. R., *et al.* (1977). *Hyperfine Interact.*, **3**, 397.
18. Alder, K. and Winther, A. (1966). *Coulomb excitation*. Academic Press, New York.
19. Newton, J. O. (1975). In *The electromagnetic interaction in nuclear spectroscopy* (ed. W. D. Hamilton), pp. 237–282. North-Holland, Amsterdam.
20. de Boer, J. and Eichler, J. (1968). *Adv. Nucl. Phys.*, **1**, 1.
21. Smilansky, U. (1970). In *Nuclear reactions induced by heavy ions* (ed. R. Bock and W. R. Hering), p. 392. North-Holland, Amsterdam.
22. Sugimoto, K., Mizobuchi, A., Nakai, K., and Matuda, K., (1966). *J. Phys. Soc. Jpn.*, **21**, 213.
23. Minaminoso, T., *et al.* (1975). *Phys. Rev. Lett.*, **34**, 1465.
24. Minaminoso, T., *et al.* (1976). *Phys. Lett.*, **B61**, 155.
25. Bodenstedt, E. (1975). In *The electromagnetic interaction in nuclear spectroscopy* (ed. W. D. Hamilton), pp. 735–74. North-Holland, Amsterdam.
26. Borchers, R. R., *et al.* (1968). *Phys. Rev. Lett.*, **20**, 424.
27. Lindhard, J. and Winther, A. (1971). *Nucl. Phys.*, **A166**, 413.
28. Eberhardt, J. L., *et al.* (1975). *Phys. Lett.*, **B56**, 329; Goldberg, M. B., *et al.* (1976). *Hyperfine Interact.*, **1**, 429.
29. van Middelkoop, G. (1977). *Hyperfine Interact.*, **4**, 238.
30. Benczer-Koller, N., Hass, M., and Sak, J. (1980). *Ann. Rev. Nucl. Part. Sci.*, **30**, 53.
31. Häusser, O., Haas, B., Ward, D., and Andrews, H. R. (1979). *Nucl. Phys.*, **A314**, 161.
32. Huber, G., *et al.* (1978). *Phys. Rev.*, **C18**, 2342.
33. Bolotin, H. H. (ed.) (1987). *Proc. int. conf. on nuclear structure through static and dynamic moments*. Conference Proceedings Press, Melbourne.
34. Towner, I. S. (ed.) (1988). *Proc. 5th int. conf. on nuclei far from stability*. American Institute of Physics, New York.
35. Otten, E. W. (1990). In *Treatise on heavy ion physics*, Vol. 8, (ed. D. A. Bromley). To be published.
36. Eberz, J., *et al.* (1987). *Nucl. Phys.*, **A464**, 9.

3
CORE POLARIZATION

It has already been pointed out that the measured magnetic moments for odd-mass nuclei differ significantly from the extreme single-particle estimates. In terms of the Schmidt diagrams, Figs. 1.1 and 1.2, experimental magnetic moments lie for the most part between the Schmidt lines. In this chapter we begin our discussion of why this is so. There are two principal reasons. One, the extreme single-particle model is too simplistic; there will be configuration admixtures which can be small in magnitude but which nonetheless can have a significant impact on the calculated magnetic moments. These corrections represent an inadequacy of the wavefunction and are the subject of the present chapter. Second, the form of the magnetic moment operator may be incomplete. There are corrections to the operator arising from the fact that nucleons in the nucleus are interacting with each other through the exchange of mesons. These processes lead to additional terms to the magnetic moment operator that are of two-body character. These meson-exchange current (MEC) processes will be discussed in the next chapter.

Consider first the situation in heavy nuclei where the jj coupling scheme is the most appropriate, and consider an odd-mass nucleus of spin a, whose wavefunction can be written symbolically as

$$\psi_a = |a\rangle + \sum_k A_k |k, a\rangle \qquad (3.1)$$

where $|a\rangle$ represents a simple shell-model configuration (e.g. the extreme single-particle configuration) and the $|k, a\rangle$ represent admixed configurations, such configurations being characterized by the index k. The magnetic moment of the nucleus is then obtained by calculating the expectation value of the magnetic moment operator using this function. For small A_k, the most important contributions to the magnetic moment will be those linear in A_k, and the condition that contributions of this kind should occur is that $|a\rangle$ and $|k, a\rangle$ differ at most by one single-particle state and that this single-particle state be diagonal in the orbital angular momentum quantum numbers. We can think of $|a\rangle$ as a Slater determinant of single-particle wavefunctions for the A occupied orbitals. Then $|k, a\rangle$ is another Slater determinant of the same dimension differing at most from $|a\rangle$ by one row. The one differing row has an occupied single-particle state ϕ_h with angular momentum $h = l + \frac{1}{2}$ in the determinant $|a\rangle$ replaced by the state ϕ_p with $p = l - \frac{1}{2}$ in $|k, a\rangle$. The index k then differentiates between all the possible angular momentum recouplings available that lead to the same resultant angular momentum a and all the possible orbits with orbital angular momentum l in which the $h = l + \frac{1}{2}$ orbital is occupied (or partially occupied) and the $p = l - \frac{1}{2}$ orbital is unoccupied (or partially unoccupied). Since, in the extreme case of a nucleus being predominantly described by a closed-shell-plus-one configuration, the admixture involves exciting an orbital from the closed-shell core to an unoccupied orbital this mechanism has come to be known as core polarization.

It remains to specify how the amplitudes A_k are determined. Since the calculation only keeps terms linear in A_k it is sufficient to use first-order perturbation theory to obtain

$$A_k = \langle a| \mathcal{V} |k, a\rangle / (E_a - E_{k,a}), \qquad (3.2)$$

where $\mathcal{V}$ is the residual interaction and E the unperturbed energies of the configurations.

3.1 First-order core polarization

The simplest situation to discuss involves a single nucleon outside a closed-shell core. The Hamiltonian is divided into a one-body Hamiltonian and a residual interaction: $H = H_0 + \mathcal{V}$ where $H_0 = T + U$, the sum of kinetic energy and one-body potential energy operators, and $\mathcal{V} = V - U$ where V is the two-body potential energy operator. The eigenfunctions of H_0 form the basis of the calculation. The closed shell is defined as the full occupancy of the lowest-energy eigenstates of the one-body Hamiltonian H_0. The valence orbitals are the next few unoccupied eigenstates. The single nucleon occupies one of these valence orbitals. For example, the ground state of ^{209}Bi in the lowest approximation would be considered as a closed-shell ^{208}Pb with a single proton in the $h_{9/2}$ orbital. Likewise its magnetic moment in the lowest-order approximation is simply the bare magnetic moment operator evaluated in the $h_{9/2}$ single-particle state. Corrections to this come from a perturbative expansion in the residual interaction, $\mathcal{V}$. In general, then, to first-order in a closed-shell-plus-one-nucleus the matrix element of a one-body operator μ is given by

$$\langle \psi_b | \mu | \psi_a \rangle = \langle b | \mu | a \rangle + \sum_{\alpha \neq a,b} \frac{\langle b| \mu |\alpha\rangle\langle \alpha| \mathcal{V} |a\rangle}{E_i - E_\alpha} + \sum_{\alpha \neq a,b} \frac{\langle b| \mathcal{V} |\alpha\rangle\langle \alpha| \mu |a\rangle}{E_f - E_\alpha}. \qquad (3.3)$$

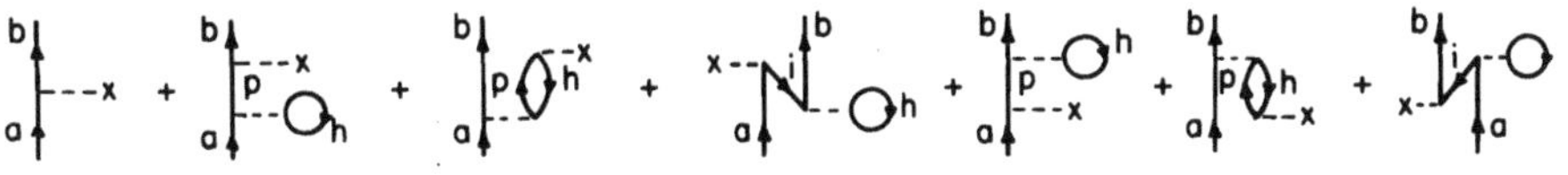

3.1 FIRST-ORDER CORE POLARIZATION

Fig. 3.1 Zeroth-order, diagram (a), and first-order, diagrams (b) to (g), perturbation corrections to the matrix element of a one-body operator in a closed-shell-plus-one nucleus.

Here a and b are single-particle valence states and α an infinite set of single-particle or two-particle one-hole (2p–1h) states constructed from the eigenfunctions of H_0. Likewise the energy denominators are constructed from the single-particle eigenenergies of H_0. Note that energy denominators are negative quantities. A graphical representation of all the possible terms is given in Fig. 3.1. Two comments are in order.

1. Consider graph (b) whose matrix element is given by the expression

$$\langle b| \text{Fig. 3.1(b)} |a\rangle = \sum_{\substack{p \neq b,a \\ h}} \frac{\langle b| \mu |p\rangle \langle ph| V - U |ah\rangle}{\varepsilon_a - \varepsilon_p}, \qquad (3.4)$$

where ε are the single-particle eigenenergies of H_0, namely $E_i = E_{\text{core}} + \varepsilon_a$ and, in this graph, $E_\alpha = E_{\text{core}} + \varepsilon_p$. This graph contains what is called a Hartree–Fock insertion. If the unperturbed Hamiltonian were to be chosen as the Hartree–Fock Hamiltonian that minimized the energy of a single Slater determinant characterizing the closed-shell core, then the potential U would be defined in the Hartree–Fock procedure by its matrix elements as

$$\langle b| U |a\rangle = \sum_h \langle bh| V |ah\rangle, \qquad (3.5)$$

where the sum over h is a sum over all the occupied states in the closed-shell core. Thus with H_0 being the Hartree–Fock Hamiltonian, graphs (b), (d), (e), and (f) are all identically zero. In practice, it is rare that calculations are done in a Hartree–Fock basis. It is more usual to use harmonic oscillator functions or in some cases Saxon–Woods functions. Then there is no reason to drop the Hartree–Fock insertion graphs except that one might argue that oscillator or Saxon–Woods functions are close to Hartree–Fock functions so the insertion graphs would be small. Explicit calculations by Ellis and Mavromatis[1] in light nuclei with oscillator functions fail to support this idea. Hartree–Fock insertion graphs give a non-negligible contribution.

Fortunately the magnetic moment operator is a special case. This operator being simply a combination of L and S, the orbital and spin angular momentum operators, is diagonal in orbital space. Thus in the single-particle matrix element, graph (a), the spatial part of the single-particle wavefunctions, $|a\rangle$ and $|b\rangle$, must be the same. In the Hartree–Fock insertion graph (b), the matrix element $\langle ph| V - U |ah\rangle$ requires that the particle state $|p\rangle$ has the same total angular momentum as $|a\rangle$, while the matrix element $\langle b|\boldsymbol{\mu}|p\rangle$ requires the spatial part of $|p\rangle$ to be the same as that of $|b\rangle$. In short, these restrictions limit $|p\rangle$ to being equal to $|a\rangle$, and this is expressly forbidden. Thus graph (b) is identically zero for the magnetic moment operator. This argument holds for all Hartree–Fock insertion graphs in all orders for the magnetic moment operator. Of course, if one looks at higher multipoles, or in magnetic electron scattering at finite momentum transfers, then these arguments break down and Hartree–Fock insertion graphs have to be considered.

2. The second comment concerns the infinite sum over intermediate states α. In principle these summations are unlimited, but in the first-order graphs shown in Fig. 3.1, the selection rules on the one-body operators restrict the number of possible intermediate states. (This limitation is not present in second order.) Again the magnetic moment operator provides a special case. Consider graph (c) given by the expression

$$\langle b| \text{Fig. 3.1(c)} |a\rangle = \sum_{p,h} \frac{\langle h|\boldsymbol{\mu}|p\rangle \langle bp| V |ah\rangle}{\varepsilon_a + \varepsilon_h - \varepsilon_p - \varepsilon_b}. \tag{3.6}$$

The restrictions require the spatial parts of $|p\rangle$ and $|h\rangle$ to be the same. This will limit orbitals p and h to being just spin–orbit partners and the number of terms in the summation to just one or two. Indeed in light nuclei where the shell closures occur at LS closed shells, spin–orbit partners are either both particles or both holes and no terms survive at all. Thus for the magnetic moment operator at closed LS-shells-plus-one-nuclei, such as ^{17}O and ^{41}Ca, all first-order corrections are identically zero. In heavy nuclei where the shell closures divide spin–orbit partners, such as ^{209}Bi, there are just two contributions to graph (c) coming from proton $h_{11/2}$ and $h_{9/2}$ and neutron $i_{13/2}$ and $i_{11/2}$ orbitals.

Thus we see that magnetic moments (and the Gamow–Teller operator of β-decay and the M1 operator of γ-decay in the long-wavelength approximation) play a rather important role in the tests of effective interaction and effective operator theory. It is convenient for calculation to rewrite eqn (3.6) using angular-momentum coupled states. Let λ be the tensorial rank of the one-body operator ($\lambda = 1$ for the magnetic moment operator) and let m_λ be the z-projection of λ, then

$$\langle h| \mu^{(\lambda)}_{m_\lambda} |p\rangle \equiv -\langle 0| \mu^{(\lambda)}_{m_\lambda} |h^{-1}p\rangle$$
$$= -(-)^{h+m_h+\lambda+m_\lambda} \hat{\lambda}^{-1} \langle h-m_h p m_p | \lambda - m_\lambda\rangle \langle 0\| \mu^{(\lambda)} \|(h^{-1}p)\lambda\rangle. \tag{3.7}$$

3.1 FIRST-ORDER CORE POLARIZATION

Our notation here is a little cryptic. On the left-hand side of the equation the letters h and p are standing for the angular momentum and z-projection quantum numbers of the single-particle states. On the right-hand side the magnetic quantum numbers are explicitly displayed in the Clebsch–Gordan coefficients and sign factors; the reduced matrix element has no dependence on magnetic quantum numbers. We also use the notation that $\hat{j} = (2j+1)^{1/2}$. Similarly the two-body matrix element is rewritten in particle–hole coupling as

$$\langle bp| V |ah\rangle = \langle h^{-1}p| V |b^{-1}a\rangle$$
$$= \sum_{IM} (-)^{j_h+m_h+j_b+m_b} \langle j_h - m_h j_p m_p | IM\rangle \langle j_b - m_b j_a m_a | IM\rangle$$
$$\times \langle (h^{-1}p)I| V |(b^{-1}a)I\rangle. \quad (3.8)$$

Inserting eqns (3.8) and (3.7) into (3.6) and summing over the magnetic quantum numbers m_h, m_p and M (orthogonality of the Clebsch–Gordan coefficients leads to the requirement $I = \lambda$) yields the following for the reduced matrix element:

$$\langle b\| \text{Fig. 3.1(c)} \|a\rangle$$
$$= -\sum_{h,p} \hat{b}^{-1} \frac{\langle 0\| \mu^{(\lambda)} \|(h^{-1}p)\lambda\rangle \langle (h^{-1}p)\lambda| V |(b^{-1}a)\lambda\rangle}{(\varepsilon_a + \varepsilon_h - \varepsilon_p - \varepsilon_b)}. \quad (3.9)$$

A similar calculation for the Hermitian diagram (f) in Fig. 3.1 yields

$$\langle b\| \text{Fig. 3.1(f)} \|a\rangle$$
$$= \sum_{h,p} (-)^{a+b-\lambda} \hat{\lambda} \hat{b}^{-1} \frac{\langle (a^{-1}b)\lambda| V |(h^{-1}p)\lambda\rangle \langle (h^{-1}p)\lambda\| \mu^{(\lambda)} \|0\rangle}{(\varepsilon_b + \varepsilon_h - \varepsilon_a - \varepsilon_p)}$$
$$= (-)^{a-b-k} \frac{\hat{a}}{\hat{b}} \langle a\| \text{Fig. 3.1(c)} \|b\rangle, \quad (3.10)$$

where k is determined from the Hermitian property of the operator $\mu^{(\lambda)}$:

$$(\mu_{m_\lambda}^{(\lambda)})^\dagger = (-)^{k-m_\lambda} \mu_{-m_\lambda}^{(\lambda)}$$

for the magnetic moment operator $k = 0$. The magnetic moment and M1 transition probabilities are defined for single particle matrix elements as

$$\mu = \left(\frac{a}{a+1}\right)^{1/2} \langle a\| g_L L + g_S S \|a\rangle$$
$$B(M1; b \to a) = \frac{3}{4\pi} |\langle b\| g_L L + g_S S \|a\rangle|^2 \quad (3.11)$$

where $g_L = 1$ and $g_S = 5.586$ for a proton, and $g_L = 0$ and $g_S = -3.826$ for a neutron using free-nucleon values. Recall we are using the definitions of Brink and Satchler[2] for the reduced matrix element with the initial state for transitions written on the left.

3.2 Zero-range residual interaction

It is useful to evaluate the first-order core-polarization correction using a zero-range form for the residual interaction. This lends itself to some algebraic simplifications and yields some insight into the calculation. We write the zero-range interaction in the following general form

$$V(r) = C_o(f + f'\boldsymbol{\tau}_1 \cdot \boldsymbol{\tau}_2 + g\boldsymbol{\sigma}_1 \cdot \boldsymbol{\sigma}_2 + g'\boldsymbol{\sigma}_1 \cdot \boldsymbol{\sigma}_2\boldsymbol{\tau}_1 \cdot \boldsymbol{\tau}_2)\delta(r_1 - r_2), \quad (3.12)$$

where C_o is the overall strength of the interaction; and f, f', g, and g', the coefficients of the different spin and isospin exchange admixtures, are known as the Landau–Migdal parameters.[3] We choose to express the strength of the interaction in terms of the pion–nucleon coupling constant, namely $C_o = 4\pi f^2_{\pi NN}/m^2_\pi = 392 \text{ MeV fm}^3$, then the Landau–Migdal parameters are said to be given in 'pionic units'. Here $f^2_{\pi NN} \simeq 0.08$ and m_π is the pion mass. The reader is warned that there are many other normalizations and parametrizations in the literature for a general central, spin- and isospin-dependent, zero-range interaction. In shell-model calculations it is generally required to calculate matrix elements of $V(r)$ between antisymmetrized two-particle states. That is, there are direct and exchange matrix elements to be evaluated. However, a special feature of a zero-range interaction is that the exchange matrix element is not independent of the direct matrix element but is proportional to it. Thus without any loss of generality the calculation can be restricted to a calculation of just direct matrix elements with an appropriate reinterpretation of the Landau–Migdal parameters. This is particularly convenient in the following sense: consider a closed-LS-shell configuration in a light nucleus and consider excited states obtained by promoting one nucleon from an occupied to an unoccupied orbital, the one particle–one hole (1p–1h) states. A standard calculation of the energy of these 1p–1h states using the direct matrix elements of a zero-range interaction produces one state significantly shifted in energy from all the others, the so-called giant resonance. Furthermore the calculated energy of the giant resonance is mainly influenced by just one of the Landau–Migdal parameters. For example, natural parity states (those for which the parity of the particle–hole state is $(-)^I$, where I is the spin of the state) are governed by the parameter f if they are isoscalar and f' if isovector, while unnatural parity states (those for which the parity is $(-)^{I+1}$) are governed by g if isoscalar and g' if isovector. Similarly core-polarization calculations in a closed-shell-plus-one nucleus based on direct matrix elements of a zero-range interaction depend only on f for isoscalar electric electromagnetic transition operators, on f' for isovector electric operators, on g for isoscalar magnetic operators, and on g' for isovector magnetic operators. This result will be become evident as we go through a detailed calculation for the magnetic moment operator.

3.2 ZERO-RANGE RESIDUAL INTERACTION

Calculations of the Landau parameters in nuclear matter by the Tubingen,[4] Jülich,[5] and Brooklyn[6] groups obtain as typical values: $f \simeq -0.2$, $f' \simeq 0.2$, $g \simeq 0.1$, $g' \simeq 0.5$ with a considerable variation between calculations of at least ± 0.1. For magnetic moments we will be mainly concerned with the parameter g' whose value typically lies in the range 0.5 to 0.6.

We will use Greek letters $\alpha, \beta \ldots$ to represent particle–hole coupled states, namely $|\alpha\rangle = |(h_\alpha^{-1} p_\alpha) I\rangle$ of angular momentum I, then the direct matrix element of the zero-range interaction eqn (3.12) is:[7]

$$\langle \beta | V | \alpha \rangle = (\lambda a_\alpha a_\beta + \mu b_\alpha b_\beta) I_{\alpha\beta}, \tag{3.13}$$

where a_α and b_α are essentially angular momentum coupling coefficients (Wigner 3j-symbols)

$$a_\alpha = (-)^{p_\alpha - 1/2} \hat{h}_\alpha \hat{p}_\alpha \begin{pmatrix} h_\alpha & p_\alpha & I \\ \tfrac{1}{2} & -\tfrac{1}{2} & 0 \end{pmatrix},$$

$$b_\alpha = (-)^{l_{h_\alpha}} \hat{h}_\alpha \hat{p}_\alpha \begin{pmatrix} h_\alpha & p_\alpha & I \\ \tfrac{1}{2} & \tfrac{1}{2} & -1 \end{pmatrix}, \tag{3.14}$$

with $\hat{h} = (2h+1)^{1/2}$ etc. The isospin dependence is contained entirely in the strength of the interaction

$$\lambda_{T=0} = C_0[(f+g) + (-)^y(f-g)]$$
$$\lambda_{T=1} = C_0[(f'+g') + (-)^y(f'-g')], \tag{3.15}$$
$$\mu_{T=0} = 2C_0 g,$$
$$\mu_{T=1} = 2C_0 g',$$

with $(-)^y = +1$ for electric states ($l_{h_\alpha} + l_{p_\alpha} + I$ = even) and $(-)^y = -1$ for magnetic states ($l_{h_\alpha} + l_{p_\alpha} + I$ = odd). Note that $\lambda = \mu$ for magnetic states. In heavy nuclei, we prefer to use a proton–neutron formalism. If the states α and β are both proton particle–proton hole states ($\pi\pi$) or both neutron particle–neutron hole states ($\nu\nu$) then

$$\lambda_{\pi\pi} = \lambda_{\nu\nu} = \tfrac{1}{2}(\lambda_{T=0} + \lambda_{T=1})$$
$$= \tfrac{1}{2} C_0[(f+f'+g+g') + (-)^y(f+f'-g-g')], \tag{3.16}$$
$$\mu_{\pi\pi} = \mu_{\nu\nu} = C_0(g+g').$$

But with α a proton particle–hole state and β a neutron particle–hole state ($\pi\nu$) or vice versa ($\nu\pi$) we have

$$\lambda_{\pi\nu} = \lambda_{\nu\pi} = \tfrac{1}{2}(\lambda_{T=0} - \lambda_{T=1})$$
$$= \tfrac{1}{2} C_0[(f-f'+g-g') + (-)^y(f-f'-g+g')], \tag{3.17}$$
$$\mu_{\pi\nu} = \mu_{\nu\pi} = C_0(g-g').$$

Finally $I_{\alpha\beta}$ is a radial integral

$$I_{\alpha\beta} = \frac{1}{4\pi} \int R_{p_\alpha}(r) R_{h_\alpha}(r) R_{p_\beta}(r) R_{h_\beta}(r) r^2 \, dr, \qquad (3.18)$$

with $R(r)$ the radial form for the single-particle wavefunction.

Likewise the one-body matrix elements of electromagnetic transition operators can be expressed in terms of the same angular momentum coupling coefficients, eqn (3.14). Consider the case of the spin part of the magnetic λ-pole operator (the term with coefficient g_S) evaluated between a particle–hole state and closed-shell configuration:

$$\langle 0 \| \mu^{(\lambda)} \| \beta \rangle = (-)^\lambda \tfrac{1}{2} g_S^{(\beta)} \lambda \left[a_\beta + (-)^\lambda \left(\frac{\lambda + 1}{\lambda} \right)^{1/2} b_\beta \right] I_\beta^{\lambda - 1}$$
$$I_\beta^{\lambda - 1} = \int R_{p_\beta}(r) r^{\lambda - 1} R_{h_\beta}(r) r^2 \, dr. \qquad (3.19)$$

In particular for the magnetic moment operator ($\lambda = 1$), the radial integral becomes unity and

$$\langle 0 \| \mu^{(1)} \| \beta \rangle = -\tfrac{1}{2} g_S^{(\beta)} (a_\beta - \sqrt{2} b_\beta), \qquad (3.20)$$

where $g_S^{(\beta)}$ is $g_S^{(\pi)}$ if β is a proton particle–hole state and $g_S^{(\nu)}$ if a neutron particle–hole state. On substituting eqns (3.20) and (3.13) into (3.9) and denoting the valence state $|(b^{-1}a)\lambda\rangle \equiv |\alpha\rangle$ and the particle–hole state $|(h^{-1}p)\lambda\rangle \equiv |\beta\rangle$ we obtain

$$\langle b \| \text{Fig. 3.1(c)} \| a \rangle$$

$$= \tfrac{1}{2} \hat{b}^{-1} C_o \sum_\beta g_S^{(\beta)} (g \pm g')(a_\beta - \sqrt{2} b_\beta)(a_\alpha a_\beta + b_\alpha b_\beta) I_{\alpha\beta} / (E_\alpha - E_\beta) \qquad (3.21)$$

with $E_\alpha = \varepsilon_a - \varepsilon_b$ and $E_\beta = \varepsilon_p - \varepsilon_h$. The upper value, $g + g'$, is used if the valence and particle–hole states are of like type (i.e. both protons or both neutrons) and the lower value, $g - g'$, if of unlike type. In summing over particle–hole states we consider the hole state always to have angular momentum $h = l + \tfrac{1}{2}$ and the particle state $p = l - \tfrac{1}{2}$. This is generally true in nuclear physics where the sign of the spin–orbit force is such that the single-particle state $l + \tfrac{1}{2}$ lies lower in energy than the $l - \tfrac{1}{2}$ state. The angular-momentum coupling coefficients (with $\lambda = 1$) are then

$$a_\beta = -\left(\frac{2l_\beta(l_\beta + 1)}{2l_\beta + 1} \right)^{1/2},$$
$$b_\beta = \frac{1}{\sqrt{2}} \left(\frac{2l_\beta(l_\beta + 1)}{2l_\beta + 1} \right)^{1/2}, \qquad (3.22)$$

3.2 ZERO-RANGE RESIDUAL INTERACTION

and the first-order core-polarization correction becomes

$$\langle b \| \text{Fig. 3.1(c)} \| a \rangle = b^{-1} C_o \left(a_\alpha - \frac{1}{\sqrt{2}} b_\alpha \right)$$

$$\times \sum_\beta g_S^{(\beta)} (g \pm g') \left(\frac{2l_\beta(l_\beta + 1)}{2l_\beta + 1} \right) I_{\alpha\beta} / (E_\alpha - E_\beta). \quad (3.23)$$

To this must be added the result (eqn (3.9)) for the Hermitian diagram (f) of Fig. 3.1. For a zero-range interaction, this contribution is identical to that from graph (c); the above result is multiplied by 2. For simplicity we will make two other small assumptions. We assume the radial integral depends only on the orbital angular momentum of the single-particle states, namely $I_{\alpha\beta} = I(n_\alpha l_\alpha, n_\beta l_\beta)$, and not on the total angular momentum. This would be true with harmonic oscillator functions. Secondly we assume the energy difference in the valence orbits $E_\alpha = \varepsilon_a - \varepsilon_b$ to be small, namely $E_\alpha = 0$. This is obviously true for diagonal matrix elements (magnetic moments), but is not strictly true for M1 transitions. Nevertheless with these two assumptions we can write

$$\langle b \| \text{Fig. 3.1(c + f)} \| a \rangle = b^{-1} \left(a_\alpha - \frac{1}{\sqrt{2}} b_\alpha \right) K,$$

$$K = -2C_o \sum_\beta g_S^{(\beta)} (g \pm g') \left(\frac{2l_\beta(l_\beta + 1)}{2l_\beta + 1} \right) I(n_\alpha l_\alpha, n_\beta l_\beta) / E_\beta,$$
(3.24)

where K is independent of the total angular momentum of the valence orbitals a and b, and depends only on their common orbital angular momentum, l_α, through the radial integral.

Next we calculate the core-polarization correction due to the orbital part of the magnetic moment operator. We follow the same steps as above but replace eqn (3.20) by

$$\langle 0 | g_L \mathbf{L} \| \beta \rangle = \langle 0 \| g_L (\mathbf{I} - \mathbf{S}) \| \beta \rangle, \quad (3.25)$$

with $\mathbf{I}$ the total angular momentum operator. Since this matrix element is off-diagonal in the total angular momentum, the contribution from the term $g_L \mathbf{I}$ is zero. Thus the calculation reduces to that already done but with $g_S^{(\beta)}$ replaced by $-g_L^{(\beta)}$. The final result, then, for the core-polarization calculation is given by eqn (3.24) with $g_S^{(\beta)}$ replaced by $g_S^{(\beta)} - g_L^{(\beta)}$.

It is convenient to cast the result of a core-polarization calculation in terms of an equivalent effective magnetic moment operator introduced in Chapter 1 in eqn (1.48). To find values for δg_L, δg_S, and δg_P a set of three linear equations need to be inverted involving the three matrix elements (eqn (1.49)):

$$\mu_> = \left(\frac{2l+1}{2l+3}\right)^{1/2} \langle b = l + \tfrac{1}{2} \| \text{Fig. 3.1(c + f)} \| a = l + \tfrac{1}{2} \rangle$$

$$= \frac{l+2}{2l+3} K,$$

$$\mu_< = \left(\frac{2l-1}{2l+1}\right)^{1/2} \langle b = l - \tfrac{1}{2} \| \text{Fig. 3.1(c + f)} \| a = l - \tfrac{1}{2} \rangle \quad (3.26)$$

$$= -\frac{l-1}{2l+1} K,$$

$$M = \left(\frac{2l+1}{l}\right)^{1/2} \langle b = l + \tfrac{1}{2} \| \text{Fig. 3.1(c + f)} \| a = l - \tfrac{1}{2} \rangle$$

$$= -\tfrac{3}{2} K.$$

Substituting these values in eqn (1.50) we obtain the final result we have been seeking:

$$\delta g_L = 0,$$

$$\delta g_S = \tfrac{4}{3} K = -\tfrac{8}{3} C_o \sum_\beta (g_S^{(\beta)} - g_L^{(\beta)})(g \pm g')$$

$$\times \left(\frac{2l_\beta(l_\beta + 1)}{2l_\beta + 1}\right) I(nl, n_\beta l_\beta)/E_\beta, \quad (3.27)$$

$$\delta g_P = -\tfrac{1}{2}(2\pi)^{1/2} \delta g_S.$$

This result is discussed by Arima and Hyuga.[8] We note the following points:

1. With a zero-range effective interaction the correction to g_L is identically zero. This is a general phenomenon for first-order (but not higher-order) core-polarization calculations. For example, for finite-range effective interactions (to be discussed shortly) the correction δg_L remains very small, but is not identically zero.

2. The correction δg_S is large, being typically of order $\delta g_S/g_S \simeq -0.4$ for the zero-range interaction. It is also almost state-independent, i.e. it does not depend on the quantum numbers of the valence orbital; the only dependence in eqn (3.27) comes in the radial integral $I(nl, n_\beta l_\beta)$. For nodeless oscillator functions this integral is

$$I(0l, 0l_\beta) = (2\pi)^{-3/2} 2^{-(l+l_\beta)} \frac{(2l + 2l_\beta + 1)!!}{(2l + 1)!! (2l_\beta + 1)!!} \alpha^3, \quad (3.28)$$

where α is the inverse-length oscillator constant: $\alpha^2 = m\omega/\hbar$, with m the nucleon mass and $\hbar\omega$ the characteristic oscillator energy. This function maximizes for $l \simeq l_\beta$. Other radial integrals $I(nl, 0l_\beta)$ with $n > 0$ are less than $I(0l, 0l_\beta)$, so the largest corrections will occur for nodeless valence states whose orbital angular momentum almost matches the orbital angular momentum l_β of the particle–hole states. For example, in the Pb region the particle–hole states involve proton 0h and neutron 0i orbitals so the largest corrections occur in the nodeless high-spin orbitals. We will show some numerical results shortly.

3. The induced tensor term, δg_P, in the effective magnetic moment operator is related, for a zero-range interaction, to the correction δg_S, eqn (3.27) independent of the orbital of the valence nucleon. Indeed this particular connection between δg_S and δg_P is such that the correction to the magnetic moment for a $p_{1/2}$ orbital is identically zero. Even with finite-range interactions, $\delta \mu$ for $p_{1/2}$ states is smaller than for other states, which explains why all $p_{1/2}$ nuclei have small deviations from the Schmidt single-particle value, an observation made in Section 1.6. However, the induced tensor term, δg_P, depends very critically on the range of the effective interaction in the core-polarization calculation. For example,[8] a central interaction of constant radial dependence, which is of infinite long range, yields $\delta g_P = 0$. Finite-range interactions produce results intermediate between the two extremes.

3.3 Finite-range interactions

The first core-polarization calculations for magnetic moments were performed in the 1950s by Arima and Horie[9] and by Blin-Stoyle.[10] The phenomenon has sometimes been called the Arima–Horie effect. Since then many calculations have been mounted, too numerous to review here. Instead we will quote some sample calculations in the Pb region. Matrix elements are calculated using harmonic oscillator wavefunctions ($\hbar\omega = 7$ MeV), while energy denominators use experimental energies: $\varepsilon_p - \varepsilon_h = 5.6$ MeV for the proton $h_{9/2} - h_{11/2}$ spin–orbit splitting, $\varepsilon_p - \varepsilon_h = 5.85$ MeV for the neutron $i_{11/2} - i_{13/2}$ splitting, and $\varepsilon_b - \varepsilon_a = 0.0$ for magnetic moments and $\varepsilon_b - \varepsilon_a = E_\gamma$, the photon energy for an M1 transition. There is an inconsistency here since we are using oscillator functions for the single-particle wavefunctions and experimental energies for the single-particle energies. Ideally both should come as the eigenfunctions and eigenenergies of the single-particle Hamiltonian H_0 used in establishing the perturbing interaction. For the illustrative purpose of the first-order calculations discussed here, this inconsistency is not serious. Some groups, for example,[11] have used Saxon–Woods potentials whose parameters are adjusted so that the potential's eigenenergies match the experimental values. The inconsistency, however, is a much more serious problem for magnetic transition operators of multipolarity higher than

Table 3.1 First-order core-polarization corrections to magnetic moments and $B(M1)$ in the Pb region

	Zero range, $g' = 0.6$				Arima, Huang-Lin[b]				OBEP[c]			
	$\delta\mu^a$	δg_L	δg_S	δg_P	$\delta\mu^a$	δg_L	δg_S	δg_P	$\delta\mu^a$	δg_L	δg_S	δg_P
$\pi h_{9/2}$	0.764	0.000	−2.802	3.512	0.79	0.01	−2.01	0.83	0.614	0.007	−1.591	0.658
$\pi i_{13/2}$	−1.031	0.000	−2.577	3.230	−0.60	0.02	−1.74	1.63	−0.597	0.007	−1.409	0.813
$\pi s_{1/2}^{-1}$	−0.803	0.000	−1.606	0.000	−0.77	0.00	−1.54	0.00	−0.611	0.000	−1.221	0.000
$\pi d_{3/2}^{-1}$	0.274	0.000	−1.825	2.287	0.33	−0.01	−1.77	1.51	0.283	0.004	−1.348	1.079
$\pi f_{5/2}^{-1} \to f_{7/2}$	−0.732	0.000	−1.762	2.208	−0.63	−0.00	−1.54	1.57	−0.583	0.001	−1.374	1.075
$\nu g_{9/2}$	0.586	0.000	1.433	−1.796	0.59	−0.02	1.58	−1.71	0.518	−0.004	1.138	−0.471
$\nu p_{1/2}^{-1}$	0.000	0.000	1.318	−1.652	−0.11	−0.01	1.32	−0.91	−0.128	−0.004	1.056	−0.382
$\nu f_{5/2}^{-1}$	−0.335	0.000	1.562	−1.957	−0.46	−0.01	1.57	−1.14	−0.403	−0.005	1.217	−0.400
$\nu i_{13/2}^{-1}$	1.031	0.000	2.577	−3.230	0.85	−0.01	1.96	−1.00	0.676	−0.007	1.413	0.139
$\nu p_{3/2}^{-1} \to p_{1/2}^{-1}$	−0.429	0.000	1.351	−1.694	−0.40	−0.01	1.32	−0.91	−0.315	−0.002	1.079	−0.376
$\nu f_{7/2}^{-1} \to f_{5/2}^{-1}$	−0.622	0.000	1.727	−2.165	−0.54	−0.01	1.57	−1.14	−0.430	−0.003	1.312	−0.295

[a] Correction to either magnetic moment or $[B(M1; i \to f)]^{1/2}$.
[b] Arima and Huang-Lin;[13] see also Arima and Hyuga.[8]
[c] One-boson-exchange potential.[14]

$\lambda = 1$, or for magnetic electron scattering where the momentum dependence of the form factors is being examined.

In Table 3.1, we show the calculated first-order core-polarization corrections to the magnetic moments and $B(M1)$ transition rates for single-particle states in the vicinity of ^{208}Pb for three choices of the effective interaction. The results are expressed in terms of the corrections to the coupling constants in the effective M1 operator. The upper half of the table refers to proton states and the lower half to neutron states.

The three effective interactions are:

1. A zero-range interaction, eqn (3.12), with typical Landau–Migdal parameters of $g = 0$, $g' = 0.6$ in pionic units.

2. A G-matrix[12] constructed from the Hamada–Johnston nucleon-nucleon potential. This was the chosen interaction in the work of Arima and Huang-Lin.[13]

3. The one-boson-exchange potential (OBEP) multiplied by a short-range correlation function. Details are to be found in the review by Towner.[14] In this application the principal ingredients of the force are the π- and ρ-exchange potentials. They differ from the Hamada–Johnston potential in having a much weaker tensor force component.

In general the calculated correction to the magnetic moments and $B(M1)$ rates is dominated by the δg_S value, which in turn depends principally on the strength of the residual interaction in the spin–isospin channel. (Recall that in evaluating expectation values the δg_P is divided by a factor of $(8\pi)^{1/2}$.) In all cases the magnetic moment is quenched, i.e. $\delta g_S/g_S$ is negative, and the resultant value lies between the Schmidt lines as required. The calculated δg_P values are more variable, the largest values coming from the zero-range force. The G-matrix from the Hamada–Johnston force gives larger δg_P values than OBEP mainly because it has a stronger tensor force. The calculated δg_L values for finite-range forces, arising mainly from the two-body spin–orbit force, is small as anticipated. We will discuss comparisons with experimental data later.

3.4 Extension to RPA

The first-order results of the last section can easily be extended to higher orders in the random phase approximation (RPA). The pertinent second-order graphs are shown in Fig. 3.2. It must be stressed that these are not the only second-order graphs; there are many more which we will discuss in Section 3.7. However, these graphs are the simplest to evaluate. The selection rules that limited the number of intermediate states in the first-order graph still apply to this particular set of

68 CORE POLARIZATION

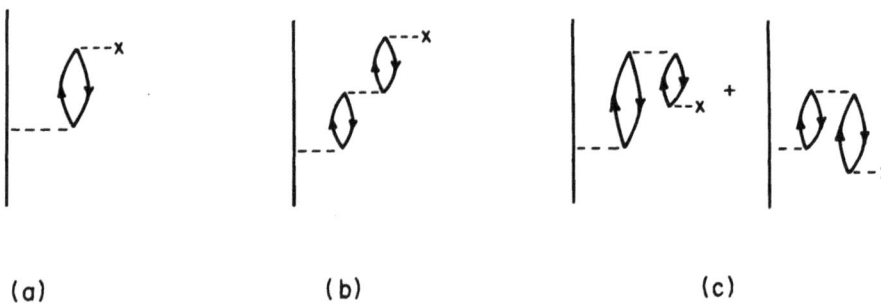

Fig. 3.2 First-order, graph (a), and second-order, graphs (b) and (c), perturbation corrections to the matrix element of a one-body operator representing the start of the RPA series in a closed-shell-plus-one nucleus. The Hermitian adjoint graphs, which must also be included, have not been drawn.

second-order graphs. Furthermore, in the Landau phenomenology these are the only second-order graphs that should be considered. All other second-order graphs are presumed to have been summed in determining the Landau parameters.

The correction to the reduced matrix element from diagrams (a), (b), and (c) in Fig. 3.2 in an angular-momentum coupled representation is given by

$$\langle b\| \text{Fig. 3.2} \|a\rangle = \sum_\alpha T_\alpha \frac{L_\alpha}{\varepsilon_\alpha} + \sum_{\alpha\beta} T_\alpha \frac{A_{\alpha\beta}}{\varepsilon_\alpha}\frac{L_\beta}{\varepsilon_\beta}$$
$$+ (-)^{\lambda+k} \sum_{\alpha\beta} T_\alpha \frac{B_{\alpha\beta}}{\varepsilon_\alpha}\frac{L_\beta}{\varepsilon_\beta}, \qquad (3.29)$$

where we are using Greek letters $\alpha, \beta \ldots$ to represent particle–hole coupled states, namely $|\alpha\rangle = |(h_\alpha^{-1}p_\alpha)\lambda\rangle$, of multipolarity λ and the following notation is being introduced:

$$T_\alpha = \langle 0\| \mu^{(\lambda)} \|(h_\alpha^{-1}p_\alpha)\lambda\rangle,$$
$$L_\alpha = -\hat{b}^{-1}\langle (h_\alpha^{-1}p_\alpha)\lambda| V |(b^{-1}a)\lambda\rangle,$$
$$A_{\alpha\beta} = \langle (h_\alpha^{-1}p_\alpha)\lambda| V |(h_\beta^{-1}p_\beta)\lambda\rangle,$$
$$B_{\alpha\beta} = \langle 0| V |(h_\alpha^{-1}p_\alpha)\lambda(h_\beta^{-1}p_\beta)\lambda\rangle \qquad (3.30)$$
$$= (-)^{h_\alpha-p_\alpha+\lambda}\langle (p_\alpha^{-1}h_\alpha)\lambda| V |(h_\beta^{-1}p_\beta)\lambda\rangle,$$
$$\varepsilon_\alpha = \varepsilon_{h_\alpha} - \varepsilon_{p_\alpha}.$$

In second and higher orders it is assumed that in the energy denominators the photon energy, $E_\gamma \simeq \varepsilon_b - \varepsilon_a$, is negligibly small. This assumption enables the two graphs in Fig. 3.2(c), which differ only in their energy denominators, to be conveniently summed to give an expression of similar form to that from Fig. 3.2(b). The matrix elements

3.4 EXTENSION TO RPA

$A_{\alpha\beta}$ and $B_{\alpha\beta}$ are exactly the matrix elements of RPA theory.[7] The sign $(-)^k$ is determined from the Hermitian property of the operator $\mu^{(\lambda)}$. For the M1 operator $(-)^{\lambda+k} = -1$. There is also a Hermitian set of graphs to be added to those displayed in Fig. 3.2. The correction from this set can be incorporated in eqn (3.29) by replacing L_α by a generalized form:

$$L_\alpha \to L_\alpha + (-)^{a-b-k} \frac{\hat{a}}{\hat{b}} L_\alpha(a \rightleftarrows b). \tag{3.31}$$

In the special case of diagonal matrix elements, namely $a = b$, the Hermitian set equals that displayed in Fig. 3.2, and the result from eqn (3.29) is simply multiplied by a factor of two.

These results, from first and second order, can be algebraically summed to all orders. In a matrix notation the all-orders expression is

$$\langle b \| \text{RPA} \| a \rangle = \sum_{\alpha\beta} T_\alpha [\mathbf{1} - (\mathbf{A} + (-)^{\lambda+k}\mathbf{B})/\varepsilon]^{-1}_{\alpha\beta} L_\beta / \varepsilon_\beta \tag{3.32}$$

where the matrix $[\mathbf{1} - (\mathbf{A} - \mathbf{B})/\varepsilon]$ is first constructed and then inverted. Here $\mathbf{1}$ is the unit matrix. Again L_β is to be interpreted in its more general form, eqn (3.31).

In Table 3.2, we give some results using this expression for the RPA correction to magnetic moments and $B(\text{M1})$ transitions in closed-shell-plus-one nuclei in the Pb region for two of the choices of residual interaction used in Table 3.1 for the first-order corrections. Again oscillator wavefunctions ($\hbar\omega = 7$ MeV) and experimental single-particle energies are used. The results show that the corrections are between 30

Table 3.2 RPA core-polarization corrections to magnetic moments and $B(\text{M1})$ in the Pb region

	Zero range, $g' = 0.6$				OBEP[b]			
	$\delta\mu^a$	δg_L	δg_S	δg_P	$\delta\mu^a$	δg_L	δg_S	δg_P
$\pi h_{9/2}$	0.445	0.000	-1.630	2.043	0.450	0.005	-1.167	0.481
$\pi i_{13/2}$	-0.599	0.000	-1.497	1.876	-0.438	0.005	-1.033	0.596
$\pi s_{1/2}^{-1}$	-0.446	0.000	-0.933	0.000	-0.448	0.000	-0.896	0.000
$\pi d_{3/2}^{-1}$	0.159	0.000	-1.060	1.329	0.208	0.003	-0.989	0.790
$\pi f_{5/2}^{-1} \to f_{7/2}$	-0.377	0.000	-0.907	1.137	-0.367	0.002	-0.911	0.793
$\nu g_{9/2}$	0.341	0.000	0.834	-1.045	0.380	-0.003	0.833	-0.344
$\nu p_{1/2}^{-1}$	0.000	0.000	0.766	-0.960	-0.094	-0.003	0.773	-0.279
$\nu f_{5/2}^{-1}$	-0.194	0.000	0.907	-1.137	-0.295	-0.004	0.891	-0.292
$\nu i_{13/2}^{-1}$	0.599	0.000	1.497	-1.876	0.495	-0.005	1.034	0.102
$\nu p_{3/2}^{-1} \to p_{1/2}^{-1}$	-0.243	0.000	0.766	-0.960	-0.227	-0.003	0.773	-0.279
$\nu f_{7/2}^{-1} \to f_{5/2}^{-1}$	-0.327	0.000	0.907	-1.137	-0.295	-0.004	0.891	-0.292

[a] Correction to either magnetic moment or $[B(\text{M1}; i \to f)]^{1/2}$.
[b] One-boson-exchange potential.[14]

and 40% smaller than the first-order calculations in Table 3.1. The core-polarization calculation for spin operators leads, in perturbation theory, to an alternating sign series; quenching in first order, enhancement in second order, and so on. Note that for the zero-range interaction, the algebraic results from eqns (3.27) still apply: $\delta g_L = 0$ and $\delta g_P = -\frac{1}{2}(2\pi)^{1/2} \delta g_S$; the value of δg_S is reduced from that of the first order. We see, for example, the calculated RPA correction for an $h_{9/2}$ proton is of order $\delta\mu = 0.45 - 0.50 \, \mu_N$. The magnetic moment of ^{209}Bi differs from the single-particle estimate by $1.49 \, \mu_N$ and the RPA correction only gives one third of this. There are other corrections, principally from meson-exchange currents, which in this case lead to a large correction. Thus we postpone for now any comparison with experiment until we have discussed meson-exchange currents.

3.5 RPA in closed-shell nuclei

Another, in principle, simple calculation involves magnetic dipole excitations in closed-shell nuclei. These excitations are described in terms of one particle–one hole states and from the M1 selection rules there are relatively few of them. For example in ^{208}Pb there are two possible configurations for the 1^+ state, namely $\pi h_{11/2}^{-1} h_{9/2}$ and $\nu i_{13/2}^{-1} i_{11/2}$. The RPA calculation, then, is to construct the matrices A and B, eqn (3.30), and solve the RPA eigenvalue problem

$$\begin{pmatrix} A & B \\ B & A \end{pmatrix} \begin{pmatrix} X \\ Y \end{pmatrix} = E \begin{pmatrix} X \\ -Y \end{pmatrix}. \quad (3.33)$$

To the diagonal elements of A, the particle–hole energies are added. With the eigenvectors, the $B(M1)$ from the ground state to the 1^+ state is evaluated using

$$\langle 0^+ \| \mu^{(\lambda)} \| 1^+ \rangle = \sum_\alpha \{X_\alpha + (-)^{\lambda+k} Y_\alpha\} T_\alpha,$$

$$B(M1; 0^+ \to 1^+) = \frac{3}{4\pi} |\langle 0^+ \| \mu^{(\lambda)} \| 1^+ \rangle|^2, \quad (3.34)$$

where $T_\alpha = \langle 0 \| \mu^{(\lambda)} \| (h_\alpha^{-1} p_\alpha)\lambda \rangle$ and the sum α is over the particle–hole basis states. For the M1 operator, the sign$(-)^{\lambda+k}$ is -1. In the limit that the matrices B are set to zero, and hence Y_α is zero, the calculation reduces to the Tamm–Dancoff approximation (TDA). As a benchmark for the calculation, it is useful to consider the summed strength, $S = \sum_f B(M1; 0^+ \to 1^+)$, to all final states in the TDA calculation. This value of S is independent of the choice of residual interaction,[7] but does depend on the number of basis states in the calculation and the coupling constants in the M1 operator, $\mu^{(\lambda)}$. In ^{208}Pb with just two particle–hole

3.5 RPA IN CLOSED-SHELL NUCLEI

basis states and free-nucleon coupling constants this summed strength is $S = 50\mu_N^2$.

There have been many TDA and RPA calculations[15-17] of the $B(M1)$ strength in ^{208}Pb. Typically in TDA there is a 1^+ state very close in energy to the unperturbed particle–hole energy that is the 'isoscalar' combination of $\pi h_{11/2}^{-1} h_{9/2}$ and $\nu i_{13/2}^{-1} i_{11/2}$ states and with a $B(M1) \simeq 1\mu_N^2$, and a second state shifted by about 2 MeV, which is the 'isovector' combination with a $B(M1) \simeq 49\mu_N^2$. Ground-state correlations, as included in the RPA, influence the energies and transition strengths; the principal effect is to reduce the isovector strength by about 20% to a value typically $B(M1) \simeq 40\mu_N^2$. In Table 3.3 we give some RPA results using two of the residual interactions discussed in the last section.

The experimental situation regarding $B(M1)$ strength in ^{208}Pb has had a chequered history, as documented in earlier[18] and recent[19a] reviews. For decades there has been a problem of 'missing' strength. This was resolved recently in experiments[19b] with highly tagged photons that enabled significant M1 strength to be identified below the neutron threshold in a region of strong E1 transitions. When these results are combined with earlier measurements above the neutron threshold, it becomes clear there is a locally fragmented M1 giant resonance in ^{208}Pb centered at 7.3 MeV having a strength of about $15.6\mu_N^2$. There is additional M1 strength below 6.4 MeV that amounts to about $1.9\mu_N^2$. The large amount of fragmentation suggested by the experimental work indicates that the prediction with just two states that carry all the transition strength is too simplified. A coupling of 1p–1h states with a background of 2p–2h states gives a mechanism for the spreading. Calculations[16,17] still yield a relative localization of M1 strength below 10 MeV, but in the work of Cha et al.[17] there is a long tail distributing a small amount of strength as far away as 60 MeV of excitation. In the main peak this spreading corresponds to another 20% reduction in the summed strength. A further reduction of 20% comes from meson-

Table 3.3 RPA calculations in ^{208}Pb for M1 and GT operators with harmonic oscillator wavefunctions and experimental single-particle energies, for two residual interactions discussed in the text

Interaction	M1			GT		
	E_x	$B(M1;\uparrow)$	% Sum	E_x	$B(GT;\uparrow)$	% Sum
Zero range, $g' = 0.6$	5.8	0.6	1	10.0	55	26
	7.5	37.5	75	15.1	134	64
OBEP	5.8	0.9	2	10.2	41	20
	7.0	43.9	88	16.2	156	75

exchange currents, and in particular isobar currents that effectively renormalize the coupling constants of the one-body operator. We will discuss this in some detail later. Laszewski and Wambach[19a] comment that, from a theoretical point of view, the most interesting observation is the cumulative importance of a series of 20% effects including RPA correlations, 2p–2h mixing, isobar couplings, and meson-exchange currents that all reduce the summed M1 strength, down to around $20\mu_N^2$ and close to recent experiments.[19b]

It is a similar story for the Gamow–Teller β-decay operator. In this case there are about a dozen proton–particle, neutron–hole 1^+ states that would be evident in the β^- decay of ^{208}Pb (if the energy systematics were right), but which are accessible in the equivalent (p, n) reaction on ^{208}Pb at forward angles. The coherent linear superposition of these 1p–1h states form the Gamow–Teller giant resonance. The (p, n) reaction[20] clearly identifies the resonance but the transition strength is found to be less than that anticipated. The one-body operator for Gamow–Teller transitions is

$$GT_\pm = \mp g_A \frac{1}{\sqrt{2}} \sigma \tau_{\pm 1}, \qquad (3.35)$$

where $\tau_{\pm 1} = \pm(1/\sqrt{2})(\tau_x \pm i\tau_y)$ are the components of a spherical tensor and τ_x, τ_y, τ_z are the usual Cartesian Pauli matrices for isospin. The upper sign is for β^+ decays and the lower sign for β^- decays. Here g_A is the axial-vector coupling constant, which from free neutron decay is $g_A \simeq 1.26$. Again it is useful to define sum rules

$$B(GT; i \rightarrow f) = |\langle i \| GT \| f \rangle|^2,$$
$$S = \sum_f B(GT; i \rightarrow f), \qquad (3.36)$$

and from the commutator algebra of isospin $[\tau_x, \tau_y] = 2i\tau_z$ obtain the Ikeda result[21]

$$S_{(p,n)} - S_{(n,p)} = 3(N - Z)g_A^2. \qquad (3.37)$$

The difference in the sum rule for the (p, n) reaction and the (n, p) reaction on a particular target nucleus such as ^{208}Pb depends only on the neutron excess in the target and is quite independent of any nuclear structure model dependence. In fact, in a heavy nucleus the $S_{(n,p)}$ is very weak so for practical purposes $3(N - Z)g_A^2$ can be viewed as a close lower bound on the summed strength $S_{(p,n)}$. Typically (p, n) reactions[20] find between 50 and 60% of this lower bound strength in the Gamow–Teller giant resonance, after using Osterfeld's RPA results[22a] for background subtraction.

In Table 3.3 we give some sample RPA calculations for the Gamow–Teller resonance in ^{208}Bi, the excitation energy, E_x, being expressed relative to the ground state of the target nucleus of the (p, n) reaction.

Harmonic oscillator wavefunctions ($\hbar\omega = 7$ MeV) and experimental single-particle energies are used, so the calculations are not truly consistent. It is preferable to be using Hartree–Fock or at least Saxon–Woods functions and energies in RPA calculations. Since we are using a sharply defined occupancy of the orbitals, all B-matrices are in fact zero and the calculations reduce to the TDA approximation. The summed $B(\text{GT}; 0^+ \to 1^+)$ strength for the dozen 1^+ states exactly saturates the $3(N-Z)g_A^2$ sum rule. The calculations show this strength is concentrated in one or two states. In Table 3.3, we list the two strongest states, the upper being associated with the Gamow–Teller giant resonance. These two states contain between 80 and 90% of the sum rule. These results are typical of the many RPA calculations in the literature.[22b,23] Experimentally[20] the observed strength is a single continuous peak at around 12.2 MeV (in ^{208}Pb) but spreading to lower energies. Simple RPA theory with 1p–1h states is not able to describe the spreading of the strength. The observed resonance[20] contains only about 50–60% of the sum rule bound. As with the M1 resonance, background mixing with 2p–2h states,[24] and inclusion of meson-exchange currents in particular isobar currents[25,26] lead to a diminution of strength. We will return to this in Chapter 4.

Note that with the use of experimental single-particle energies the calculated energy of the Gamow–Teller resonance is too low compared with its experimental location of 19.2 MeV. This is a well known phenomenon.[27,28] As already mentioned we should be using, in RPA theory, single-particle energies evaluated in Hartree–Fock (HF) theory in the closed-shell nucleus. Typically the HF level spacings are wider apart than the experimental observed spacings. This is because the coupling to vibrational states compresses the levels near the Fermi surface, leading to the observed single-particle energies. Many of the RPA calculations in the literature are based on HF calculations using the Skyrme III interaction.[23] This interaction is characterized by an effective mass $m^*/m \simeq 0.75$. Very crudely, we can correct the experimental single-particle energies by multiplying them by m/m^*. In RPA this raises the location of the Gamow–Teller resonance by about 4 MeV and reduces the transition strength in the resonance by about 10%.

3.6 Core-polarization blocking

We discussed in Section 3.1 first-order core polarization calculations in closed-shell-plus-one nuclei. We now want to move further away from closed shells and start with the next simplest case namely that of closed-shell-plus-two nuclei. Our purpose is to demonstrate that additional two-body core-polarization graphs occur as soon as extra valence nucleons are added. Consider the graphs in Fig. 3.3. Diagram (a) is the zeroth-order graph with the one-body operator interacting with one of

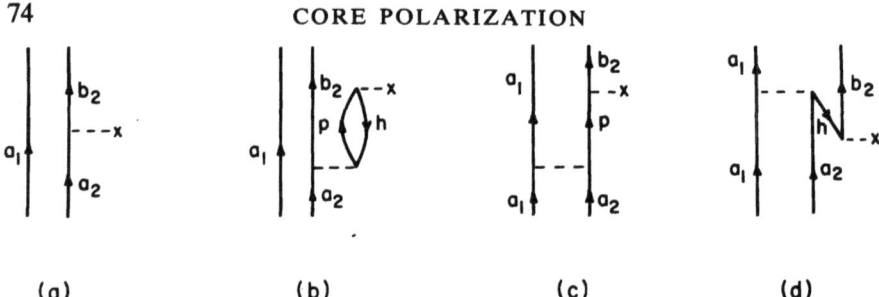

Fig. 3.3 Zeroth-order, diagram (a), and first-order, diagrams (b) to (d), perturbation corrections to the matrix element of a one-body operator in a closed-shell-plus-two nucleus. Hermitian and topologically equivalent graphs have been omitted.

the two valence nucleons. (Naturally there should be a second graph with the one-body operator interacting with the other valence nucleon. This has not been drawn, but its presence is implied. Hermitian and topologically equivalent graphs have been omitted in Fig. 3.3.) Diagram (b) is a first-order core-polarization graph but only one of the two valence nucleons is actively involved. We could proceed in this case by first calculating the graph in closed-shell-plus-one and finding, in the example of an M1 operator, an equivalent effective operator: $\mu_{\text{eff}} = g_{L,\text{eff}} L + g_{S,\text{eff}} S + g_{P,\text{eff}}[Y_2, S]$. With the effective coupling constants determined, it is then only necessary to evaluate diagram (a) with μ_{eff} rather than μ and core polarization would be included in the calculation. This, then, is the method of attacking open-shell nuclei in the nuclear shell model: use effective one-body operators in the shell-model computation whose effective coupling constants are determined from the closed-shell-plus-one situation. However, there is a flaw in this argument. Consider again graph (b). It is possible that the upgoing particle line in the bubble could be in the same quantum state as the non-interacting particle line, i.e. $p = a_1$. If this were so then the Pauli exclusion principle would require this contribution to be zero. But graph (b) would not give zero because nucleon a_1 is not antisymmetrized with respect to nucleon p in forming the 3p–1h intermediate state. This error, however, is corrected in perturbation theory by graphs (c) and (d). Thus it is essential if one wants to maintain full antisymmetry in the theory that graphs (b), (c), and (d) be taken together. In first order in the residual interaction there are no further core-polarization graphs. Therefore in open-shell nuclei the correct way to proceed is to allow for effective one-body and two-body operators to be evaluated between the many-body wavefunctions where the effective coupling constants have been obtained from the closed-shell-plus-one and closed-shell-plus-two situations. In all but the simplest cases, this is rarely done.

3.6 CORE-POLARIZATION BLOCKING

Let us consider the particular case where all the valence nucleons occupy a single j shell. We will use the seniority scheme to classify the many-body states, $|j^n v \alpha I M\rangle$, where I is the total angular momentum, M its magnetic projection, v the seniority quantum number (essentially the number of unpaired nucleons), and α any other quantum number necessary to complete the specification of the state. Then the matrix elements of the one-body magnetic multipole operator connecting states $|j^n v \alpha I M\rangle$ and $|j^n v' \alpha' I' M\rangle$ vanish unless $v = v'$, and are independent of n. Proofs can be found in standard shell-model textbooks.[29,30] In particular, for the one-body magnetic moment operator, the many-body matrix element can be related to the single-nucleon matrix element

$$\frac{\langle j^n v I \| \mu^{(1)} \| j^n v I \rangle}{\langle j^n v I \| I \| j^n v I \rangle} = \frac{\langle j \| \mu^{(1)} \| j \rangle}{\langle j \| I \| j \rangle}. \tag{3.38}$$

This result holds for any one-body operator, $\mu^{(1)}$, of tensorial rank $\lambda = 1$. Thus the magnetic moment of a many-body state is

$$\mu(j^n v I) \equiv \left(\frac{I}{I+1}\right)^{1/2} \langle j^n v I \| \mu^{(1)} \| j^n v I \rangle = I \mu(j)/j \tag{3.39}$$

and we arrive at the well-known result that the g-factor (being the magnetic moment divided by the spin, $g = \mu/I$) for the state $|j^n v \alpha I M\rangle$ is identical to the g-factor of the single-particle $|jm\rangle$. This result has been tested in many cases and found to be true to within 10%. However, there is a systematic departure that grows with n. This departure cannot be due to any renormalization of the one-body operator, such as given by the one-body core-polarization graph of Fig. 3.3(b), but is indeed a signal of the presence of two-body graphs, such as those in Fig. 3.3(c) and Fig. 3.3(d).

There are two cases to be considered. In the first case valence nucleons occupy a $j = l + \frac{1}{2}$ orbital, while the spin–orbit partner orbital $j = l - \frac{1}{2}$ remains empty. Examples[31] would be the calcium isotopes where neutrons are filling the $f_{7/2}$ orbital or the $N = 28$ isotones where protons are filling the $f_{7/2}$ orbital; in either case the $f_{5/2}$ orbital is unoccupied. In this case the graph of Fig. 3.3(c) contributes to the two-body core polarization with the intermediate particle line being the spin–orbit partner orbital, $j = l - \frac{1}{2}$. The graph of Fig. 3.3(d) gives no contribution. In the second case valence nucleons occupy a $j = l - \frac{1}{2}$ orbital, while the spin–orbit partner orbital $j = l + \frac{1}{2}$ is fully occupied. Examples[32,33] would be the $N = 126$ isotones where protons are filling the $h_{9/2}$ orbital. In this case it is Fig. 3.3(d) that contributes to the core polarization, while Fig. 3.3(c) gives no contribution.

For two particles in the valence orbit j, $|j^2 v I M\rangle$, the expression for the normalized and antisymmetrized reduced matrix element corresponding to Fig. 3.3(c) is

$$\langle j^2 vI \| \text{Fig. 3.3(c)} \| j^2 vI \rangle$$
$$= \sqrt{2}\,(1+(-)^k) \sum_{j'} U(\lambda j' I j; jI) \langle j \| \mu^{(\lambda)} \| j' \rangle \langle jj'; I | V | j^2; I \rangle / (\varepsilon_j - \varepsilon_{j'}), \tag{3.40}$$

where k relates to the Hermitian property of the one-body operator $\mu^{(\lambda)}$. For the particular case of the M1 operator, $k = 0$ and j' is restricted to being just the spin–orbit partner orbital to j. Likewise for Fig. 3.3(d), we have

$$\langle j^2 vI \| \text{Fig. 3.3(d)} \| j^2 vI \rangle$$
$$= -\sqrt{2}\,(1+(-)^k) \sum_{j'} U(\lambda j' I j; jI) \langle j \| \mu^{(\lambda)} \| j' \rangle \langle jj'; I | V | j^2; I \rangle / (\varepsilon_{j'} - \varepsilon_j). \tag{3.41}$$

We will write these expressions, eqns (3.40) and (3.41), as $\langle j^2 vI \| G^{(1)} \| j^2 vI \rangle$ where G represents a two-body operator. What we are interested in is the number dependence when this operator is evaluated between many-body states of configuration, $j^n v\alpha IM$. Nomura[34] has given the general result for two-body operators of odd tensorial rank, $G^{(\lambda)}$ with $\lambda = $ odd, which simplifies for the diagonal case with $\lambda = 1$ to be:

$$\langle j^n v\alpha I \| G^{(1)} \| j^n v\alpha I \rangle = \frac{2j+1-2n}{2j+1-2v} \langle j^v vI \| G^{(1)} \| j^v vI \rangle$$
$$+ \frac{n-v}{2j+1-2v} [I(I+1)]^{1/2} M_0, \tag{3.42}$$

where

$$M_0 = \frac{1}{[j(j+1)]^{1/2}} \sum_{K,\,\text{even}} \frac{2K+1}{2j+1} \left(\frac{K(K+1)}{j(j+1)}\right)^{1/2} \langle j^2; K \| G^{(1)} \| j^2; K \rangle. \tag{3.43}$$

Note the very simple linear n-dependence. Contrast this with one-body magnetic operators, $\mu^{(\lambda)}$ with λ-odd, which are independent of n. For the particular case of low seniority, $v = 1$ or 2, the expression simplifies even further and the contribution to the g-factor (being the magnetic moment divided by spin) from two-body operators reduces to:[34]

$$g(j^n vI) = \frac{\langle j^n v\alpha I \| G^{(1)} \| j^n v\alpha I \rangle}{[I(I+1)]^{1/2}}$$
$$= \frac{n-1}{2j-1} M_0 + \tfrac{1}{2}(1+(-)^n)\frac{2j+1-2n}{2j-3} Z(I), \tag{3.44}$$

where

$$Z(I) = \frac{\langle j^2 I \| G^{(1)} \| j^2 I \rangle}{[I(I+1)]^{1/2}} - \frac{1}{2j-1} M_0. \tag{3.45}$$

3.6 CORE-POLARIZATION BLOCKING

Note that from its construction $Z(I)$ satisfies $\sum I(I+1)(2I+1)Z(I) = 0$, the sum being over even values of I. Nomura[34] gives arguments why $Z(I)$ is expected to be small. Thus on examining the magnetic moments of odd-mass nuclei, with assumed seniority $v = 1$ ground states, and of excited states in even–even nuclei with assumed seniority $v = 2$, a linear dependence on the number of nucleons in the single valence shell is to be expected in first-order core-polarization theory. We examine examples of the two cases cited above.

1. *The $N = 28$ isotones.* Magnetic moments of ^{51}V, ^{53}Mn, and ^{55}Co are known[35] together with the magnetic moments of the 2^+ and 6^+ states in ^{54}Fe. We assume the configurations for these states are $\pi f_{7/2}^n$ with seniority, $v = 1$ for the odd-mass cases, and seniority $v = 2$ in ^{54}Fe. The contribution to the g-factors from one-body operators will be written as g_{sp} and is independent of n. Thus for one-body and two-body operators taken together, we expect

$$g(7/2^n vI) = g_{sp} + \tfrac{1}{6}(n-1)M_0 + \tfrac{1}{4}(1+(-)^n)(4-n)Z(I). \quad (3.46)$$

In principle, g_{sp} should be determined from the magnetic moment of ^{49}Sc, but since this moment is not known, we will treat g_{sp} as a parameter to be fitted in the analysis to ^{51}V. We have calculated M_0 and $Z(I)$ using eqns (3.43) and (3.45) with the one-boson-exchange potential for V, the free-nucleon M1 operator for $\mu^{(1)}$, and the spin–orbit splitting between the proton $f_{7/2}$ and $f_{5/2}$ orbitals taken as 5 MeV. Harmonic oscillator wavefunctions ($\hbar\omega = 11$ MeV) are used. The results are given in Table 3.4. The theoretical results show a reduction in the g-factor with n, as expected, although the calculated fall-off is too fast, namely $\Delta \equiv g(^{55}\text{Co}) - g(^{51}\text{V})$ has the experimental value $\Delta = -0.093$ and theoretical value $\Delta = -0.236$. This is a characteristic feature of first-order calculations; the discrepancy would be reduced in higher orders.

Table 3.4 Experimental and calculated g-factors for many-body states whose configuration is presumed to be $j^n vI$ with lowest seniority, v.

$N = 28$ isotones			$N = 126$ isotones		
State	Expt.[a]	Theory	State	Expt.[a]	Theory
^{51}V(7/2$^-$)	1.471	1.471[b]	^{209}Bi(9/2$^-$)	0.913	0.913[b]
^{53}Mn(7/2$^-$)	1.435(2)	1.353	^{210}Po(8$^+$)	0.919(5)	0.904
^{54}Fe(2$^+$)	1.38(30)	1.339	^{210}Po(6$^+$)	0.913(6)	0.897
^{54}Fe(6$^+$)	1.37(3)	1.284	^{212}Rn(8$^+$)	0.894(2)	0.874
^{55}Co(7/2$^-$)	1.378(1)	1.235	^{214}Ra(8$^+$)	0.885(4)	0.844

[a] Raghavan.[35]
[b] fitted.

2. *The $N = 126$ isotones.* Magnetic moments of the ground state of ^{209}Bi and the 8^+ states in ^{210}Po, ^{212}Rn, and ^{214}Ra[35] are known. We assume the configurations for these states are $\pi h_{9/2}^n$ with lowest seniority. Again the contribution from one-body operators, g_{sp}, is fixed to the known experimental g-factor in ^{209}Bi, but with one additional comment. Consider the one-body graph from first-order core polarization given in Fig. 3.3(b). There is a contribution to this graph with the particle line in the bubble being the valence orbital, $p \equiv h_{9/2}$, and the hole line being the spin–orbit partner orbital, $h = h_{11/2}$. We will single out this contribution and evaluate it to get

$$\langle j \| \text{Fig. 3.3(b)} \| j \rangle = -\sqrt{2}(1 + (-)^k) \sum_K \frac{(2K+1)}{[(2j+1)(2j'+1)]^{1/2}} U(\lambda j j K; j'j)$$

$$\times \langle j \| \mu^{(\lambda)} \| j' \rangle \langle jj'; K | V | j^2; K \rangle / (\varepsilon_{j'} - \varepsilon_j)$$

$$= -[j(j+1)]^{1/2} M_0, \qquad (3.47)$$

where $j = h_{9/2}$ and $j' = h_{11/2}$, and where M_0 is defined in eqn (3.43) and obtained using Fig. 3.3(d) for the two-body operator. Thus this particular contribution to the one-body core polarization is intimately connected with the two-body contribution. Indeed the g-factor from this one-body contribution is just $g(\text{Fig. 3.3(b)}) = -M_0$. We will separate this piece out from g_{sp}; that is, we define g_{sp} such that for ^{209}Bi: $g(^{209}\text{Bi}) = g_{sp} - M_0$. With this definition, the g-factor for other $N = 126$ isotones is

$$g(9/2^n vI) = g_{sp} - \tfrac{1}{8}(9-n)M_0 + \tfrac{1}{6}(1 + (-)^n)(5-n)Z(I). \qquad (3.48)$$

Note that at the closed-shell-minus-one configuration, $n = 9$, the contribution from the term proportional to M_0 is zero; that is, the contribution from Fig. 3.3(b) is exactly cancelled from the contribution Fig. 3.3(d). This is what is required to satisfy the Pauli exclusion principle. This phenomenon is called core-polarization blocking. Again we have calculated M_0 and $Z(I)$ using eqns (3.42), (3.43), and (3.45) with the one-boson-exchange potential for V and the spin–orbit splitting between the proton $h_{9/2}$ and $h_{11/2}$ orbitals taken as 5.6 MeV. Harmonic osciallator wavefunctions ($\hbar\omega = 7$ MeV) are used. The results are given in Table 3.4. Again the theoretical g-factors show a characteristic fall-off with n that is about a factor of two larger than experiment. Namely $\Delta \equiv g(^{214}\text{Ra}; 8^+) - g(^{210}\text{Po}; 8^+)$ has the experimental value $\Delta = -0.034$ and theoretical value $\Delta = -0.060$. Furthermore the difference $\Delta_1 = g(^{210}\text{Po}; 8^+) - g(^{209}\text{Bi}; 9/2^-)$ as calculated has the wrong sign as has been noted before.[32]

As is seen here, and in previous works,[31,32] first-order core-polarization calculations universally overestimate the core-polarization blocking. This deficiency is largely removed when the calculation is taken beyond first order. Towner et al.[33] in all-orders calculation involving

1p–1h configurations: $|h_{9/2}^{n+1}h_{11/2}^{-1}\rangle$, $|h_{9/2}^{n}(\nu i_{11/2}, i_{13/2}^{-1})\rangle$, and 2p–2h configurations: $|h_{9/2}^{n+2}h_{11/2}^{-2}\rangle$, $|h_{9/2}^{n+1}h_{11/2}^{-1}(\nu i_{11/2}, i_{13/2}^{-1})\rangle$, $|h_{9/2}^{n}(\nu i_{11/2}^{2} i_{13/2}^{-2})\rangle$ obtained an acceptable fit to the data on the $N = 126$ isotones. The conclusion is that two-body core-polarization graphs have an important role to play in understanding the state-dependence of magnetic moments of many-body systems, but as was the case discussed in Section 3.3, first-order calculations tend to overestimate the effect. There is also the possibility that two-body operators coming from meson-exchange currents will also give a contribution to $\langle j^2 \| G^{(1)} \| j^2 \rangle$ and hence to M_0 and $Z(J)$. We have evaluated these using the meson-exchange current operators to be discussed shortly and find that their influence on the measure of blocking, Δ, is small. Typically Δ changed by about 3%. The blocking effect is further masked by the impurity of the configuration. The analysis here is based on an assumed configuration, $j^n \nu \alpha I M$, with low seniority. Mixing with other orbitals also has a role to play, and in this instance is responsible for understanding the sign of $\Delta_1 = g(^{210}\text{Po}; 8^+) - g(^{209}\text{Bi}; 9/2^-)$. Thus the quantitative understanding of the blocking requires more than just first-order core-polarization calculations.

3.7 Second-order core polarization

In eqn (3.3) in Section 3.1, we gave the expression for the first-order core-polarization correction to the expectation value of a one-body operator in a closed-shell-plus-one nucleus. Extending this expression, now, to second order the following additional terms are to be added

$$\sum_{\alpha,\beta \neq a,b} \langle b|\mu|\alpha\rangle \frac{\langle a|V|\beta\rangle\langle\beta|V|a\rangle}{(E_a - E_\alpha)(E_a - E_\beta)} - \sum_{\alpha \neq a,b} \langle b|\mu|\alpha\rangle \frac{\langle \alpha|V|a\rangle\langle a|V|a\rangle}{(E_a - E_\alpha)^2}$$

$$+ \sum_{\alpha,\beta \neq a,b} \frac{\langle b|V|\beta\rangle\langle\beta|V|\alpha\rangle}{(E_b - E_\beta)(E_b - E_\alpha)} \langle \alpha|\mu|a\rangle$$

$$- \sum_{\alpha \neq a,b} \frac{\langle b|V|b\rangle\langle b|V|\alpha\rangle}{(E_b - E_\alpha)^2} \langle \alpha|\mu|a\rangle$$

$$+ \sum_{\beta,\gamma \neq a,b} \frac{\langle b|V|\beta\rangle}{E_b - E_\beta} \langle\beta|\mu|\gamma\rangle \frac{\langle\gamma|V|a\rangle}{E_a - E_\gamma}$$

$$- \tfrac{1}{2} \sum_{\beta \neq a,b} \frac{\langle b|V|\beta\rangle\langle\beta|V|b\rangle}{(E_b - E_\beta)^2} \langle b|\mu|a\rangle$$

$$- \tfrac{1}{2} \sum_{\beta \neq a,b} \langle b|\mu|a\rangle \frac{\langle a|V|\beta\rangle\langle\beta|V|a\rangle}{(E_a - E_\beta)^2}, \tag{3.49}$$

where, as before, a and b are single-particle valence states, α an infinite set of 2p–1h states, and β, γ an infinite set of 2p–1h and 3p–2h states. We are not considering any Hartree–Fock insertions. When μ is the M1 magnetic moment operator, the selection rules on $\langle b | \mu | \alpha \rangle$ severely limit the number of intermediate states of type α. However, there is no such restriction on the intermediate states of type β and γ. Second-order calculations, therefore, are computationally quite time consuming even for the M1 operator. If we limit the discussion to light nuclei and closed LS shells, then there is no first-order correction to the magnetic moment operator as already discussed in Section 3.1. The reason is that the one-body M1 operator cannot create (or annihilate) a particle–hole state at LS closed shells because the M1 selection rules require the particle and the hole state to have the same orbital structure. The same reasoning eliminates the first four terms of eqn (3.49) for the second-order core-polarization correction. This leaves the fifth, sixth, and seventh terms, each of which carries intermediate state summations over 2p–1h and 3p–2h states. Note that the last two terms represent simply a normalization correction to the single-particle matrix element $\langle b | \mu | a \rangle$. In Fig. 3.4, we give a graphical representation of these three terms for a closed-shell-plus-one nucleus. The normalization corrections are represented by folded graphs.[36,37] These particular groups of graphs are sometimes called the number conserving set.[38] This is because if the one-body operator were to be replaced by the number operator the sum from these graphs would be zero. This is quickly seen if we rewrite the last three terms in eqn (3.49) for the diagonal matrix element with $a = b$

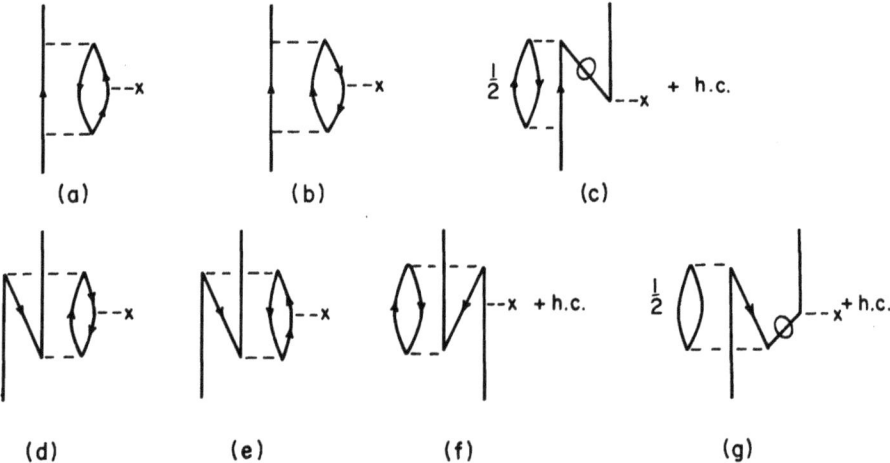

Fig. 3.4 Second-order core-polarization graphs that give a correction to the magnetic moment of a closed-LS-shell-plus-one nucleus. Graphs (a), (b), and (c) involve 2p–1h intermediate states, graphs (d), (e), (f), and (g) involve 3p–2h intermediate states. Graphs (c) and (g) are 'folded' graphs that correct (to second order) the normalization of the single-particle state.

3.7 SECOND-ORDER CORE-POLARIZATION

in the following equivalent form

$$\left\langle a \left| V \frac{Q}{e} \left[\mu, \frac{Q}{e} V \right] \right| a \right\rangle. \quad (3.50)$$

Here the operator Q/e represents $\Sigma_\beta |\beta\rangle\langle\beta|/(E_\alpha - E_\beta)$. Thus if the one-body operator, μ, commutes with both the energy denominator, e, and the residual interaction, V, then the second-order correction to the diagonal matrix element will vanish. The number operator does just that.

In second-order calculations the unperturbed Hamiltonian is usually the harmonic oscillator and the unperturbed energies simple multiples of the characteristic oscillator energy, $\hbar\omega$. Thus the one-body operator will always commute with e. The residual interaction, V, can be decomposed into tensors in orbital and spin space:

$$V = \sum_k \mathcal{L}^{(k)} \mathcal{S}^{(k)}, \quad (3.51)$$

where $k = 0$, 1, or 2 corresponds to central, spin–orbit, and tensor components respectively. Consider, first, the isoscalar magnetic moment operator comprising operators L and S. This operator clearly commutes with the central component of the residual interaction. Thus only non-central, and in particular tensor, interactions can generate a second-order correction to the single-particle isoscalar moment. A similar argument holds for the spin component of the isovector magnetic moment operator, $S\tau$, which commutes with any spin-independent central interaction such as Wigner and Majorana exchange forces. If the spin dependence of the residual interaction is weak, then one can expect the second-order configuration mixing due to the central interaction to produce only a small correction to the matrix element of $S\tau$. The same argument does not apply to the orbital operator, $L\tau$, which does not commute with any central interaction. From these considerations it is clear that the tensor force will play an important role in the second-order core-polarization corrections to magnetic moments.

There have been several calculations of second-order core polarization reported in the literature. Ichimura and Yazaki[39] using only central interactions obtained no correction to the isoscalar magnetic moments and only a small correction to $S\tau$. Mavromatis and Zamick[40] used the G-matrix of Kuo and Brown,[41] which contains non-central components, but limited the intermediate-state summation to states whose oscillator energy were just $2\hbar\omega$ above that of the valence nucleon. Shimizu et al.[42] were the first to point out the importance of extending the intermediate-state summation beyond $2\hbar\omega$ since the tensor force, in particular, has a strong coupling to highly excited intermediate states. In that work the Hamada–Johnston potential was used (with a short-range correlation function). This potential has a 'strong' tensor force in that the short-range phenomenological additions to the one-pion-exchange potential in the Hamada–Johnston parametrization enhances the tensor force over that of pion exchange. Towner and Khanna[43] have repeated these earlier calculations using the one-boson exchange potential. In this case, the potential is said to have a 'weak' tensor force because the short-range additions, coming from ρ exchange, have the opposite sign and weaken the pion tensor force. Despite this there is still a need for extensive intermediate-state summations. In the following, we will discuss some numerical results from Towner.[14]

Again it is convenient to express the results of calculations such as these in terms of an equivalent effective one-body operator. For M1, we write

$$\mu_{\text{eff}} = g_{L,\text{eff}} L + g_{S,\text{eff}} S + g_{P,\text{eff}}[Y_2, S] \qquad (3.52)$$

where as before $g_{L,\text{eff}} = g_L + \delta g_L$ etc. Here g_L is the free-nucleon single-particle coupling constant and δg_L is the calculated correction. We will also introduce a superscipt, e.g. $\delta g_L^{(0)}$ and $\delta g_L^{(1)}$, to distinguish isoscalar and isovector coupling constants. The free-nucleon values are $g_L^{(0)} = 0.5$, $g_S^{(0)} = 0.88$, $g_P^{(0)} = 0.0$, and $g_L^{(1)} = 0.5$, $g_S^{(1)} = 4.706$, $g_P^{(1)} = 0.0$. For the Gamow–Teller β-decay operator, we write

$$\text{GT}_{\text{eff}} = g_{LA,\text{eff}} L + g_{A,\text{eff}} \sigma + g_{PA,\text{eff}}[Y_2, \sigma]. \qquad (3.53)$$

Note the traditional use of the spin operator $\sigma = 2S$. The free-nucleon values are $g_{LA} = 0.0$, $g_A = 1.26$, and $g_{PA} = 0.0$. The computed second-order core-polarization corrections were obtained using harmonic oscillator wavefunctions with $\hbar\omega = 20.4$ MeV for closed-shell core $A = 4$, $\hbar\omega = 13.3$ MeV for $A = 16$, and $\hbar\omega = 10.8$ MeV for $A = 40$. These values are chosen such that the mean square radius of the nuclear charge density of the closed-shell nucleus calculated with oscillator functions reproduces the experimental value.

In Table 3.5, we summarize the results of second-order core-polarization calculations in light nuclei for valence orbitals at the closed LS shells $A = 4$, 16, and 40. Quoted there are the corrections to the magnetic moment expressed as a percentage of the single-particle matrix element and the corrections to the coupling constants, δg, in the equivalent effective one-body operator. These latter values enable the correction to magnetic moments in the single-particle states $\mu(j = l + \frac{1}{2})$ and $\mu(j = l - \frac{1}{2})$ and the off-diagonal matrix element $(j = l + \frac{1}{2} \| \mu \| j = l - \frac{1}{2})$ to be obtained since, by construction, these δg values only depend on the orbital quantum numbers of the state nl, and not the total angular momentum j. Note the results show only a weak state dependence for δg_L and δg_S with a tendency for the hole orbits to have a larger correction than the particle orbits. The δg_P values are more variable, but the matrix elements $\delta g_P \langle Y_2, S \rangle$ are generally small when compared with

3.7 SECOND-ORDER CORE-POLARIZATION

Table 3.5 Second-order core-polarization corrections in closed-shell-plus (or minus)-one configurations to magnetic moments and Gamow–Teller matrix elements expressed in terms of the coefficients of the effective operator and as a percentage of the single-particle matrix element.

		Isovector				Isoscalar				Gamow–Teller			
		δg_L	δg_S	δg_P	%	δg_L	δg_S	δg_P	%	δg_{LA}	δg_A	δg_{PA}	%
$A=4$	$0s_{1/2}^{-1}$		−0.468		−10.0		−0.103		−11.7		−0.119		−9.5
$A=4$	$0p_{3/2}$	−0.072	−0.329	0.141	−8.1	0.015	−0.065	−0.042	−2.0	0.011	−0.088	0.038	−5.9
$A=16$	$0p_{1/2}^{-1}$	−0.183	−0.699	0.177	6.5	0.013	−0.138	−0.022	18.5	0.013	−0.181	0.052	−13.1
$A=16$	$0d_{5/2}$	−0.101	−0.508	0.146	−13.3	0.010	−0.097	−0.012	−2.1	0.011	−0.136	0.005	−9.0
$A=16$	$1s_{1/2}$		−0.562		−11.9		−0.103		−11.7		−0.149		−11.9
$A=40$	$0d_{3/2}^{-1}$	−0.198	−0.845	0.286	26.9	0.010	−0.158	−0.007	10.3	0.009	−0.221	0.048	−18.1
$A=40$	$1s_{1/2}^{-1}$		−0.855		−18.2		−0.154		−17.5		−0.217		−17.2
$A=40$	$0f_{7/2}$	−0.127	−0.646	0.225	−17.9	0.006	−0.121	0.002	−2.1	0.009	−0.173	−0.005	−11.8
$A=40$	$1p_{3/2}$	−0.123	−0.693	−0.141	−16.6	0.007	−0.122	−0.003	−5.8	0.006	−0.183	0.006	−14.0

those of $\delta g_S \langle S \rangle$. For isoscalar M1 and Gamow–Teller operators the calculated δg_L and δg_P are negligible. For the other coupling constants we find $\delta g_L^{(1)} \simeq -0.13 \pm 0.06$, $\delta g_S^{(1)} \simeq -0.6 \pm 0.2$, $\delta g_S^{(0)} \simeq -0.12 \pm 0.02$, and $\delta g_A \simeq -0.16 \pm 0.04$ span the range of results tabulated in Table 3.5. Expressed as a fraction of the free-nucleon coupling constant, the three spin operators are all in the range $\delta g_S / g_S = -0.13 \pm 0.03$; that is, second-order core polarization gives a quenching to spin operators of around $13 \pm 3\%$.

We postpone a comparison with experiment until after a discussion of meson-exchange currents in the next chapter.

3.8 Summary

In this chapter we have discussed the microscopic analysis of the corrections required in the spherical shell model due to the necessity of having to work in a truncated model space.

These calculations are based on a perturbation expansion in which one or several nucleons are excited from closed-shell configurations. We have given results for both first- and second-order perturbation theory for magnetic moments. With a zero-range force we were able to get algebraic results obtaining, for example, a 40% quenching of the spin operator. We also discussed briefly an extension of these calculations in the Tamm–Dancoff and random phase approximations and considered the long-standing problem of the excitation of the 1^+ state in ^{208}Pb.

In this chapter we have also introduced the topic of valence polarization. With more than one nucleon outside closed shell, the mechanism of core polarization can be partly blocked by the action of the Pauli principle. Generally first-order calculations overestimate this blocking.

Finally, before a comparison between theory and experiment can be effected, a further round of corrections associated with higher-order terms must be considered. These are the meson-exchange corrections which we now discuss in Chapter 4.

3.9 References

1. Ellis, P. J. and Mavromatis, H. A. (1971). *Nucl. Phys.*, **A175**, 309.
2. Brink, D. M. and Satchler, G. R. (1968). *Angular momentum*. Clarendon, Oxford.
3. Landau, L. D. (1956). *Sov. Phys.–JETP*, **3**, 920; (1957). **5**, 101; (1959). **8**, 70; Migdal, A. B. (1967). *Theory of finite Fermi systems and applications to atomic nuclei*. Interscience, London.
4. Dickhoff, W. H., Faessler, A., Müther, H., and Wu, S. S. (1983). *Nucl. Phys.*, **A405**, 534.
5. Nakayama, K., Krewald, S., and Speth, J. (1984). *Phys. Lett.*, **B145**, 310.

6. Celenza, L. S., Pong, W. S., and Shakin, C. M. (1982). *Phys. Rev.*, **C25,** 3115; Anastasio, M. R., Celenza, L. S., Pong, W. S., and Shakin, C. M. (1983). *Phys. Rep.*, **100,** 327.
7. Towner, I. S. (1977). *A shell model description of light nuclei.* Clarendon, Oxford.
8. Arima, A. and Hyuga, H. (1979). In *Mesons in nuclei* (ed. D. H. Wilkinson and M. Rho), p. 685. North-Holland, Amsterdam.
9. Arima, A. and Horie, H. (1954). *Prog. Theor. Phys.*, **11,** 504; (1954). **12,** 623.
10. Blin-Stoyle, R. J. (1953). *Proc. Phys. Soc.*, **A66,** 1158.
11. Cha, D. and Speth, J. (1984). *Phys. Lett.*, **B143,** 297; Osterfeld, F., Cha, D., and Speth, J. (1985). *Phys. Rev.*, **C31,** 372.
12. Kuo, T. T. S. (1968). *Nucl. Phys.*, **A122,** 325; and private communication to reference.[13]
13. Arima, A. and Huang-Lin, L. J. (1972). *Phys. Lett.*, **B41,** 429 and 435.
14. Towner, I. S. (1987). *Phys. Rep.*, **155,** 263.
15. Broglia, R. A., Molinari, A., and Sorensen, B. (1968). *Nucl. Phys.*, **A109,** 353; Vergados, J. D. (1971). *Phys. Lett.*, **36B,** 12; Ring, P. and Speth, J. (1973). *Phys. Lett.*, **44B,** 447; Anastasio, M. R. and Brown, G. E. (1977). *Nucl. Phys.*, **A285,** 516; Cwiok, S. and Wygonowska, M. (1973). *Acta Phys. Pol.*, **B4,** 233; Knupfer, W., et al. (1978). *Phys. Lett.*, **77B,** 367; Frey, R., et al. (1978). *Phys. Lett.*, **74B,** 45; Brown, G. E., Dehesa, J. S., and Speth, J. (1979). *Nucl. Phys.*, **A330,** 290; Knupfer, W., Dillig, M., and Richter, A. (1980). *Phys. Lett.*, **95B,** 349; Speth, J., Klemt, V., Wambach, J., and Brown, G. E. (1980). *Nucl. Phys.*, **A343,** 382; Lipparini, E. and Richter, A. (1984). *Phys. Lett.*, **144B,** 13.
16. Lee, T. S. H. and Pittel, S. (1975). *Phys. Rev.*, **C11,** 607; Dehesa, J. S., Speth, J., and Faessler, A. (1977). *Phys. Rev. Lett.*, **38,** 208.
17. Cha, D., Schwesinger, B., Wambach, J., and Speth, J. (1984). *Nucl. Phys.*, **A430,** 321.
18. Brown, G. E. and Raman, S. (1980). *Comm. Nucl. Part. Phys.*, **9,** 79; Holt, R. J., Jackson, H. E., Laszewski, R. M., and Specht, J. R. (1979). *Phys. Rev.*, **C20,** 93; Raman, S. (1979). In *Neutron capture γ-ray spectroscopy* (ed. R. E. Chrien and W. R. Kane) p. 193. Plenum, New York.
19a. Laszewski, R. M. and Wambach, J. (1985). *Comm. Nucl. Part. Phys.*, **14,** 321.
19b. Laszewski, R. M., Alarcon, R., Dale, D. S., and Hoblit, S. D. (1988). *Phys. Rev. Lett.*, **61,** 1710.
20. Gaarde, C., et al. (1981). *Nucl. Phys.*, **A369,** 258; Bainum, D. E., et al. (1980). *Phys. Rev. Lett.*, **44,** 1751; Horen, D. J., et al. (1980). *Phys. Lett.*, **95B,** 27; Rapaport, J. (1983). *American Institute of Physics Conference Proceedings,* No. 42 (ed. M. O. Meyer) p. 365. American Institute of Physics, New York.
21. Ikeda, K., Fujii, S., and Fujita, J. I. (1963). *Phys. Lett.*, **3,** 271.
22a. Osterfeld, F. (1982). *Phys. Rev.*, **C26,** 762; Osterfeld, F. and Schulte, A. (1984). *Phys. Lett.*, **138B,** 23.
22b. Ikeda, K. (1964). *Prog. Theor. Phys.*, **31,** 434; Bertsch, G., Cha, D., and Toki, H. (1981). Phys. Rev., **C24,** 533; Auerbach, N., Zamick, L., and Klein, A. (1982). *Phys. Lett.*, **118B,** 256; Grotz, K., et al. (1983). *Phys. Lett.*, **126B,** 417; Bertsch, G. F., Bortignon, P. F., and Broglia, R. A.

(1983). *Rev. Mod. Phys.*, **55**, 287; Bortignon, P. F. and Broglia, R. A. (1981). *Nucl. Phys.*, **A371**, 405.

23. van Giai, N., and Sagawa, H. (1981). *Phys. Lett.*, **106B**, 379.
24. Bertsch, G. F. and Hamamoto, I. (1982). *Phys. Rev.*, **C26**, 1323; Fiebig, H. R. and Wambach, J. (1982). *Nucl. Phys.*, **A386**, 381; Muto, K., Oda, T., and Horie, H. (1982). *Phys. Lett.*, **118B**, 261; Schwesinger, B. and Wambach, J. (1984). *Phys. Lett.*, **134B**, 29; (1984). *Nucl. Phys.*, **A426**, 253; Cha, D., Schwesinger, B., Wambach, J., and Speth, J. (1984). *Nucl. Phys.*, **A430**, 321; Takayanagi, K., Shimizu, K., and Arima, A. (1985). *Nucl. Phys.*, **A444**, 436.
25. Cha, D. and Speth, J. (1984). *Phys. Lett.*, **B143**, 297; Osterfeld, F., Cha, D., and Speth, J. (1985). *Phys. Rev.*, **C31**, 372.
26. Sagawa, H. and van Giai, N. (1982). *Phys. Lett.*, **113B**, 119; Grotz, K., et al. (1983). *Phys. Lett.*, **132B**, 22.
27. Brown, G. E. and Rho, M. (1981). *Nucl. Phys.*, **A372**, 397.
28. Krewald, S., Osterfeld, F., Speth, J., and Brown, G. E. (1981). *Phys. Rev. Lett.*, **46**, 103.
29. de Shalit, A. and Talmi, I. (1963). *Nuclear shell theory*. Academic, New York.
30. Lawson, R. D. (1980). *Theory of the nuclear shell model*. Clarendon, Oxford.
31. Arima, A. (1971). *The structure of $1f_{7/2}$ nuclei* (ed. R. A. Ricci), p. 385. Editorice Compositori, Bologna; Horie, H. (1973). *J. Phys. Soc. Japan Suppl.*, **34**, 461; Zamick, L. (1973). *J. Phys. Soc. Japan Suppl.*, **34**, 470; Arita, K. (1973). *J. Phys. Soc. Japan Suppl.*, **34**, 516; Goode, P. and Zamick, L. (1972). *Part. and Nucl.*, **3**, 125.
32. Arima, A. (1973). *J. Phys. Soc. Japan Suppl.*, **34**, 205; Tonozuka, I., Sasaki, K., and Harada, K. (1973). *J. Phys. Soc. Japan Suppl.*, **34**, 475; Arita, K. (1973). *J. Phys. Soc. Japan Suppl.*, **34**, 516; Arita, K. (1973). *Proc. int. conf. on nuclear physics, Munich, 1973*, Vol. 1, (ed. J. de Boer and H. J. Mang), p. 264. North-Holland, Amsterdam.
33. Towner, I. S., Khanna, F. C., and Häusser, O. (1977). *Nucl. Phys.*, **A277**, 285.
34. Nomura, M. (1972). *Phys. Lett.*, **40B**, 522.
35. Raghavan, P. (1989). *At. Data and Nucl. Data Tables*, **42**, 189.
36. Brandow, B. H. (1967). *Rev. Mod. Phys.*, **39**, 771.
37. Ellis, P. J. and Osnes, E. (1977). *Rev. Mod. Phys.*, **49**, 777.
38. Ellis, P. J. and Siegel, S. (1971). *Phys. Lett.*, **34B**, 177.
39. Ichimura, M. and Yazaki, K. (1965). *Nucl. Phys.*, **63**, 401.
40. Mavromatis, H. A. and Zamick, L. (1966). *Phys. Lett.*, **20**, 191; (1967). *Nucl. Phys.*, **A104**, 19.
41. Kuo, T. T. S. and Brown, G. E. (1966). *Nucl. Phys.*, **85**, 40.
42. Shimizu, K., Ichimura, M., and Arima, A. (1974). *Nucl. Phys.*, **A226**, 282.
43. Towner, I. S. and Khanna, F. C. (1983). *Nucl. Phys.*, **A399**, 334; (1979). *Phys. Rev. Lett.*, **42**, 51.

4

MESON-EXCHANGE CURRENTS

To introduce the topic of meson-exchange currents it is necessary to review some elements of quantum electrodynamics. We have relegated most of this to Appendix A, which serves mainly as a way of introducing our notation. We will assume the reader has some familiarity with the topics of field theory, Lagrangians, Dirac equation, and Feynman diagrams; if not, there are many textbooks that can be consulted. Our notation for four-vectors is as follows. If $A_\mu = (\mathbf{A}, iA_0)$ and $B_\mu = (\mathbf{B}, iB_0)$ are two Lorentz four-vectors then their scalar product is

$$A_\mu B_\mu \equiv A \cdot B = \mathbf{A} \cdot \mathbf{B} + A_4 B_4 = \mathbf{A} \cdot \mathbf{B} - A_0 B_0,$$

with $A_4 = iA_0$. Note only subscripts are used in this notation (in contrast to the introduction of covariant and contravariant vectors with subscripts and superscripts). A discussion of this notation is to be found in de Wit and Smith.[1]

4.1 Minimal coupling

Let $A_\mu(\mathbf{r}, t) = A_\mu(x)$ be a four-vector potential describing a given set of classical electric and magnetic fields. The coupling of a lepton of charge q ($q = -e$ for an electron) to these fields is found by the substitution

$$\partial_\mu \to \partial_\mu - iqA_\mu \qquad (4.1)$$

in the particle's equations of motion. This is the so-called 'minimum coupling' rule. If we apply this prescription to the free Dirac Lagrangian, $\mathscr{L}_D$:

$$\mathscr{L}_D = -\bar{\psi}(x)(\gamma_\mu \partial_\mu + m)\psi(x) \qquad (4.2)$$

an additional term is obtained in the Lagrangian

$$\mathscr{L}_{em} = iq\bar{\psi}(x)\gamma_\mu \psi(x) A_\mu(x)$$
$$\equiv j_\mu(x) A_\mu(x). \qquad (4.3)$$

This equation defines the electromagnetic current, $j_\mu(x)$. In particular the fourth component represents the charge density and the integral over all space the charge operator, Q:

$$Q = \int d^3x\, j_0(x) = q \int d^3x\, \bar{\psi}(x)\gamma_4 \psi(x).$$

Expanding the field operators, as given in Appendix A, we obtain

$$Q = q \sum_s \mathrm{d}^3 p \frac{E}{m} [a_s^\dagger(p) a_s(p) - b_s^\dagger(p) b_s(p)].$$

We now calculate one-particle matrix elements of Q using the anticommutation rules. For particles created by $a_s^\dagger(p)$ we find

$$Q a_s^\dagger(p) |0\rangle = +q a_s^\dagger(p) |0\rangle,$$

while for particles created by $b_s^\dagger(p)$:

$$Q b_s^\dagger(p) |0\rangle = -q b_s^\dagger(p) |0\rangle.$$

These results show that particles created by a-operators and particles created by b-operators have opposite 'charge'. This 'charge' can be the electric charge, but it can also be any other charge-like quantum number that respects the fermion's equations of motion. (We discuss this further in Section 4.4.) Thus a-particles and b-particles are charge conjugates to each other. We arbitrarily decide (in keeping with normal terminology) that a-operators create particles and b-operators create antiparticles.

The electromagnetic coupling, $j_\mu(x)$ of eqn (4.3), when used in the standard Feynman–Dyson perturbation theory, gives splendid agreement with electromagnetic experiments (for leptons). It leads to the result that the g-factor of a lepton is 2. Any deviation from this so-called Dirac value comes about from higher-order terms in the perturbation theory (radiative corrections).

The proof that $g = 2$ for leptons is as follows.[2] Consider the interaction of a lepton with a stationary magnetic field, which is described by a three-vector potential A: $B = \nabla \times A$. For simplicity consider the decay of a lepton with an initial momentum $p_1 \equiv (p_1, iE_1)$ to a final lepton of momentum $p_1' \equiv (p_1', iE_1')$ emitting a photon of momentum $k = (k, iE_k)$. Then from Appendix A, the Feynman rules for the decay process governed by the Lagrangian, $iq \bar\psi \gamma_\mu \psi A_\mu$, the T-matrix element is

$$T_{\mathrm{fi}} = -iq \bar{u}_{s'}(p_1') \gamma_\mu u_s(p_1) \left(\frac{m}{E_1'} \frac{m}{E_1} \right)^{1/2} A_\mu. \qquad (4.4)$$

We evaluate T_{fi} for the case μ = space-like choosing the Breit coordinate system defined by the requirement $p_1 + p_1' = 0$. This system of reference is particularly convenient since the limit of taking the squared four-momentum transfer $k^2 = (p_1 - p_1')^2 \to 0$ to zero leads automatically to the rest system of the particle. Using the expressions in Appendix A for the plane wave spinors $u_{s'}(-p_1)$ and $u_s(p_1)$ and the standard representation for γ we obtain for the particular case of an electron with $q = -e$

$$T_{\mathrm{fi}} = -2i \frac{e}{2m} \chi_{s'}^\dagger \cdot \boldsymbol{\sigma} \cdot \boldsymbol{p}_1 \times \boldsymbol{A} \chi_s \qquad (4.5)$$

4.1 MINIMAL COUPLING

in the limit $E_1 = m(1 + \ldots)$ and $E_1' = m(1 + \ldots)$. This result is then compared with a non-relativistic expression obtained from the well known interaction of a magnetic moment μ is a constant external field B

$$|T_{\text{fi}}|_{\text{nr}} = (\psi_{-p_1}, H_{\text{int}} \psi_{p_1}), \tag{4.6}$$

with

$$H_{\text{int}} = -\mu \cdot B = -g \frac{e}{2m} S \cdot B,$$

$$\psi_{p_1} = (2\pi)^{-3/2} \exp(i p_1 \cdot x_1) \chi_s, \tag{4.7}$$

where μ is the g-factor times the spin in magneton units, and $B = \nabla \times A$. Note, also, that $\sigma = 2S$. Thus

$$|T_{\text{fi}}|_{\text{nr}} = \int d^3 x_1 \psi^*_{-p_1}(x_1) H_{\text{int}} \psi_{p_1}(x_1)$$

$$= g \frac{e}{2m} \tfrac{1}{2} \chi_s^\dagger \cdot \sigma \chi_s (2\pi)^{-3} \int d^3 x_1 \exp(2i p_1 \cdot x_1) \nabla \times A$$

$$= -i g \frac{e}{2m} \chi_s^\dagger \sigma \cdot p_1 \times A \chi_s \tag{4.8}$$

where we have integrated by parts dropped the surface term, and, because A represents a constant external field, obtained a delta function. Comparing $|T_{\text{fi}}|_{\text{nr}}$ with T_{fi} then identifies $g = 2$. In very much the same way one can show that the corresponding antiparticle has the opposite magnetic moment.

However, this minimal coupling prescription is not unique. This has to do with the ambiguity in $\mathscr{L}_D$. Two Lagrange densities that differ by a term of the form $\partial_\mu O_\mu(x)$ yield the same equations of motion providing that $\int d^4x \, \partial_\mu O_\mu(x) = 0$, since they both give rise to the same Lagrange function, $L = \int \mathscr{L}(x) \, d^4x$. Thus such a term can be arbitrarily added to $\mathscr{L}_D$ with impunity; but in the minimal coupling prescription it will generate new terms in $\mathscr{L}_{\text{em}}$ and lead to an entirely different electromagnetic interaction to the one written above. To take a specific example, let us write

$$\mathscr{L}_D' = -\bar{\psi}(x)(\gamma_\mu \partial_\mu + m) \psi(x) - i \frac{K}{2m} \partial_\mu (\bar{\psi} \sigma_{\mu\nu} \partial_\nu \psi) \tag{4.9}$$

where K is an arbitrary constant and the factor $1/2m$ introduced so that K remains dimensionless. This Lagrangian will generate an additional electromagnetic interaction of the form

$$-e \frac{K}{2m} \bar{\psi} \sigma_{\mu\nu} \partial_\nu \psi A_\mu(x).$$

This added coupling is often called an intrinsic Pauli-moment coupling, 'intrinsic' since when it is put into the theory in this way K is a fundamental constant intrinsic to the lepton and not connected in any specific way with the electric charge. There is something unattractive about adding such terms, since one would like to keep the number of fundamental constants in the theory down to an absolute minimum. This additional term modifies the g-factor of a lepton to be $2(1 + K)$. For the electron certainly, and for the rest of the leptons most probably, such terms seem to be entirely irrelevant, since the magnetic properties of these particles can be computed or estimated without them. Thus we would like to have a principle that eliminates these terms from the beginning. This principle usually goes under the name of 'minimal electromagnetic coupling' which essentially states that all Lagrangian densities should contain a minimum number of derivatives.

For strongly interacting fermions we proceed in an analogous way and determine the electromagnetic current by minimal substitution in the Lagrangian. Consider the case of the nucleon, which has two possible states distinguished by the electric charge that can be either $+e$ or 0. In field theory, the nucleon field will now be described by an eight-component spinor operator

$$\psi = \begin{pmatrix} \psi_\pi \\ \psi_\nu \end{pmatrix},$$

with ψ_π and ψ_ν each being four-component Dirac spinor fields for the proton and neutron respectively. In an isospin notation, three operators τ_x, τ_y, τ_z forming the components of a vector, $\boldsymbol{\tau}$, can act upon ψ. The operators are represented by the usual Pauli matrices. In particular acting with τ_z on a pure proton state or on a pure neutron state leads to

$$\tau_z \begin{pmatrix} \psi_\pi \\ 0 \end{pmatrix} = \begin{pmatrix} \psi_\pi \\ 0 \end{pmatrix}; \qquad \tau_z \begin{pmatrix} 0 \\ \psi_\nu \end{pmatrix} = -\begin{pmatrix} 0 \\ \psi_\nu \end{pmatrix}.$$

It is convenient to introduce a projection operator, P_τ, that selects a pure proton part from a composite ψ:

$$P_\tau \psi \equiv \tfrac{1}{2}(1 + \tau_z)\psi = \begin{pmatrix} 1 & 0 \\ 0 & 0 \end{pmatrix} \begin{pmatrix} \psi_\pi \\ \psi_\nu \end{pmatrix} = \begin{pmatrix} \psi_\pi \\ 0 \end{pmatrix}.$$

Note that $P_\tau^2 = P_\tau$. Now the Lagrangian for a system of proton and neutron fields is simply the sum of the two fields

$$\mathscr{L}_{NN} = -\bar{\psi}_\pi(\gamma_\mu \partial_\mu + M)\psi_\pi - \bar{\psi}_\nu(\gamma_\mu \partial_\mu + M)\psi_\nu, \qquad (4.10)$$

neglecting the known mass difference between the physical proton and the physical neutron. The electromagnetic current is given by the minimal

4.1 MINIMAL COUPLING

substitution: $\partial_\mu \psi_\pi \to \partial_\mu \psi_\pi - ieA_\mu \psi_\pi$ and $\partial_\mu \psi_\nu \to \partial_\mu \psi_\nu$, with the charge on the proton being $q = +e$ and on the neutron $q = 0$. The additional term this substitution gives to the Lagrangian is

$$\mathscr{L}_{em} = ie\bar{\psi}_\pi \gamma_\mu \psi_\pi A_\mu$$

and hence a current (for protons) essentially the same as eqn (4.3). Inserting the isospin projection operator we can write this current in terms of composite nucleon fields:

$$j_\mu(x) = ie\bar{\psi}(x)\gamma_\mu \tfrac{1}{2}(1 + \tau_z)\psi(x). \tag{4.11}$$

This current leads to the Dirac value of $g = 2$ for the g-factor of a proton and $g = 0$ for the neutron. Unfortunately, experimental measurements of the magnetic moment of the proton and neutron show they differ significantly from the Dirac values, indicating presumably that the proton and neutron are composite particles. Nevertheless we will continue to describe nucleons as elementary fermions described by Dirac fields but bear in mind that such principles as minimal coupling might have to be modified. Indeed experiment dictates there will be significant Pauli terms. Thus the approach is more heuristic. In it we construct the electromagnetic current by writing down all possible ways of making a Lorentz vector out of the vectors at our disposal: $(p_1)_\mu$, the momentum of the incoming nucleon; $(p_1')_\mu$ the momentum of the outgoing nucleon; and γ_μ, the spin matrices. In all there are twelve such possible linearly independent terms. Note, however, that j_μ is taken in the matrix element between free Dirac spinors, so that we can use the Dirac equation to reduce the number of independent vectors to three. Thus the current is written

$$j_\mu(x) = ie\bar{\psi}(x)\left(F_1 \gamma_\mu + i\frac{F_2}{2M}\sigma_{\mu\nu}\partial_\nu + F_3(p_1' - p_1)_\mu\right)\psi(x) \tag{4.12}$$

where F_1, F_2, and F_3 are arbitrary functions of a Lorentz scalar made from the momenta. These functions are also isospin dependent: $F_i = \tfrac{1}{2}(F_i^{(0)} + F_i^{(1)}\tau_z)$, where $F_i^{(0)}$ and $F_i^{(1)}$ are the isoscalar and isovector components respectively. We choose the Lorentz scalar to be k^2 where $k_\mu = (p_1' - p_1)_\mu$. Any other scalar can be expressed in terms of k^2 and the nucleon's mass, M. Recall $p_1^2 = -M^2$, $p_1'^2 = -M^2$, and hence $2p_1' \cdot p_1 = -k^2 - 2M^2$ and $(p_1' + p_1)^2 = -k^2 - 4M^2$. The number of arbitrary functions can be reduced one further by specifying the current be Hermitian and invariant under time reversal. These conditions lead to the requirement (see Bernstein[3]) that $F_1(k^2)$ and $F_2(k^2)$ be real functions, and that $F_3(k^2) = 0$. This latter condition also follows from the requirement that j_μ be a conserved current, $\partial_\mu j_\mu = 0$. Thus we write

$$j_\mu(x) = ie\bar{\psi}(x)\left(F_1(k^2)\gamma_\mu + i\frac{F_2(k^2)}{2M}\sigma_{\mu\nu}\partial_\nu\right)\psi(x), \tag{4.13}$$

where $F_1(k^2)$ is known as the Dirac and $F_2(k^2)$ the Pauli form factor. Next we take matrix elements of this current between free-nucleon one-particle states. Again the field operators are expanded in plane waves, as given in Appendix A, leading to the result

$$\langle p'_1, s' | j_\mu(0) | p_1, s \rangle = ie\bar{u}_{s'}(p'_1)\left(F_1\gamma_\mu - \frac{F_2}{2M}\sigma_{\mu\nu}k_\nu\right)u_s(p_1). \quad (4.14)$$

Note that for economy of notation we have dropped a factor $(2\pi)^{-3}$. This factor is cancelled when the expression is multiplied by $(2\pi)^3$ times the momentum-conserving delta function that comes from the density of states; see the rules for evaluating Feynman diagrams. Note also we choose to evaluate the matrix element at the space–time point $x = 0$. Translational invariance enables us to shift to any other point, namely $\langle p'_1 | j_\mu(x) | p_1 \rangle = \exp[i(p_1 - p'_1)x]\langle p'_1 | j_\mu(0) | p_1 \rangle$.

4.2 Magnetic moment operator

Next with the expression, eqn (4.14), for the electromagnetic current of a nucleon we show how, in the non-relativistic limit, it leads to the standard magnetic moment operator. Note that we will not make this reduction in the Breit frame of reference. This is because we envisage this nucleon being embedded in a nucleus so that the rest frame of interest is the rest frame of the whole nucleus not that of a nucleon. Thus we continue with a general frame of coordinates. (The magnetic moment operator deduced in the Breit frame would only contain the intrinsic spin terms; all orbital terms would be lost.) The three steps in the reduction are: (1) a non-relativistic approximation to the current, (2) a Fourier transformation from a momentum to a coordinate space representation, and (3) a multipole decomposition to pick out the M1 multipole.

1. *Non-relativistic reduction.* Using the expressions in Appendix A for the plane-wave spinors $\bar{u}_{s'}(p'_1)$ and $u_s(p_1)$ and the standard representation for the γ matrices, we multiply out the matrices to obtain for the space part of the current

$$\langle p'_1, s' | j(0) | p_1, s \rangle = \frac{e}{2M}\chi^\dagger_{s'}[F_1(p'_1 + p_1) + (F_1 + F_2)i\sigma \times k]\chi_s + O(1/M^3), \quad (4.15)$$

with $k = p'_1 - p_1$. Here $\chi^\dagger_{s'}$ and χ_s are two-component Pauli spinors and $e/2M$ is the unit of nuclear magneton, μ_N. Equation (4.15) represents the leading term in an expansion in powers of p/M, where M is the nucleon mass and p some typical momentum. Note that the nucleon's energy

4.2 MAGNETIC MOMENT OPERATOR

associated with the spinor $E_1 = (p_1^2 + M^2)^{1/2}$ is expanded binomially $E_1 = M(1 + p_1^2/2M^2 + \ldots)$ in this development. The structure of eqn (4.15) is of an operator sandwiched between Pauli spinors. We write the operator as

$$j(k, p_1', p_1) = \frac{e}{2M}[F_1(p_1' + p_1) + (F_1 + F_2)i\sigma \times k] + O(1/M^3) \quad (4.16)$$

2. *Fourier transformation.* To transform to coordinate space we multiply the spinors by plane-wave factors, $\exp(ip_1' \cdot r_1' - ip_1 \cdot r_1)$ and the photon field by $\exp(-ik \cdot x)$, introduce a momentum-conserving delta function, $(2\pi)^3\delta(p_1 + k - p_1')$, and integrate:

$$j(x, r_1', r_1) = \frac{1}{(2\pi)^9}\int d^3k\, d^3p_1'\, d^3p_1$$
$$\times \exp(ip_1' \cdot r_1' - ip_1 \cdot r_1 - ik \cdot x)(2\pi)^3\delta(p_1 + k - p_1')j(k, p_1', p_1).$$

It is more convenient to introduce relative and centre-of-mass coordinates:

$$p = p_1' - p_1, \quad 2P = p_1' + p_1$$

$$j(x, r_1', r_1) = \frac{1}{(2\pi)^6}\int d^3k\, d^3p\, d^3P \exp[ip \cdot (r_1' + r_1)/2]$$
$$\times \exp[iP \cdot (r_1' - r_1)]\exp(-ik \cdot x)\delta(k - p)j(k, p, P)$$
$$= \frac{1}{(2\pi)^6}\int d^3k\, d^3P \exp[ik \cdot (r_1' + r_1 - 2x)/2$$
$$\times \exp[iP \cdot (r_1' - r_1)]j(k, P). \quad (4.17)$$

Note that if the current is independent of P, then it is said to be a local current since the integrations over P lead to a delta function, $\delta(r_1' - r_1)$. To handle non-local terms we use the replacement $P \exp(iP \cdot r_1) = -i\nabla_1 \exp(iP \cdot r_1)$ and interchange the order of differentiation and integration in the Fourier transform. In this way the dependence on P can be transformed away to become derivatives on the delta function. Thus

$$j(x, r_1) = \frac{1}{(2\pi)^3}\int d^3k \exp[ik \cdot (r_1 - x)]j(k, -i\nabla_1)$$
$$= \frac{1}{(2\pi)^3}\int d^3k \exp[ik \cdot (r_1 - x)][F_1(-2i\nabla_1) + (F_1 + F_2)i\sigma \times k]e/2M, \quad (4.18)$$

where F_1 and F_2 are functions of the scalar, k^2.

3. *Multipole decomposition.* There remains the practical step of projecting out the required multipole operator. For transverse magnetic

multipole moments the projection (see[4]) is:

$$T_{LM}^{\text{mag}}(q, r_1) = \int d^3x \, j_L(qx)[Y_L(\hat{x}), j(x, r_1)]_{LM} \quad (4.19)$$

where $j_L(qx)$ is a spherical Bessel function and the square bracket represents a Clebsch–Gordan coupling of a spherical harmonic of rank L with the current (tensor of rank 1) to form a resultant tensor of rank L. Inserting eqn (4.18) in eqn (4.19)

$$T_{LM}^{\text{mag}}(q, r_1) = \frac{1}{(2\pi)^3} \int d^3k \, \exp(i\mathbf{k} \cdot \mathbf{r}_1) \int d^3x$$
$$\times \exp(-i\mathbf{k} \cdot \mathbf{x}) j_L(qx)[Y_L(\hat{x}), j(k, -i\nabla_1)]_{LM},$$

and expanding the plane-wave factor, $\exp(-i\mathbf{k} \cdot \mathbf{x})$, we obtain

$$T_{LM}^{\text{mag}}(k, r_1) = \frac{1}{4\pi} i^{-L} \int d^2\hat{k} \, \exp(i\mathbf{k} \cdot \mathbf{r}_1)[Y_L(\hat{k}), j(k, -i\nabla_1)]_{LM}, \quad (4.20)$$

where an orthogonality relation on the spherical Bessel functions has been used

$$\int_0^\infty dx \, x^2 j_L(kx) j_L(qx) = \tfrac{1}{2}\pi \delta(k - q)/k^2.$$

With the explicit form of the current given in eqn (4.18) we reach our desired result that

$$T_{LM}^{\text{mag}}(k, r_1) = \frac{e}{2M} \Big\{ 2F_1(k^2) j_L(kr_1)[Y_L(\hat{r}_1), p_1]_{LM}$$
$$+ 2(F_1(k^2) + F_2(k^2)) ik \Big[\Big(\frac{L+1}{2L+1}\Big)^{1/2} j_{L-1}(kr_1)[Y_{L-1}(\hat{r}_1), S]_{LM}$$
$$+ \Big(\frac{L}{2L+1}\Big)^{1/2} j_{L+1}(kr_1)[Y_{L+1}(\hat{r}_1), S]_{LM} \Big] \Big\}, \quad (4.21)$$

where $p_1 = -i\nabla_1$ is the conjugate momentum to r_1, and $\sigma = 2S$. In the long-wavelength approximation, $k^2 \to 0$, this operator reduces to the magnetic multipole operator introduced in eqn (1.36), namely

$$-i T_{LM}^{\text{mag}}(k^2 \to 0, r_1) = \frac{k^L}{(2L+1)!!} \Big(\frac{2L+1}{4\pi}\Big)^{1/2} \Big(\frac{L+1}{L}\Big)^{1/2} T_{LM}(M)$$

and in particular for the magnetic moment operator, $(L = 1)$,

$$-i T_{10}^{\text{mag}}(k^2 \to 0, r_1) = k(6\pi)^{-1/2} \mu, \quad (4.22)$$

where we have identified $g_L = \tfrac{1}{2} F_1(0)$ and $g_S = F_1(0) + F_2(0)$. It is convenient to introduce isospin components as defined in eqn (1.10) by

4.2 MAGNETIC MOMENT OPERATOR

writing $F_1 = \frac{1}{2}(F_1^{(0)} + F_1^{(1)}\tau_z) = g_L^{(0)} + g_L^{(1)}\tau_z$ and $F_2 = \frac{1}{2}(F_2^{(0)} + F_2^{(1)}\tau_z)$ where τ_z is $+1$ for a proton and -1 for a neutron. Then the values of the form factors in the limit $k^2 \to 0$ obtained from the free-nucleon magnetic moments are

$$F_1^{(0)}(k^2 \to 0) = 1.0, \qquad F_1^{(1)}(k^2 \to 0) = 1.0,$$
$$F_2^{(0)}(k^2 \to 0) = -0.12, \qquad F_2^{(1)}(k^2 \to 0) = 3.706. \qquad (4.23)$$

Finally there is one other expression for the magnetic moment operator that we derive now for later use by inserting the defining equation, eqn (4.19), into eqn (4.22) and take the long-wavelength approximation

$$\boldsymbol{\mu}(r_1) = \frac{1}{2} \int d^3x \, \boldsymbol{x} \times \boldsymbol{j}(\boldsymbol{x}, r_1). \qquad (4.24)$$

To obtain the electric quadrupole moment we proceed in an analogous way starting with the time component rather than the space component of the electromagnetic current. Recall that the time component of a four-vector current corresponds to a charge density. The equation analogous to eqn (4.14) is

$$\langle \boldsymbol{p}_1', s' | j_0(0) | \boldsymbol{p}_1, s \rangle = e \bar{u}_{s'}(\boldsymbol{p}_1') \left(F_1 \gamma_4 - \frac{F_2}{2M} \sigma_{4\nu} k_\nu \right) u_s(\boldsymbol{p}_1). \qquad (4.14a)$$

The non-relativistic reduction produces the operator

$$j_0(\boldsymbol{k}, \boldsymbol{p}_1', \boldsymbol{p}_1) = eF_1 + O(1/M^2), \qquad (4.16a)$$

to leading order in p/M. The Fourier transform is straightforward

$$j_0(\boldsymbol{x}, r_1) = \frac{1}{(2\pi)^3} \int d^3k \, \exp[i\boldsymbol{k} \cdot (r_1 - \boldsymbol{x})] eF_1(k^2), \qquad (4.18a)$$

while the projection operator[4] is

$$M_{LM}^{\text{Coul}}(q, r_1) = \int d^3x \, j_L(qx) Y_{LM}(\hat{\boldsymbol{x}}) j_0(\boldsymbol{x}, r_1). \qquad (4.19a)$$

Again substituting eqn (4.18a) in eqn (4.19a) and using orthogonality produces

$$M_{LM}^{\text{Coul}}(k, r_1) = \frac{1}{4\pi} i^{-L} \int d^2\hat{k} \, \exp(i\boldsymbol{k} \cdot r_1) Y_{LM}(\hat{\boldsymbol{k}}) eF_1(k^2) \qquad (4.20a)$$

$$= eF_1(k^2) j_L(kr_1) Y_{LM}(\hat{r}_1). \qquad (4.21a)$$

In the long-wavelength approximation, $k^2 \to 0$, this operator reduces to the electric multipole operator introduced in eqn (1.36), namely

$$M_{LM}^{\text{Coul}}(k^2 \to 0, r_1) = \frac{k^L}{(2L+1)!!} \left(\frac{2L+1}{4\pi} \right)^{1/2} T_{LM}(E),$$

and in particular for the quadrupole moment, $Q = 2T_{20}(E)$ (see eqn (1.37)),

$$M^{\text{Coul}}_{20}(k^2 \to 0, r_1) = \frac{k^2}{15}\left(\frac{5}{16\pi}\right)^{1/2} Q. \qquad (4.22a)$$

Some of the steps in this development may seem to be a little pedantic and unnecessary. That is true. However, this is the framework that will have to be followed in developing operators from the more complicated two-body meson-exchange currents.

4.3 Relativistic corrections

The first step we took in deriving the magnetic moment operator was a non-relativistic reduction of the current operator in which only the leading orders in p/M were retained. A rough estimate of the relativistic correction is obtained, then, by carrying out this reduction to the next higher order. To do this, the normalization factor of the free-nucleon spinor (multiplied by $(M/E)^{1/2}$) is expanded to get $(M/E)^{1/2}[(E+M)/2M]^{1/2} \simeq 1 - p^2/8M^2$ and the factor $(E+M)^{-1}$ in the lower component of the Dirac spinor likewise expanded $(E+M)^{-1} \simeq (2M)^{-1}(1-p^2/4M^2)$, using the on-mass-shell relation $E^2 = p^2 + M^2$ for each spinor. Then, on following the same steps as before, the magnetic moment operator becomes

$$\boldsymbol{\mu} = g_L[\boldsymbol{L}(1-p^2/2M^2) - p^2/2M^2(\boldsymbol{S}-(\boldsymbol{S}\cdot\hat{\boldsymbol{p}})\hat{\boldsymbol{p}})] + g_S\boldsymbol{S}(1-p^2/2M^2) \qquad (4.25)$$

in units of the nuclear magneton, $e/2M$. It is convenient to express the correction this operator gives to the lowest-order result, eqn (4.22), in terms of effective coupling constants introduced in eqn (1.48); namely

$$\boldsymbol{\mu}_{\text{eff}} = g_{L,\text{eff}}\boldsymbol{L} + g_{S,\text{eff}}\boldsymbol{S} + g_{P,\text{eff}}[Y_2, \boldsymbol{S}]$$

where $g_{L,\text{eff}} = g_L + \delta g_L$ etc. In this parametrization the relativistic correction is

$$\begin{aligned}
\delta g_L &= -\tfrac{1}{2}g_L\langle p^2/M^2\rangle, \\
\delta g_S &= (-\tfrac{1}{3}g_L - \tfrac{1}{2}g_S)\langle p^2/M^2\rangle, \\
\delta g_P &= -\tfrac{1}{6}(8\pi)^{1/2}g_L\langle p^2/M^2\rangle,
\end{aligned} \qquad (4.26)$$

where $\langle p^2/M^2\rangle$ is an expectation value for a nucleon in a nucleus. If this nucleon is in the shell-model orbital with quantum numbers n and l, and if its radial wavefunction is given by a harmonic oscillator function, then $\langle p^2/M^2\rangle$ is trivially evaluated

$$\langle p^2/M^2\rangle = (2n + l + \tfrac{3}{2})\hbar\omega/M, \qquad (4.27)$$

where $\hbar\omega$ is the characteristic oscillator energy and M the nucleon mass. Note that this expectation value involves the second derivative and so is very sensitive to the choice of the single-particle radial wavefunction. Estimates of this relativistic correction can be found in the literature.[5] The main result is that the free-nucleon values of g_L and g_S are reduced by roughly 2 to 3%.

We can proceed in the same way to find a relativistic correction to the quadrupole moment operator. Firstly the time component of the electromagnetic current is evaluated to order $1/M^2$:

$$j_0(k, p'_1, p_1) = eF_1 - e\frac{(F_1 + 2F_2)}{8M^2}[k^2 + i\sigma \times (p_1 + p'_1) \cdot k], \quad (4.28)$$

and the Fourier transform is

$$j_0(x, r_1) = \frac{e}{(2\pi)^3} \int d^3k \, \exp[ik \cdot (r_1 - x)]\left(F_1 - \frac{(F_1 + 2F_2)}{8M^2}(k^2 + 2\sigma \times \nabla_1 \cdot k)\right). \quad (4.29)$$

There are two terms to the correction: the first in k^2/M^2 is called the Darwin–Foldy and the second in $\sigma \times \nabla_1 \cdot k/M^2$ the spin–orbit term. See Friar[6] and earlier references cited there for a discussion of these terms. To find the correction to the quadrupole moment operator we need to make a multipole decomposition, eqn (4.22a), for multipolarity $L = 2$, and take the long-wavelength approximation. The Darwin–Foldy term gives no contribution. In lowest order, $O(1/M^0)$, the standard quadrupole operator

$$Q = eF_1\left(\frac{16\pi}{5}\right)^{1/2} r_1^2 Y_{20}(\hat{r}_1)$$

is obtained (to within a constant) and a correction term of order $1/M^2$,

$$\Delta Q = e\frac{(F_1 + 2F_2)}{(2M)^2}\left(\frac{2}{5}\right)^{1/2} T_{20}(\sigma \times p_1, r_1) \quad (4.30)$$

where T_{20} is a spherical tensor of rank 2 formed from the coupling of the vectors $\sigma \times p_1$ and r_1. Here $p_1 = -i\nabla_1$ is the conjugate momentum to r_1.

These relativistic corrections are rough estimates. We have evaluated the expectation values of the current operators in plane-wave Dirac spinors without a nuclear potential, $V(r)$, present. If such a term is included in the Dirac Hamiltonian, the $V(r)$ term comes into the charge operator, j_0, like V/M^3 and into the current operator, j, like V/M^2. Since in the weak binding limit the potential is considered to be of order $1/M$, this will affect the corrections to the current in order $1/M^3$ (comparable to corrections already included in eqn (4.25)) and to the charge operator in order $1/M^4$ (negligible compared to eqn (4.30)). A discussion of this and other problems of relativistic corrections is to be found in Friar.[6]

4.4 Current conservation

One very important property of the electromagnetic current is that it is a conserved current: $\partial_\mu j_\mu = 0$. To see how this comes about we briefly review some results of the Lagrange formalism.[1] The Lagrangian is a function of fields and their derivatives and independently a function of adjoint fields and derivatives

$$\mathscr{L} = \mathscr{L}(\psi, \partial_\mu \psi; \bar\psi, \partial_\mu \bar\psi). \tag{4.31}$$

Each field is a function of the space–time variable x. An integral, d^4x, of the Lagrangian from space–time point x_1 to space–time point x_2 is called the action. The action is a functional in that it depends on the path of integration from x_1 to x_2. Hamilton's variational principle states that the trajectory followed by the system is the one for which the action is an extremum. Applying this principle to the general form of the Lagrangian, eqn (4.31), leads to the so-called Euler–Lagrange equations which are indeed the equations of motion for the system:

$$\partial_\mu \frac{\partial \mathscr{L}}{\partial(\partial_\mu \psi)} - \frac{\partial \mathscr{L}}{\partial \psi} = 0, \qquad \partial_\mu \frac{\partial \mathscr{L}}{\partial(\partial_\mu \bar\psi)} - \frac{\partial \mathscr{L}}{\partial \bar\psi} = 0. \tag{4.32}$$

For example the Euler–Lagrange equations applied to the Dirac Lagrangian, $\mathscr{L}_D$,

$$\begin{aligned}\mathscr{L}_D &= -\tfrac{1}{2}\bar\psi(\gamma_\mu \partial_\mu + m)\psi + \text{h.c.} \\ &= -\tfrac{1}{2}\bar\psi\gamma_\mu(\partial_\mu \psi) + \tfrac{1}{2}(\partial_\mu \bar\psi)\gamma_\mu \psi - m\bar\psi\psi,\end{aligned} \tag{4.33}$$

leads to the Dirac equation

$$(\gamma_\mu \vec\partial_\mu + m)\psi = 0, \qquad \bar\psi(-\vec\partial_\mu \gamma_\mu + m) = 0. \tag{4.34}$$

Now consider the phase transformation $\psi \to \psi' = e^{ie\Lambda}\psi$ and $\bar\psi \to \bar\psi' = e^{-ie\Lambda}\bar\psi$, where Λ is a real constant, independent of space–time. It is clear the Lagrangian, $\mathscr{L}_D$ is invariant under this transformation because each term in $\mathscr{L}_D$ contains $\bar\psi$ and ψ linearly. Consider an infinitesimal version of this transformation, keeping terms linear in Λ, but not Λ^2 or higher powers. Then the infinitesimal variations in the fields are

$$\delta\psi = ie\Lambda\psi, \qquad \delta\bar\psi = -ie\Lambda\bar\psi,$$

and the corresponding variation in the Lagrangian, eqn (4.31)

$$\delta\mathscr{L} = \frac{\partial \mathscr{L}}{\partial \psi}\delta\psi + \frac{\partial \mathscr{L}}{\partial(\partial_\mu \psi)}\delta(\partial_\mu \psi) + \frac{\partial \mathscr{L}}{\partial \bar\psi}\delta\bar\psi + \frac{\partial \mathscr{L}}{\partial(\partial_\mu \bar\psi)}\delta(\partial_\mu \bar\psi).$$

4.4 CURRENT CONSERVATION

Integrating the second and fourth terms by parts leads to

$$\delta\mathcal{L} = \left(\frac{\partial\mathcal{L}}{\partial\psi} - \partial_\mu\frac{\partial\mathcal{L}}{\partial(\partial_\mu\psi)}\right)\delta\psi + \partial_\mu\left(\frac{\partial\mathcal{L}}{\partial(\partial_\mu\psi)}\delta\psi\right)$$

$$+ \left(\frac{\partial\mathcal{L}}{\partial\bar\psi} - \partial_\mu\frac{\partial\mathcal{L}}{\partial(\partial_\mu\bar\psi)}\right)\delta\bar\psi + \partial_\mu\left(\frac{\partial\mathcal{L}}{\partial(\partial_\mu\bar\psi)}\delta\bar\psi\right)$$

$$= \partial_\mu\frac{\partial\mathcal{L}}{\partial(\partial_\mu\psi)}ie\Lambda\psi - \partial_\mu\frac{\partial\mathcal{L}}{\partial(\partial_\mu\bar\psi)}ie\Lambda\bar\psi$$

because of the Euler–Lagrangian equations. Since the Lagrangian is invariant under the transformation, the infinitesimal variation is zero. That is

$$\delta\mathcal{L} = ie\Lambda\,\partial_\mu\left(\frac{\partial\mathcal{L}}{\partial(\partial_\mu\psi)}\psi - \bar\psi\frac{\partial\mathcal{L}}{\partial(\partial_\mu\bar\psi)}\right) = 0 \tag{4.35}$$

or that $\partial_\mu S_\mu = 0$ where, because Λ is an arbitrary constant $\Lambda \neq 0$,

$$S_\mu = ie\left(\frac{\partial\mathcal{L}}{\partial(\partial_\mu\psi)}\psi - \bar\psi\frac{\partial\mathcal{L}}{\partial(\partial_\mu\bar\psi)}\right). \tag{4.36}$$

In particular, for the Dirac Lagrangian eqn (4.33), the current S_μ reduces to $ie\bar\psi\gamma_\mu\psi$ and is exactly the electromagnetic current, j_μ, determined in eqn (4.3) from minimal coupling. Two comments are in order:

1. The proof above demonstrates that for every transformation that leaves the Lagrangian invariant there is a corresponding conserved current. This is known as Noether's theorem. In particular, if the Lagrangian is invariant with respect to a set of constant gauge transformations Λ^α, with Λ^α independent of the space–time variable x, then for every α there is an associated divergenceless current, $\partial_\mu j_\mu^\alpha = 0$, and a 'generalized charge', Q^α, that is time independent. This latter statement also assumes that the space part of the current, $\boldsymbol{j}^\alpha$, vanishes at large distances such that Gauss' theorem is applicable

$$\int d^3x\,\boldsymbol\nabla\cdot\boldsymbol{j}^\alpha = -\int_S \boldsymbol{j}\cdot d\boldsymbol{S} = 0. \tag{4.37}$$

Then the generalized charge defined as

$$Q^\alpha = \int d^3x\,j_0(x) \tag{4.38}$$

is time independent because

$$\frac{d}{dt}Q^\alpha = \int d^3x \frac{\partial}{\partial x_0} j_0^\alpha(x)$$
$$= \int d^3x \left(\frac{\partial}{\partial x_0} j_0^\alpha(x) - \nabla \cdot j^\alpha(x)\right)$$
$$= -\int d^3x \, \partial_\mu j_\mu^\alpha(x)$$
$$= 0 \qquad (4.39)$$

for a conserved current. Thus Q^α is a constant of motion and commutes with the Hamiltonian. Examples of charges that are constants of motion because they are derivable from conserved currents are electric charge, lepton, and baryon number.

2. Note that the current deduced from the invariance of the Langrangian under a constant gauge transformation is identical to the current deduced from the minimal coupling with the electromagnetic field. This shows that the replacement $\partial_\mu \to \partial_\mu - ieA_\mu$ guarantees the invariance of the total Lagrangian, $\mathcal{L}_D + \mathcal{L}_{em}$, under simultaneous gauge transformations

$$\psi \to e^{ie\Lambda}\psi,$$
$$A_\mu \to A_\mu + \partial_\mu \Lambda. \qquad (4.40)$$

However, it must be remembered that the current determined by either method has not been determined unambiguously. One can also add terms to the current such that these additional terms in themselves are divergenceless.

Let us consider strongly interacting fermions and the postulated form of their current given in eqns (4.12) and (4.13). Note that this is a conserved current, $\partial_\mu j_\mu = 0$, provided $F_3(k^2) = 0$. There is no experimental evidence to date to suggest that $F_3(k^2) \neq 0$ for strongly interacting fermions; indeed the conservation of charge in strong interactions is a consequence of the current being conserved.

Consider next the matrix elements of the divergence of the current taken between free-nucleon one-particle states, namely

$$\langle p_1', s' | \partial_\mu j_\mu(x) | p_1, s \rangle = 0,$$
$$\langle p_1', s' | \nabla \cdot j(x) | p_1, s \rangle = -\left\langle p_1', s' \left| \frac{\partial}{\partial t} j_0(x) \right| p_1, s \right\rangle.$$

Since the matrix element of $j_0(x)$ is time independent (see comment (1) above), the time derivative can be replaced by the commutator with the

Hamiltonian. Furthermore from translational invariance we can rewrite the left-hand side of the equation to get

$$\nabla \exp[i(\boldsymbol{p}_1 - \boldsymbol{p}_1') \cdot \boldsymbol{x}] \cdot \langle \boldsymbol{p}_1', s' | \boldsymbol{j}(0) | \boldsymbol{p}_1, s \rangle = -i \langle \boldsymbol{p}_1', s' | [H, j_0] | \boldsymbol{p}_1, s \rangle,$$
(4.41)
$$\langle \boldsymbol{p}_1', s' | \boldsymbol{k} \cdot \boldsymbol{j}(0) | \boldsymbol{p}_1, s \rangle = \langle \boldsymbol{p}_1', s' | [H, j_0] | \boldsymbol{p}_1, s \rangle,$$

where $\boldsymbol{k} = \boldsymbol{p}_1' - \boldsymbol{p}_1$. We make a non-relativistic reduction of both sides, and keep just the leading order terms. First for the left-hand side

$$\langle \boldsymbol{p}_1', s' | \boldsymbol{k} \cdot \boldsymbol{j}(0) | \boldsymbol{p}_1, s \rangle = \frac{e}{2M} F_1 \chi_{s'}^\dagger [\boldsymbol{k} \cdot (\boldsymbol{p}_1' + \boldsymbol{p}_1)] \chi_s$$
$$\boldsymbol{k} \cdot \boldsymbol{j} = e F_1 (p_1'^2 - p_1^2)/2M$$
(4.42)

using eqn (4.15). Note that the Pauli term in the current, the term proportional ot F_2, is entirely transverse and gives no contribution to the current conservation equation. Second for the right-hand side and using only the kinetic energy part of the Hamiltonian, T:

$$\langle \boldsymbol{p}_1', s' | [T, j_0] | \boldsymbol{p}_1, s \rangle = \frac{1}{2M}(p_1'^2 - p_1^2) \langle \boldsymbol{p}_1', s' | j_0 | \boldsymbol{p}_1, s \rangle$$
$$[T, j_0] = e F_1 (p_1'^2 - p_1^2)/2M$$
(4.42a)

using eqn (4.16a). Thus the matrix element of the current, eqn (4.16), satisfies the relation

$$\boldsymbol{k} \cdot \boldsymbol{j} = [T, j_0],$$
(4.43)

rather than the relationship $\boldsymbol{k} \cdot \boldsymbol{j} = [H, j_0]$ deduced from current conservation. Thus, there has to be an additional term to the current, $\boldsymbol{j}^{ex}$, such that

$$\boldsymbol{k} \cdot \boldsymbol{j}^{ex} = [V, j_0],$$
(4.44)

where V is the potential energy part of the Hamiltonian. This extra current, $\boldsymbol{j}^{ex}$, the meson-exchange current, is two-body in character since V is a two-body potential. Furthermore, on identifying V with the exchange of an isovector meson, such as the pion, with an isospin dependence $V = v_{12} \boldsymbol{\tau}_1 \cdot \boldsymbol{\tau}_2$ then the commutator with the isovector part of the charge density, $j_0^{(1)}$, does not vanish:

$$[V, j_0] = e[v_{12} \boldsymbol{\tau}_1 \cdot \boldsymbol{\tau}_2, \tfrac{1}{2} F_1^{(1)} \tau_{1z}] = ie(\boldsymbol{\tau}_1 \times \boldsymbol{\tau}_2)_z F_1^{(1)} v_{12}.$$
(4.45)

Recall we have written $j_0 = j_0^{(0)} + j_0^{(1)}$ where $j_0^{(0)} = \tfrac{1}{2} e F_1^{(0)}$ and $j_0^{(1)} = \tfrac{1}{2} e F_1^{(1)} \tau_{1z}$ are the isoscalar and isovector parts of the charge density operator to leading order in the non-relativistic reduction. *Thus current continuity and the isospin dependence of the nuclear force guarantee the existence of exchange currents.* The magnetic moment due to the exchange current, from eqn (4.24), is

$$\mu^{ex}(r_1, r_2) = \frac{1}{2}\int d^3x\, x \times j^{ex}(x, r_1, r_2), \qquad (4.46)$$

which can be broken into two pieces[7,8] in the following way

$$\mu^{ex}(r_1, r_2) = \tfrac{1}{2}R \times \int d^3x\, j^{ex}(x, r_1, r_2)$$
$$+ \frac{1}{2}\int d^3x\, [x - \tfrac{1}{2}(r_1 + r_2)] \times j^{ex}(x, r_1, r_2) \qquad (4.47)$$

where $R = (r_1 + r_2)/2$. Since j^{ex} depends only on the relative coordinates $x - r_1$, $x - r_2$ the second term in eqn (4.47) is translationally invariant. By contrast, the first term depends on the choice of origin and is not translationally invariant. Using Green's theorem and the fact that currents j^{ex} vanish exponentially at large distances, the integral in the first term of eqn (4.47) becomes

$$\int d^3x\, j^{ex} = -\int d^3x\, x(\nabla \cdot j^{ex}), \qquad (4.48)$$

and hence through the equation of continuity is known unambiguously. Therefore the magnetic moment terms, which are translationally non-invariant are determined uniquely:

$$\mu^{ex}(r_1, r_2) = -\tfrac{1}{2}e(R \times r)(\tau_1 \times \tau_2)_z F_1^{(1)} v_{12}, \qquad (4.49)$$

where $r = r_1 - r_2$. This is known as the Sachs moment and is completely model independent inasmuch as it is determined by the nucleon–nucleon potential, v_{12}. The Sachs moment is the dominant contribution to the meson-exchange correction to the magnetic moment. Note, in particular, that this correction is isovector in character.

This approach, however, tells us nothing about the translationally invariant and hence model dependent second term in eqn (4.47). For this reason the method of constructing the exchange-current operator directly from the continuity equation has largely been superseded by an *ab initio* approach rather similar to the way the one-boson-exchange model is constructed for the nuclear force: all Feynman diagrams in which one meson is exchanged between two nucleons are drawn and an external photon line is linked in all possible places. Graphs involving two meson exchanges are generally too complicated to consider, but the essence of the one-boson-exchange model is that the influence of such graphs can be simulated by the exchange of a boson, for example two pions in a singlet state are described by a fictitious scalar σ-meson. When carried out consistently, the exchange current from this approach will automatically satisfy the continuity equation, with the potential V in the commutator

4.5 Pion-exchange graphs

with the charge density, eqn (4.44), being the one-boson-exchange potential. This Feynman graph approach was pioneered by Chemtob and Rho[9] and is called the S-matrix method.

We start by considering graphs involving pion exchange. These are the most important graphs because the pion, being the lightest-mass meson, generates the longest-range exchange current operator. Heavy-meson exchange will lead to short-range operators, which in nuclear physics are less important since the nuclear wavefunction goes rapidly to zero at short distances. This is a consequence of the short-range repulsion in the nucleon–nucleon interaction.

The Lagrangian for a system of nucleons and pions is a sum of three terms

$$\mathscr{L} = \mathscr{L}_{\pi\pi} + \mathscr{L}_{NN} + \mathscr{L}_{\pi NN}$$

where $\mathscr{L}_{\pi\pi}$ is the free-pion Lagrangian, $\mathscr{L}_{NN}$ the free nucleon Lagrangian, and $\mathscr{L}_{\pi NN}$ an interaction Lagrangian describing the emission and absorption of a pion by a nucleon. Recall that the pion is a spin-0 boson that has three different charge states, and therefore requires three independent Hermitian field operators for its description. We write these operators as ϕ_x, ϕ_y, and ϕ_z, envisaging them as three components of a vector, $\boldsymbol{\phi}$. It is more convenient to define

$$\phi_{+1} = -(1/\sqrt{2})(\phi_x + i\phi_y); \qquad \phi_{-1} = +(1/\sqrt{2})(\phi_x - i\phi_y). \quad (4.50)$$

Note that $\phi^\dagger_{+1} = -\phi_{-1}$, $\phi^\dagger_{-1} = -\phi_{+1}$, and $\phi^\dagger_0 = \phi_0 \,(= \phi_z)$. The interpretation of these field operators is: $\phi^\dagger_{+1}$ creates a π^+ meson or destroys a π^- meson, while $\phi^\dagger_{-1}$ creates a π^- meson or destroys a π^+ meson. The Lagrangian density for the free π meson field is simply the superposition of three one-component field Lagrangians

$$\mathscr{L}_{\pi\pi} = -\tfrac{1}{2}\sum_r [(\partial_\mu \phi_r)^2 + m_\pi^2 \phi_r^2] \quad (4.51)$$

where r sums over the Cartesian labels x, y, z. Each ϕ_r field operator must be a pseudoscalar.

The free-nucleon Lagrangian, $\mathscr{L}_{NN}$, is given in eqn (4.10) in terms of its two field operators ψ_π and ψ_ν for the proton and neutron charge states respectively. Lastly the interaction Lagrangian must itself be a Lorentz scalar constructed from pseudoscalar pion field operators, ϕ_r, and pseudoscalar combinations of nucleon field operators. Assuming the simplest form of direct coupling there are two possibilities:

$$\mathscr{L}^{PS}_{\pi NN} = ig_{\pi NN}\bar\psi(x)\gamma_5\tau_r\psi(x)\phi_r(x), \quad (4.52a)$$

$$\mathscr{L}^{PV}_{\pi NN} = -i\frac{g_{\pi NN}}{2M}\bar\psi(x)\gamma_\mu\gamma_5\tau_r\psi(x)\,\partial_\mu\phi_r(x), \quad (4.52b)$$

in each case summed over the isospin label r. The first expression is known as the pseudoscalar πNN coupling and the second as the pseudovector πNN coupling. Note the use of the isospin Pauli matrix with the pion field: $\sum_r \tau_r \phi_r = \boldsymbol{\tau} \cdot \boldsymbol{\phi}$. Written out explicitly this scalar product is

$$\boldsymbol{\tau} \cdot \boldsymbol{\phi} = \tau_x \phi_x + \tau_y \phi_y + \tau_z \phi_z$$

$$= \begin{pmatrix} 1 & 0 \\ 0 & -1 \end{pmatrix} \phi_0 + \sqrt{2} \begin{pmatrix} 0 & 1 \\ 0 & 0 \end{pmatrix} \phi_{-1} - \sqrt{2} \begin{pmatrix} 0 & 0 \\ 1 & 0 \end{pmatrix} \phi_{+1}$$

and hence in terms of proton and neutron fields the pseudoscalar Lagrangian becomes

$$\mathcal{L}_{\pi NN}^{PS} = ig_{\pi NN} \{ \bar{\psi}_\pi \gamma_5 \psi_\pi \phi_0 - \bar{\psi}_v \gamma_5 \psi_v \phi_0 + \sqrt{2} \, \bar{\psi}_\pi \gamma_5 \psi_v \phi_{-1}$$
$$- \sqrt{2} \, \bar{\psi}_v \gamma_5 \psi_\pi \phi_{+1} \}.$$

There is a similar expression for the pseudovector Lagrangian. Note the probability for emission of a charged meson by a nucleon will contain a supplementary factor $(\sqrt{2})^2$ or 2 when compared to the probability for emission of a neutral meson. There is no *a priori* reason to choose one form of the interaction Lagrangian over the other. In lowest-order processes they both lead to the same expressions, so the choice is in most cases not important. Only in higher orders will differences appear, and without imposing any supplementary conditions (such as chiral invariance) these differences remain unresolved. They become a measure of the model dependence in the chosen Lagrangian. We will not pursue this topic further. Should it become necessary to select one form of the Lagrangian over the other, we will use the pseudovector form.

As a brief aside and as an example of the use of Feynman rules, let us calculate the one-pion-exchange potential as depicted in the Feynman diagram in Fig. 4.1. The momenta on the nucleon lines are defined in the

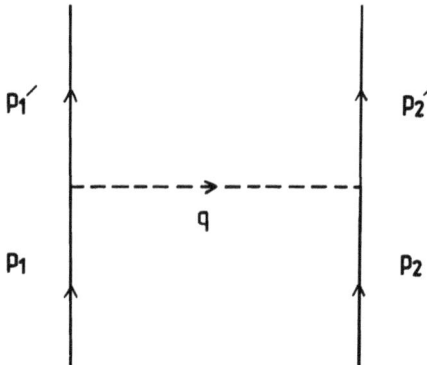

Fig. 4.1 Feynman diagram for the exchange of a pion from nucleon 1 to nucleon 2.

4.5 PION-EXCHANGE GRAPHS

diagram; the momentum transfer on the pion is $q = p'_2 - p_2$. Note the direction of the arrow on the pion line. From the rules in Appendix A, we can write down an expression for the T-matrix for the case of pseudovector coupling:

$$T_{\text{fi}} = \frac{g_{\pi NN}^2}{(2M)^2} \boldsymbol{\tau}_1 \cdot \boldsymbol{\tau}_2 \left(\frac{M}{E'_1} \frac{M}{E'_2} \frac{M}{E_1} \frac{M}{E_2} \right)^{1/2} \bar{u}_{s'_1}(\boldsymbol{p}'_1)[\not{q} \gamma_5] u_{s_1}(\boldsymbol{p}_1)$$
$$\times (q^2 + m_\pi^2)^{-1} u_{s'_2}(\boldsymbol{p}'_2)[\not{q} \gamma_5] u_{s_2}(\boldsymbol{p}_2),$$

where ∂_μ acting on the pion field is replaced by $-iq_\mu$ when the direction of the arrow points out of the vertex and by $+iq_\mu$ when the arrow points into the vertex. Here $\not{q}$ is $\gamma_\mu q_\mu$. Using the plane-wave spinor solutions, and multiplying out the Dirac matrices keeping only terms to leading order in p/M, we obtain a non-relativistic version

$$T_{\text{fi}} = (\chi_{s'_1} \chi_{s'_2}, V \chi_{s_1} \chi_{s_2}),$$
$$V(q) = -\frac{g_{\pi NN}^2}{(2M)^2} \boldsymbol{\tau}_1 \cdot \boldsymbol{\tau}_2 \frac{(\boldsymbol{\sigma}_1 \cdot \boldsymbol{q})(\boldsymbol{\sigma}_2 \cdot \boldsymbol{q})}{q^2 + m_\pi^2} + O(1/M^4), \quad (4.53)$$

where χ_s are two-component Pauli spinors. The same expression, to leading order in p/M, is obtained with pseudoscalar coupling as the reader can easily confirm.

The meson-exchange currents of interest are obtained from the basic pion-exchange graph, Fig. 4.1, by linking one external photon line in all possible ways. We consider for the moment only graphs involving nucleons and pions. This subset are known as the Born graphs. Graphs in which the photon can excite the nucleon to a resonant state (such as the isobar Δ) or the pion to a heavy meson (such as the ρ or ω) are deferred until later and are known as the non-Born graphs. In Fig. 4.2 are drawn the three basic Born graphs: (a) the seagull graph, (b) the pair graph, and (c) the current graph. We consider each one in turn.

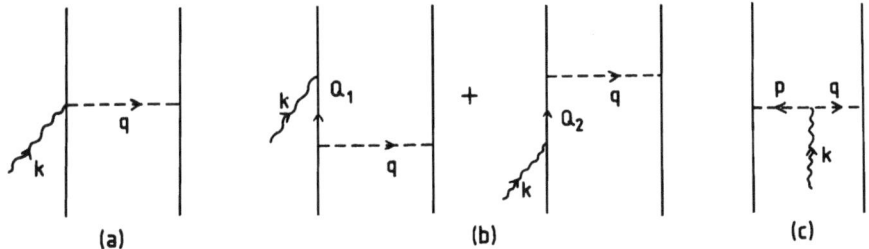

Fig. 4.2 MEC Born graphs of pion range: (a) seagull graph, (b) pair graph, and (c) current graph. The momenta on the four external lines are p_1, p_2, p'_1, p'_2 as in Fig. 4.1. Graphs in which the photon interacts with nucleon 2 rather than nucleon 1 have not been drawn, but must be included in the mathematical evaluations.

4.5.1 The seagull graph

The seagull graph, in Fig. 4.2(a), has a four-point vertex involving $\bar{\psi}$, ψ, ϕ, and A, the nucleon, pion, and photon fields. To find the Lagrangian for this vertex, we use the minimal substitution prescription, eqn (4.1), in the pseudovector form of the πNN Lagrangian. With the charge $q = +e$ for the π^+ meson, $q = -e$ for the π^- meson, the substitution in terms of the creation operators becomes

$$\partial_\mu \phi^\dagger_{+1} \to \partial_\mu \phi^\dagger_{+1} - ie\phi^\dagger_{+1} A_\mu,$$
$$\partial_\mu \phi^\dagger_{-1} \to \partial_\mu \phi^\dagger_{-1} + ie\phi^\dagger_{-1} A_\mu,$$
$$\partial_\mu \phi^\dagger_0 \to \partial_\mu \phi^\dagger_0.$$

In terms of the annihilation operators, these expressions become

$$\partial_\mu \phi_{-1} \to \partial_\mu \phi_{-1} - ie\phi_{-1} A_\mu,$$
$$\partial_\mu \phi_{+1} \to \partial_\mu \phi_{+1} + ie\phi_{+1} A_\mu,$$
$$\partial_\mu \phi_0 \to \partial_\mu \phi_0,$$

which can be combined into one expression in the Cartesian labels:

$$\partial_\mu \phi_r \to \partial_\mu \phi_r - e\varepsilon_{rsz} \phi_s A_\mu. \tag{4.54}$$

Here ε_{rst} is the antisymmetric tensor in three indices: $\varepsilon_{rst} = +1$ if rst is a cyclic permutation of xyz, $\varepsilon_{rst} = -1$ if rst is an odd permutation of xyz, $\varepsilon_{rst} = 0$ if any two indices are equal. Inserting eqn (4.54) into the Lagrangian, eqn (4.52b), generates a new term linear in the photon field

$$\mathcal{L}_{\pi NN\gamma} = ie\varepsilon_{rsz} \frac{g_{\pi NN}}{2M} \bar{\psi}(x) \gamma_\mu \gamma_5 \tau_r \psi(x) \phi_s(x) A_\mu(x). \tag{4.55}$$

Thus for the meson-exchange graph, Fig. 4.2(a), there are three ingredients: a photo-pion production amplitude governed by eqn (4.55), a pion absorption amplitude governed by eqn (4.52b), and a pion propagator. Using the Feynman rules from Appendix A, we obtain for the T-matrix

$$T_{fi} = -ie \frac{g^2_{\pi NN}}{(2M)^2} (\boldsymbol{\tau}_1 \times \boldsymbol{\tau}_2)_z \left(\frac{M}{E'_1} \frac{M}{E'_2} \frac{M}{E_1} \frac{M}{E_2} \right)^{1/2} \bar{u}_{s'_1}(\boldsymbol{p}'_1)[\gamma_\mu \gamma_5] u_{s_1}(\boldsymbol{p}_1)$$
$$\times (q^2 + m_\pi^2)^{-1} \bar{u}_{s'_2}(\boldsymbol{p}'_2)[\slashed{q}\gamma_5] u_{s_2}(\boldsymbol{p}_2) \varepsilon_\mu(\boldsymbol{k}, \lambda) + (1 \rightleftarrows 2)$$

where $\varepsilon_\mu(\boldsymbol{k}, \lambda)$ is the polarization vector for the photon. Note that a second graph, not drawn in Fig. 4.2(a), in which the photon interacts with nucleon 2, has been included by adding a term in which indices 1 and 2 are interchanged. This leads to a T-matrix that is symmetric under the interchange. Using the plane-wave spinor solutions and multiplying

4.5 PION-EXCHANGE GRAPHS

out the Dirac matrices keeping only terms to leading order in p/M, we obtain a non-relativistic version for the case when the Lorentz index μ is space like

$$T_{fi} = (\chi_{s_1'}\chi_{s_2'}, -j_\mu \varepsilon_\mu \chi_{s_1}\chi_{s_2}),$$

$$j = -ie\frac{g_{\pi NN}^2}{(2M)^2}(\tau_1 \times \tau_2)_z \left(\frac{\sigma_1(\sigma_1 \cdot q)}{q^2 + m_\pi^2} - \frac{\sigma_2(\sigma_1 \cdot p)}{p^2 + m_\pi^2}\right), \quad (4.56)$$

and for when μ is time like

$$j_0 = -ie\frac{g_{\pi NN}^2}{(2M)^3}(\tau_1 \times \tau_2)_z \left(\frac{\sigma_1 \cdot (p_1' + p_1)(\sigma_2 \cdot q)}{q^2 + m_\pi^2} - \frac{\sigma_2 \cdot (p_2' + p_2)(\sigma_1 \cdot p)}{p^2 + m_\pi^2}\right), \quad (4.57)$$

where $p = p_1' - p_1$ and from momentum conservation $k = p + q$. (Note that the T-matrix is a representation of the scattering matrix and its matrix elements are the off-diagonal matrix elements of the Hamiltonian for the system. Thus when discussing potential scattering, the T-matrix element is naturally the matrix element of the potential. And when discussing electromagnetic interactions, the T-matrix element is the matrix element of the interaction Hamiltonian: $H_{int} = -j \cdot A$, where A is the vector potential of the electromagnetic field.) We observe that the deduced MEC current, j, is of order $(1/M^2)$, and is one order smaller than the single-nucleon one-body current given in eqn (4.16). On the other hand the MEC charge density, j_0, is of order $(1/M^3)$ and is three orders smaller than the single-nucleon density of eqn (4.16a). The MEC corrections to the charge density are generally very small. Note as well that the seagull graph leads to purely isovector MEC current and charge density. There are no isoscalar terms.

For the pseudoscalar πNN Lagrangian there are no seagull graphs at all. This is because there are no derivatives in the Lagrangian and the minimal substitution rule cannot be applied.

4.5.2 The pair graph

The pair graph, Fig. 4.2(b), comprises photon–nucleon vertices and pion–nucleon vertices governed by Lagrangians eqns (4.13) and (4.52). We must not forget the isospin dependence of the form factors: $F_1 = \frac{1}{2}(F_1^{(0)} + F_1^{(1)}\tau_z)$ and $F_2 = \frac{1}{2}(F_2^{(0)} + F_2^{(1)}\tau_z)$, because they do not commute with the isospin matrices associated with the pion–nucleon vertices.

First for pseudovector coupling, Feynman diagram rules lead to the following expression for the T-matrix:

$$\begin{aligned}
T_{fi} = i\frac{g_{\pi NN}^2}{(2M)^2}&\left(\frac{M\,M\,M\,M}{E_1'E_2'E_1E_2}\right)^{1/2}\bar{u}(p_1')\\
&\times [e(F_1\gamma_\mu - F_2\sigma_{\mu\nu}k_\nu/2M)S_F(Q_1)\rlap{/}{q}\gamma_5\tau_{1r}\\
&+ \rlap{/}{q}\gamma_5\tau_{1r}S_F(Q_2)e(F_1\gamma_\mu - F_2\sigma_{\mu\nu}k_\nu/2M)]u(p_1)\\
&\times (q^2 + m_\pi^2)^{-1}\bar{u}(p_2')[\rlap{/}{q}\gamma_5]u(p_2)\tau_{2r}\varepsilon_\mu(k,\lambda) + (1 \rightleftarrows 2). \quad (4.58)
\end{aligned}$$

For pseudoscalar πNN coupling, we replace the factor $\not{q}\gamma_5$ by $2Mi\gamma_5$. Here $S_F(Q)$ is the nucleon propagator (without the factor $-i/(2\pi)^4$)

$$S_F(Q) = \frac{1}{i\not{Q} + M}.$$

It is convenient to divide this propagator into positive-frequency and negative-frequency components corresponding to the internal fermion line representing a particle or an antiparticle, namely

$$S_F(Q) = S_F^{(+)}(Q) + S_F^{(-)}(Q),$$

$$S_F^{(+)}(Q) \simeq \frac{1}{2M(M - Q_0)}(-i\boldsymbol{\gamma}\cdot\boldsymbol{Q} + M\gamma_4 + M), \quad (4.59)$$

$$S_F^{(-)}(Q) \simeq \frac{1}{(2M)^2}(-i\boldsymbol{\gamma}\cdot\boldsymbol{Q} - M\gamma_4 + M).$$

From the labelling of the lines in Fig. 4.2(b) we have $Q_1 = p_1' - k$ and hence $M - (Q_1)_0 = M - E_1' + k_0 \simeq k_0$, and $Q_2 = p_1 + k$ with $M - (Q_2)_0 = M - E_1 - k_0 \simeq -k_0$. Similarly the derived current from eqn (4.58) is likewise divided into $j_\mu = j_\mu^{(+)} + j_\mu^{(-)}$. Then a non-relativistic reduction for the space part of the current leads to the following expressions:

$$j^{(+)} = \frac{1}{k_0}\left(V_\pi(q^2)\frac{e}{2M}[F_1(\boldsymbol{Q}_2 + \boldsymbol{p}_1) + (F_1 + F_2)i\boldsymbol{\sigma}_1 \times \boldsymbol{k}]\right.$$

$$\left. - \frac{e}{2M}[F_1(\boldsymbol{Q}_1 + \boldsymbol{p}_1') + (F_1 + F_2)i\boldsymbol{\sigma}_1 \times \boldsymbol{k}]V_\pi(q^2)\right) + O(1/M^4) + (1\rightleftarrows 2),$$

$$= \frac{1}{k_0}[V_\pi(q^2)j(\boldsymbol{k}, \boldsymbol{Q}_2, \boldsymbol{p}_1) - j(\boldsymbol{k}, \boldsymbol{p}_1', \boldsymbol{Q}_1)V_\pi(q^2)] + O(1/M^4) + (1\rightleftarrows 2),$$

$$j^{(-)} = O(1/M^4), \quad (4.60)$$

where $V_\pi(q^2)$ is the pion-exchange potential, eqn (4.53), and $j(\boldsymbol{k}, \boldsymbol{Q}_2, \boldsymbol{p}_1)$ is the nucleon one-body current operator, eqn (4.16), evaluated between nucleon states of momentum $\boldsymbol{Q}_2$ and $\boldsymbol{p}_1$. This result shows that the two-body current is singular in the limit that the photon energy tends to zero, $k_0 \to 0$. More importantly, the positive-frequency part of the propagator has led to a current that is not a true two-body exchange current, but is a product of a one-body current and a potential. This term is already included in the standard calculation where the matrix elements of the one-body current are evaluated with shell-model wavefunctions obtained from the solution of the Schrödinger equation that contains the one-pion exchange potential. Therefore to avoid double counting this term is not included in the exchange-current operator.

4.5 PION-EXCHANGE GRAPHS

Similarly for the time component of the electromagnetic current, a non-relativistic reduction leads to terms

$$j_0^{(+)} = \frac{1}{k_0}[V_\pi(q^2)j_0(k, Q_2, p_1) - j_0(k, Q_2, p_1)V_\pi(q^2)]$$

$$+ \frac{g_{\pi NN}^2}{(2M)^3} e\left(\frac{\sigma_1 \cdot k \sigma_2 \cdot q}{q^2 + m_\pi^2}(F_1^{(1)}\tau_{2z} + F_1^{(0)}\tau_1 \cdot \tau_2)\right.$$

$$\left. + \frac{\sigma_1 \cdot (p_1 + p_1')\sigma_2 \cdot q}{(q^2 + m_\pi^2)} F_1^{(1)} i(\tau_1 \times \tau_2)_z\right) + O(1/M^5) + (1 \rightleftarrows 2),$$

$$j_0^{(-)} = O(1/M^5).$$

(4.61)

Again in the low-energy photon limit there is a singular term that is recognizable as the product of a one-body current density with the pion potential. As before, this term is not considered part of the exchange-current operator. However, to order $1/M^3$, there are additional non-singular terms from the positive-frequency propagator. Note in particular the term in $\sigma_1 \cdot (p_1 + p_1')$. This term, on Fourier transforming to coordinate space, leads to a derivative operator, $-2i\sigma_1 \cdot \nabla_1$, and is a non-local operator. Yet this term exactly cancels a corresponding term from the seagull graph, eqn (4.57), in the low-energy limit $F_1^{(1)}(k^2 \to 0) = 1$. Thus the non-local term is eliminated to order $1/M^3$.

For the pseudoscalar πNN coupling the corresponding non-relativistic reductions yield again

$$j^{(+)} = \frac{1}{k_0}[V_\pi(q^2)j(k, Q_2, p_1) - j(k, p_1', Q_1)V_\pi(q^2)] + O(1/M^4) + (1 \rightleftarrows 2),$$

$$j^{(-)} = -i\frac{g_{\pi NN}^2}{(2M)^2} eF_1^{(1)}\frac{\sigma_1(\sigma_2 \cdot q)}{q^2 + m_\pi^2}(\tau_1 \times \tau_2)_z + O(1/M^4) + (1 \rightleftarrows 2),$$

$$j_0^{(+)} = \frac{1}{k_0}[V_\pi(q^2)j_0(k, Q_2, p_1) - j_0(k, p_1', Q_1)V_\pi(q^2)] + O(1/M^5) + (1 \rightleftarrows 2),$$

(4.62)

$$j_0^{(-)} = \frac{g_{\pi NN}^2}{(2M)^3}\frac{\sigma_1 \cdot k \sigma_2 \cdot q}{q^2 + m_\pi^2} e[(F_1^{(0)} + F_2^{(0)})\tau_1 \cdot \tau_2 + (F_1^{(1)} + F_2^{(1)})\tau_{2z}]$$

$$+ O(1/M^5) + (1 \rightleftarrows 2).$$

In lowest order, the terms from the positive-frequency part of the propagator are singular and are discarded as not being part of the exchange-current operator. The terms from the negative-frequency part, on the other hand, are immediately identifiable. The current term, $j^{(-)}$, is exactly the term obtained from the seagull graph for pseudovector πNN coupling in the limit $F_1^{(1)}(k^2 \to 0) = 1$. Thus in lowest order the same expression is obtained for the exchange current irrespective of whether pseudoscalar or pseudovector pion coupling is used. In the literature this contribution is often called the 'pair' current, because it comes from the negative-frequency propagator with pseudoscalar coupling. For pseudovector coupling, the origin is quite different but the same name is often used. It is almost the same story for the charge density, j_0, except that for the pseudovector coupling the contribution is proportional to F_1 while for the pseudoscalar coupling it is proportional to $F_1 + F_2$. This difference is of the order of 10% and is negligible for the isoscalar term but is quite significant for the isovector term. These differences cannot be resolved; they give a measure of the model dependence in the chosen Lagrangians. Only by inserting additional theoretical constraints can any progress be made. For example, Adam and Truhlik[10] start with a Lagrangian exhibiting pseudovector coupling of pions with nucleons and transform it to one exhibiting pseudoscalar coupling by a chiral transformation on the nucleon fields of

$$\psi \to \psi' = \exp\left(\frac{ig_A}{2f_\pi}\gamma_5(\tau\cdot\phi)\right)\psi.$$

Here g_A is the axial-vector coupling constant and f_π the pion decay constant, which is related by the Goldberger–Trieman relation to the pion–nucleon coupling constant: $f_\pi = Mg_A/g_{\pi NN}$. The point is that the original Lagrangian and the transformed Lagrangian yield identical pion photoproduction amplitudes which at low energy are exactly the same amplitudes as those derived from a model-independent low-energy theorem.[11,12] However, the transformed Lagrangian, besides containing pseudoscalar pion–nucleon coupling, also contains an additional four-point vertex (seagull graph) that involves the electromagnetic F_2 form factor. Indeed this extra graph when added to the pseudoscalar pair graphs is just what is needed to cancel the F_2 dependence in the expression for meson-exchange charge density operator and restore the equivalence with the pseudovector result. Another hint along the same lines comes from the constituent quark model. Beyer et al.[13] consider a photon impinging on an individual quark rather than on a nucleon. Because the quark is considered a point-like fermion with Dirac coupling any anomalous Pauli terms in its Lagrangian are quite small. Writing f_1 and f_2 as the Dirac and Pauli electromagnetic couplings to a quark, Beyer et al. fix the f_2 terms such that a composite three-quark structure has the same electromagnetic couplings as a nucleon, namely

$$F_1^{(0)} + F_2^{(0)} = 3(f_1^{(0)} + f_2^{(0)}) = 0.88,$$
$$F_1^{(1)} + F_2^{(1)} = 5(f_1^{(1)} + f_2^{(1)}) = 4.706.$$

Then in a calculation of the pair graph, assuming a pion is exchanged between quarks rather than between nucleons and assuming the pion–quark coupling is pseudoscalar, Beyer et al. obtain the same expression for the exchange-current charge density operator eqn (4.62) with the same magnitude for the isoscalar term but with reduced magnitude, $(F_1^{(1)} + F_2^{(1)}) \rightarrow (9/25)(F_1^{(1)} + F_2^{(1)})$, in the isovector term. This reduction goes a long way towards reconciling pseudoscalar pion–nucleon coupling with pseudovector coupling. Ultimately one might appeal to experiment to differentiate between the two possible couplings. Lina and Goulard[14] considering electron scattering data on ^{3}H and ^{3}He find the computed isoscalar charge form factor little altered in switching from pseudoscalar to pseudovector coupling as might have been expected. For the isovector charge form factor, pseudoscalar coupling puts the first diffraction minimum in the electron scattering form factor at too small a momentum transfer but agrees with data at higher momentum transfer, while pseudovector coupling correctly reproduces the diffraction minimum but underestimates the data at higher momentum transfer. The problem remains basically unsolved, although the prejudices of most practitioners leans towards pseudovector coupling.

There is one final comment we wish to make concerning the pair graph, and this has to do with the choice of the photon–nucleon coupling to be used in evaluating Fig. 4.2(b). Recall that in writing down the expression for the current in eqn (4.13) in terms of F_1 and F_2 we used a number of heuristic arguments to reduce the number of possible Lorentz vectors that could make up the most general expression for the current from twelve to two. In particular we used the fact that the matrix element of the current would be evaluated between Dirac spinors satisfying the free Dirac equation. However, in the pair graph the internal fermion line with momentum Q does not represent an on-mass-shell fermion, i.e. $E_Q^2 \equiv Q^2 + M^2 \neq Q_0^2$. Therefore it is not obvious what form of the photon–nucleon coupling should be used. For example there is another version of eqn (4.14), known as the Sachs form, written as

$$\langle p_1', s' | j_\mu(0) | p_1, s \rangle = \left(1 + \frac{k^2}{4M^2}\right)^{-1} e\bar{u}_{s'}(p_1')\left(G_E \frac{P_\mu}{M} + \frac{G_M}{4M^2} r_\mu\right) u_s(p_1),$$

with

$$r_\mu = -i(\gamma_\mu \not{P}\not{k} - \not{k}\not{P}\gamma_\mu),$$

$$G_E(k^2) = F_1(k^2) - \frac{k^2}{4M^2} F_2(k^2), \qquad (4.63)$$

$$G_M(k^2) = F_1(k^2) + F_2(k^2),$$

where $2P_\mu = (p_1' + p_1)_\mu$ and $k_\mu = (p_1' - p_1)_\mu$. For on-mass-shell nucleons this form is equivalent to eqn (4.14). Here $G_E(k^2)$ is known as the electric and $G_M(k^2)$ the magnetic form factor. Indeed this parametrization is preferred in the analysis of electron scattering experiments on free nucleons, because it allows a simple interpretation of G_E and G_M as charge and magnetic form factors of the nucleon in contrast to the Dirac form factors F_1 and F_2.

In evaluating pair graphs, we are faced with simplifying expressions such as: $\bar{u}(p'_1)\Gamma_\mu S_F(Q)\Gamma_\pi u(p_1)$, where Γ_μ is the photon–nucleon coupling, either eqn (4.14) involving F_1 and F_2 or eqn (4.63) involving G_E and G_M, and Γ_π is the pion-nucleon coupling, either $\displaystyle{\not}q\gamma_5$ or $2Mi\gamma_5$. The positive-frequency part of the nucleon propagator, $S_F^{(+)}(Q)$, is proportional to $\sum_s u_s(Q)\bar{u}_s(Q)$, and so the photon vertex becomes $\bar{u}(p'_1)\Gamma_\mu u(Q)$ which in leading order gives the same result for the F_1, F_2 and the G_E, G_M versions of the coupling. The negative-frequency part of the propagator, $S_F^{(-)}(Q)$, on the other hand, is proportional to $\sum_s v_s(-Q)\bar{v}_s(-Q)$ and the photon vertex $\bar{u}(p'_1)\Gamma_\mu v(-Q)$ does not lead to the same result in the two versions of Γ_μ. With the pseudovector pion coupling this is not serious since the pair graph anyway gives no contribution in the leading order, and this remaims true for the G_E, G_M version. For the pseudoscalar coupling, on the other hand, there are different expressions in leading order, which is not at all satisfactory. This then is another reason for preferring pseudovector over pseudoscalar coupling.

4.5.3 *The current graph*

The current graph is given in Fig. 4.2(c) and involves a photon–pion vertex. To find the Lagrangian for this vertex, we start with the free-pion Lagrangian and use minimal substitution, eqn (4.54). This produces a term in the Lagrangian linear in the photon field, A_μ, of

$$\mathscr{L}_{\pi\pi\gamma} = e\varepsilon_{rsz}[(\partial_\mu\phi_r)\phi_s - \phi_r(\partial_\mu\phi_s)]A_\mu. \tag{4.64}$$

Then from diagram rules in Appendix A the T-matrix corresponding to Fig. 4.2(c) is

$$T_\text{fi} = ie\frac{g_{\pi NN}^2}{(2M)^2}(\tau_1 \times \tau_2)_z \left(\frac{M}{E'_1}\frac{M}{E'_2}\frac{M}{E_1}\frac{M}{E_2}\right)^{\frac{1}{2}}$$
$$\times \bar{u}_{s'_1}(p'_1)[\not{p}\gamma_5]u_{s_1}(p_1)(p^2 + m_\pi^2)^{-1}(p-q)_\mu(q^2 + m_\pi^2)^{-1}$$
$$\times \bar{u}_{s'_2}(p'_2)[\not{q}\gamma_5]u_{s_2}(p_2)\varepsilon_\mu(k, \lambda),$$

where $p = p'_1 - p_1$ and $q = p'_2 - p_2$. Using the plane-wave spinor solutions and multiplying out the Dirac matrices keeping only terms to leading order in p/M, we obtain a non-relativistic version of the current operator, j_μ, for the case when the Lorentz index μ is space like

$$\boldsymbol{j} = ie\frac{g_{\pi NN}^2}{(2M)^2}(\tau_1 \times \tau_2)_z\frac{\sigma_1 \cdot \boldsymbol{p}}{p^2 + m_\pi^2}(\boldsymbol{p} - \boldsymbol{q})\frac{\sigma_2 \cdot \boldsymbol{q}}{q^2 + m_\pi^2}, \tag{4.65}$$

and for the case when μ is time like

$$j_0 = ie\frac{g_{\pi NN}^2}{(2M)^2}(\tau_1 \times \tau_2)_z\frac{\sigma_1 \cdot \boldsymbol{p}}{p^2 + m_\pi^2}(p-q)_0\frac{\sigma_2 \cdot \boldsymbol{q}}{q^2 + m_\pi^2}. \tag{4.66}$$

4.5 PION-EXCHANGE GRAPHS

In this second case, $p_0 = E_1' - E_1$ and is approximately given by $p_0 \simeq (p_1'^2 - p_1^2)/2M \simeq \boldsymbol{P} \cdot \boldsymbol{p}/M$ where $2\boldsymbol{P} = (\boldsymbol{p}_1' + \boldsymbol{p}_1)$. Similarly $q_0 = E_2' - E_2 \simeq \boldsymbol{Q} \cdot \boldsymbol{q}/M$ with $2\boldsymbol{Q} = (\boldsymbol{p}_2' + \boldsymbol{p}_2)$. Note the MEC charge operator, eqn (4.66), is therefore of order $(1/M^3)$. The operator depends on $\boldsymbol{P}$ and $\boldsymbol{Q}$, which under Fourier transforms become $-i\boldsymbol{\nabla}_1$ and $-i\boldsymbol{\nabla}_2$ respectively in coordinate space, and hence is of non-local character.

4.5.4 Current conservation

A requirement of the MEC exchange-current operator is that it satisfy the equation of continuity, eqn (4.44),

$$\boldsymbol{k} \cdot \boldsymbol{j}^{ex} = [V, j_0]$$

that follows from the statement that the electromagnetic current is a conserved current. For the pion Born graphs, the leading terms in the current come from the seagull, eqn (4.56), and current, eqn (4.65), terms. With $\boldsymbol{k}$ from momentum conservation in these two-body graphs given by $\boldsymbol{k} = \boldsymbol{p} + \boldsymbol{q}$, we have

$$\boldsymbol{k} \cdot \boldsymbol{j}^{ex} = \boldsymbol{k} \cdot (\boldsymbol{j}_{\text{seagull}} + \boldsymbol{j}_{\text{current}})$$

$$= -ie \frac{g_{\pi NN}^2}{(2M)^2} (\boldsymbol{\tau}_1 \times \boldsymbol{\tau}_2)_z \left(\frac{\boldsymbol{k} \cdot \boldsymbol{\sigma}_1 \boldsymbol{\sigma}_2 \cdot \boldsymbol{q}}{q^2 + m_\pi^2} \right.$$

$$\left. - \frac{\boldsymbol{k} \cdot \boldsymbol{\sigma}_2 \boldsymbol{\sigma}_1 \cdot \boldsymbol{p}}{p^2 + m_\pi^2} - \frac{\boldsymbol{\sigma}_1 \cdot \boldsymbol{p} \boldsymbol{k} \cdot (\boldsymbol{p} - \boldsymbol{q}) \boldsymbol{\sigma}_2 \cdot \boldsymbol{q}}{(p^2 + m_\pi^2)(q^2 + m_\pi^2)} \right)$$

$$= -ie \frac{g_{\pi NN}^2}{(2M)^2} (\boldsymbol{\tau}_1 \times \boldsymbol{\tau}_2)_z \frac{\boldsymbol{\sigma}_1 \cdot \boldsymbol{q} \boldsymbol{\sigma}_2 \cdot \boldsymbol{q}}{q^2 + m_\pi^2} + (1 \rightleftarrows 2). \quad (4.67)$$

Using the identity $i(\boldsymbol{\tau}_1 \times \boldsymbol{\tau}_2)_z = [\boldsymbol{\tau}_1 \cdot \boldsymbol{\tau}_2, \tfrac{1}{2}\tau_{1z}]$ and identifying

$$V_\pi = -\frac{g_{\pi NN}^2}{(2M)^2} \frac{\boldsymbol{\sigma}_1 \cdot \boldsymbol{q} \boldsymbol{\sigma}_2 \cdot \boldsymbol{q}}{q^2 + m_\pi^2} \boldsymbol{\tau}_1 \cdot \boldsymbol{\tau}_2, \qquad j_0^{(1)} = \tfrac{1}{2} e F_1^{(1)}(k^2) \tau_{1z},$$

we obtain the desired result, $\boldsymbol{k} \cdot \boldsymbol{j}^{ex} = [V, j_0^{(1)}]$, in the low-energy limit $F_1^{(1)}(k^2 \to 0) = 1$. Thus the pion Born terms alone satisfy the equation of continuity. Note the pion current, $\boldsymbol{j}^{ex}$, is purely isovector and the charge density in the continuity equation is the isovector component of j_0. This result then places a restriction on the non-Born terms, namely that in the low-energy limit they must lead to purely transverse currents satisfying $\boldsymbol{k} \cdot \boldsymbol{j} = 0$. Note also that these expressions for pion-exchange currents from the Born terms can be said to be model independent in the sense that they involve no parameters not already present in the one-pion-exchange potential.

4.5.5 Non-Born terms of pion range

We turn now to the non-Born graphs in which either the nucleon is raised to an excited state (Δ resonance) or the pion is converted to a heavy

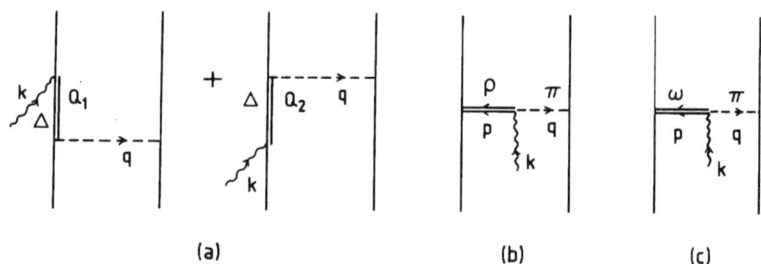

Fig. 4.3 MEC non-Born graphs of pion range: (a) isobar excitation, (b) ρ–π graph, and (c) ω–π graph. Comments in the caption to Fig. 4.2 also apply here.

meson (ρ or ω) in the photo-absorption process. The relevant Feynman graphs are shown in Fig. 4.3. We will be rather brief, just writing down the relevant vertex Lagrangians and the deduced expressions for the MEC exchange currents. Further details can be found in Towner.[15]

4.5.5a Isobar currents We treat the isobar as an elementary spin-$\frac{3}{2}$, isospin-$\frac{3}{2}$ fermion governed by the Rarita–Schwinger equations of motion.[16] The Lagrangian for the pion–nucleon-isobar vertex is

$$\mathcal{L}_{\pi N\Delta} = \frac{g_{\pi N\Delta}}{2M} \bar{\psi}_\mu T_r \psi \, \partial_\mu \phi_r + \text{h.c.}, \tag{4.68}$$

where $\bar{\psi}_\mu$, ψ, and ϕ_r are respectively isobar, nucleon, and pion field operators with the subscript r representing a Cartesian isospin index. Here T_r is a transition isospin operator,[17] (a generalization of the Pauli isospin operator τ_r.) that connects an isospin-$\frac{1}{2}$ nucleon with an isospin-$\frac{3}{2}$ isobar. It is defined through its reduced matrix elements: $\langle \frac{3}{2} \| T \| \frac{1}{2} \rangle = 1$ and $\langle \frac{1}{2} \| T^\dagger \| \frac{3}{2} \rangle = -\sqrt{2}$ (in the conventions for reduced matrix elements of Brink and Satchler, see eqn (1.4)). Lastly a coupling constant $g_{\pi N\Delta}$ is introduced whose value is a source of some uncertainty in isobar calculations. For example, the constituent quark model gives a simple prediction for the coupling constant, namely $g_{\pi N\Delta}/g_{\pi NN} = 6\sqrt{2}/5$. The Skyrme soliton model gives $3\sqrt{2}/2$. On the other hand one can appeal to experimental information. The Chew–Low theory[18] for example predicts π–nucleon scattering will proceed principally through the P_{33} channel. Using an isobar model to describe π–nucleon scattering, a value of $g_{\pi N\Delta}/g_{\pi NN} = 2$ will yield the same scattering phase shifts as the static Chew–Low theory. Alternatively, fitting the resonance width in the isobar model[19] determines $g_{\pi N\Delta}/g_{\pi NN} \simeq 2.15$, which is some 25% larger than the quark model value.

4.5 PION-EXCHANGE GRAPHS

The photo-excitation of the isobar is given by a Lagrangian

$$\mathscr{L}_{\gamma N\Delta} = \frac{1}{2} \frac{G^{(1)}_{M,\gamma N\Delta}}{2M} e\bar{\psi}_\mu \gamma_5 \gamma_\nu T_z \psi (\partial_\mu A_\nu - \partial_\nu A_\mu) + \text{h.c.}, \quad (4.69)$$

where again there is some uncertainty in the choice of coupling constant. In the constituent quark model $G^{(1)}_{M,\gamma N\Delta}/G^{(1)}_{M,\gamma NN} = 6\sqrt{2}/5$ where $G^{(1)}_{M,\gamma NN}$ is the isovector Sachs magnetic form factor, eqn (4.63): $G^{(1)}_M = F^{(1)}_1 + F^{(1)}_2$. With the Lagrangians specified, it is a straightforward but tedious exercise to apply the Feynman rules to Fig. 4.3(a) to find the T-matrix and then the corresponding electromagnetic current. The result is

$$j_\mu = -ie \frac{G^{(1)}_{M,\gamma N\Delta}}{2M} \frac{g_{\pi N\Delta} g_{\pi NN}}{(2M)^2} \left(\frac{M}{E'_1} \frac{M}{E'_2} \frac{M}{E_1} \frac{M}{E_2}\right)^{1/2}$$
$$\times \bar{u}(p'_1)[(k_\alpha \gamma_\mu - K\delta_{\alpha\mu})\gamma_5 S^\Delta_{F,\alpha\beta}(Q_1) q_\beta \tfrac{1}{2} T^\dagger_{1z} T_{1r}$$
$$+ q_\beta S^\Delta_{F,\beta\alpha}(Q_2)\gamma_5 (k_\alpha\gamma_\mu - K\delta_{\alpha\mu})\tfrac{1}{2}T^\dagger_{1r} T_{1z}]u(p_1)$$
$$\times (q^2 + m_\pi^2)^{-1} \bar{u}(p'_2)[\slashed{q}\gamma_5]u(p_2)\tau_{2r} + (1\rightleftarrows 2), \quad (4.70)$$

where $S^\Delta_{F,\alpha\beta}(Q)$ is the isobar propagator:

$$S^\Delta_{F,\alpha\beta}(Q) = \frac{-i}{i\slashed{Q} + m_\Delta}[\delta_{\alpha\beta} - \tfrac{1}{3}\gamma_\alpha\gamma_\beta + (i/3m_\Delta)$$
$$\times (\gamma_\alpha Q_\beta - \gamma_\beta Q_\alpha) + (2/3m_\Delta^2) Q_\alpha Q_\beta] + \dots. \quad (4.71)$$

The additional terms not shown all vanish when the isobar is on its mass shell.[20] Unfortunately in our application the isobar is far off the mass shell but we ignore this complication. We have $Q_1 = p'_1 - k$ and $Q_2 = p_1 + k$ such that in the static limit, $\mathbf{Q} \to 0$, the fourth component becomes $Q_4 \to iM$, the nucleon mass rather than the isobar mass, m_Δ. We use for the propagator the approximate form

$$S^\Delta_{F,\alpha\beta}(Q) = \frac{-i}{m_\Delta^2 - M^2}(-i\boldsymbol{\gamma}\cdot\mathbf{Q} + M\gamma_4 + m_\Delta)$$
$$\times [\delta_{\alpha\beta} - \tfrac{1}{3}\gamma_\alpha\gamma_\beta + (i/3m_\Delta)(\gamma_\alpha Q_\beta - \gamma_\beta Q_\alpha) + (2/3m_\Delta^2) Q_\alpha Q_\beta].$$

The leading term in the non-relativistic reduction of the current for the case that the Lorentz index μ is space like is

$$\mathbf{j} = ie \frac{G^{(1)}_{M,\gamma N\Delta}}{2M} \frac{g_{\pi N\Delta} g_{\pi NN}}{(2M)^2} \frac{1}{m_\Delta - M} \frac{\boldsymbol{\sigma}_2 \cdot \mathbf{q}}{q^2 + m_\pi^2}$$
$$\times [\tfrac{4}{9}\tau_{2z}\mathbf{q} \times \mathbf{k} + \tfrac{1}{9}(\boldsymbol{\tau}_1 \times \boldsymbol{\tau}_2)_z (\mathbf{k}\cdot\mathbf{q}\boldsymbol{\sigma}_1 - \boldsymbol{\sigma}_1\cdot\mathbf{k}\mathbf{q})] + (1\rightleftarrows 2). \quad (4.72)$$

In deriving this result, we have used the property of transition spin operators

$$T^\dagger_r T_s = \tfrac{2}{3}\delta_{rs} - \tfrac{1}{3}i\varepsilon_{rsz}\tau_z, \quad (4.73)$$

which is easily proven[2] by taking matrix elements of $T^\dagger T$ between nucleon spinors and using a completeness relation. Note that the derived current is transverse, $\mathbf{k} \cdot \mathbf{j} = 0$, as required by the continuity equation.

4.5.5b ρ–π graph Strong interactions are invariant under arbitary rotations in isospin and under charge conjugation. Thus mesons can be categorized according to their transformation properties under the product operator known as G-parity defined as

$$G = C \exp(i\pi T_y).$$

Here C is the charge conjunction operator, and $\exp(i\pi T_y)$ a rotation of $180°$ about the isospin y-axis. For example, the pion has negative G-parity, i.e. $G|\pi\rangle = -|\pi\rangle$. We will use the properties of mesons under G-parity transformations to discuss the isospin structure of meson–photon vertices. First the proof that a pion has negative G-parity is as follows. From angular momentum algebra (Brink and Satchler[21] and eqn (2.18) therein) we have the result

$$\exp(i\pi T_y)|T, T_z\rangle = (-)^{T+T_z}|T, -T_z\rangle,$$

indicating that a state of isospin T, magnetic projection T_z is transformed into a state $T, -T_z$ under the rotation. Writing the pion field operators in a spherical basis, $\phi_{\pm 1} = \mp(2)^{-1/2}(\phi_x \pm i\phi_y)$, isospin rotations give

$$\exp(i\pi T_y)\phi_{\pm 1} \exp(-i\pi T_y) = \phi_{\mp 1},$$
$$\exp(i\pi T_y)\phi_0 \exp(-i\pi T_y) = -\phi_0.$$

Under charge conjugation the pion field operators transform as

$$C\phi_r C^{-1} = \phi_r^\dagger.$$

Remembering that in the Cartesian representation, the pion fields are Hermitian, e.g. $\phi_x^\dagger = \phi_x$, we obtain the result

$$G\phi_i G^{-1} = -\phi_i,$$

where i is the spherical index, $+1, -1$, or 0. Hence for a state of n pions with any charge,

$$G|n\pi\rangle = (-)^n|n\pi\rangle.$$

Moreover the electromagnetic currents, separated into their isoscalar and isovector components, have the following transformation properties under G[3]:

$$Gj_\mu^{(0)}G^{-1} = -j_\mu^{(0)}, \qquad Gj_\mu^{(1)}G^{-1} = j_\mu^{(1)}.$$

Currents with the opposite sign under G-parity transformations are known as second-class currents. There is no experimental evidence to

4.5 PION-EXCHANGE GRAPHS

date for the existence of any second-class currents. Since $G|0\rangle = |0\rangle$ it follows that

$$\langle 0| j_\mu^{(0)} + j_\mu^{(1)} |n\pi\rangle = (-)^n \langle 0| -j_\mu^{(0)} + j_\mu^{(1)} |n\pi\rangle.$$

That is, states with an odd number of pions can be generated from the vacuum by an isoscalar current and states with an even number of pions generated by an isovector current. For example, consider the π-current graph discussed in the last section. Here the electromagnetic current generates two pions, so the $\pi\pi\gamma$ vector is isovector in structure as would be expected from these G-parity arguments. For the ρ–π graph, the ρ meson can decay to two pions and so under G-parity has the same transformation properties as two pions. Therefore the $\rho\pi\gamma$ vertex is isoscalar in structure. Similarly the $\omega\pi\gamma$ vertex is isovector, since the ω meson decays to three pions.

The Lagrangian for the $\rho\pi\gamma$ vertex is given in de Alfaro et al.[12] as

$$\mathscr{L}_{\rho\pi\gamma} = \mathrm{i} e \frac{g_{\rho\pi\gamma}}{m_\rho} \varepsilon_{\nu\mu\rho\sigma} \partial_\nu A_\mu \partial_\rho \rho_{\sigma,r} \phi_r \tag{4.74}$$

where A_μ, $\rho_{\sigma,r}$, and ϕ_r are the photon, ρ meson, and pion field operators, the subscript r being a Cartesian isospin index. The coupling constant can be determined from the radiative decay of the ρ, namely $\rho \to \pi\gamma$. The width is given by the expression[22]

$$\Gamma(\rho \to \pi\gamma) = \frac{1}{24} \frac{e^2}{\hbar c} g_{\rho\pi\gamma}^2 m_\rho (1 - m_\pi^2/m_\rho^2)^3.$$

A recent determination[23] gives $\Gamma = 71 \pm 7$ keV and hence

$$|g_{\rho\pi\gamma}| = 0.578 \pm 0.028.$$

It is amusing to note that de Alfaro et al.[12] using SU(3) relations determine $g_{\rho\pi\gamma} = 2/3$, which is within 15% the value deduced from the radiative width.

The ρNN Lagrangian is given by a form analogous to the γNN interaction (both ρ and γ are vector fields), namely

$$\mathscr{L}_{\rho\mathrm{NN}} = \mathrm{i} g_{\rho\mathrm{NN}} \bar\psi \left(\gamma_\mu + \mathrm{i} \frac{K_\rho}{2M} \sigma_{\mu\nu} \partial_\nu \right) \psi \rho_{\mu,r} \tau_r. \tag{4.75}$$

In vector dominance models, in which the electromagnetic form factors of a nucleon $F_1^{(1)}(k^2)$ and $F_2^{(1)}(k^2)$ are given in terms of the ρ-meson propagator, namely $F_1^{(1)}(k^2) = m_\rho^2/(m_\rho^2 + k^2)$ and $F_2^{(1)}(k^2) = K_\rho m_\rho^2/(m_\rho^2 + k^2)$, the coupling constant K_ρ is naturally identified as $K_\rho = F_2^{(1)}(0)/F_1^{(1)}(0)$ and equals $K_\rho = 3.7$ from the known isovector anomalous magnetic moment. There is growing evidence that K_ρ, in fact, should be larger than the vector dominance value. Höhler and Pietarinen[24] examining the isovector $J = 1^-$ channel in πN scattering suggest $K_\rho = 6.6$

with $g_{\rho NN} = 2.63$. Large values of K_ρ are also found in the Bonn one-boson-exchange potential[25] in fits to nucleon–nucleon scattering data.

With the Lagrangians of eqns (4.74) and (4.75), diagram Fig. 4.3(b) can be evaluated to obtain for the current

$$j_\mu = -eg_{\rho NN} \frac{g_{\rho\pi\gamma} g_{\pi NN}}{m_\rho \ 2M} \tau_1 \cdot \tau_2 \left(\frac{M}{E_1'}\frac{M}{E_2'}\frac{M}{E_1}\frac{M}{E_2}\right)^{1/2}$$

$$\times \bar{u}(p_1')\left(\gamma_\sigma - \frac{K_\rho}{2M}\sigma_{\sigma\tau}p_\tau\right)u(p_1)(p^2 + m_\rho^2)^{-1}\varepsilon_{\nu\mu\rho\sigma}k_\nu p_\rho$$

$$\times (q^2 + m_\pi^2)^{-1}\bar{u}(p_2')[\slashed{q}\gamma_5]u(p_2) + (1 \rightleftarrows 2). \tag{4.76}$$

Multiplying out the Dirac spinors with Dirac matrices and keeping terms to leading order in $1/M$, we obtain a non-relativistic expression for the current

$$j = -ieg_{\rho NN}\frac{g_{\rho\pi\gamma} g_{\pi NN}}{m_\rho \ 2M}\tau_1 \cdot \tau_2 \frac{(k \times q)(\sigma_2 \cdot q)}{(p^2 + m_\rho^2)(q^2 + m_\pi^2)} + (1 \rightleftarrows 2). \tag{4.77}$$

Note this current is transverse, $k \cdot j = 0$, as required.

4.5.5c ω–π graph The evaluation of this graph follows analogously that of the ρ–π graph. The $\omega\pi\gamma$ Lagrangian is

$$\mathcal{L}_{\omega\pi\gamma} = ie\frac{g_{\omega\pi\gamma}}{m_\omega}\varepsilon_{\nu\mu\rho\sigma}\partial_\nu A_\mu \partial_\rho \omega_\sigma \phi_z, \tag{4.78}$$

where ω_σ is the ω-meson field operator. As before the coupling constant is determined from the decay width for the radiative decay $\omega \to \pi\gamma$. The expression for the width is[22]

$$\Gamma(\omega \to \pi\gamma) = \frac{1}{24}\frac{e^2}{\hbar c}g_{\omega\pi\gamma}^2 m_\omega(1 - m_\pi^2/m_\omega^2)^3.$$

The Particle Data Group's fit[26] to the partial decay widths of the three principal decay modes of the ω meson yields for the radiative width $\Gamma = 852 \pm 52$ keV and hence

$$|g_{\omega\pi\gamma}| = 1.98 \pm 0.06.$$

Again SU(3) relations yield a simple result of $g_{\omega\pi\gamma} = 2$, in excellent agreement with the value deduced from the radiative widths. Indeed with SU(3) we have $g_{\omega\pi\gamma} = 3g_{\rho\pi\gamma}$ which is approximately satisfied by the experimental widths.

The ωNN Lagrangian is

$$\mathcal{L}_{\omega NN} = ig_{\omega NN}\bar{\psi}\left(\gamma_\mu + i\frac{K_\omega}{2M}\sigma_{\mu\nu}\partial_\nu\right)\psi\omega_\mu. \tag{4.79}$$

4.5 PION-EXCHANGE GRAPHS

In contrast to the ρNN Lagrangian, the tensor coupling for the ω meson is known to be small,[22] thus it is sufficient to use the vector dominance value of $K_\omega = F_2^{(0)}(0)/F_1^{(0)}(0) = -0.12$. In SU(3), the ω-meson coupling constant is related to that of the ρ meson: $g_{\omega NN} = 3g_{\rho NN} = 7.89$, and we adopt this value.

The evaluation of the Feynman graph, Fig. 4.3(c), follows analogously that for the ρ-π graph. The resulting current in the non-relativistic limit is

$$j = -ie\tau_{2z}g_{\omega NN} \frac{g_{\omega\pi\gamma}g_{\pi NN}}{m_\omega} \frac{(\mathbf{k} \times \mathbf{q})(\mathbf{\sigma}_2 \cdot \mathbf{q})}{2M (p^2 + m_\omega^2)(q^2 + m_\pi^2)} + (1 \rightleftarrows 2). \quad (4.80)$$

The current is transverse as required.

4.5.6 Form factors and heavy mesons

A general form for the electromagnetic current at the photon–nucleon vertex is written down in eqn (4.13) in terms of two Lorentz vectors multiplied by form factors $F_1(k^2)$ and $F_2(k^2)$. These are known as the Dirac and Pauli form factors. In the Sachs parametrization, eqn (4.63), two different Lorentz vectors are selected and two different form factors $G_E(k^2)$ and $G_M(k^2)$ introduced. For on-mass-shell nucleons the two forms are equivalent and the two sets of form factors are related. One deficiency in the S-matrix method of constructing the MEC exchange current, and in the use of a minimal substitution prescription, is that no information on the momentum dependence of the coupling constant at the photon vertex is supplied. Indeed the proper choice of the electromagnetic form factor for the MEC in a non-relativistic description has been the subject of some controversy because agreement or disagreement with experimental data at high momentum transfer (in for example electron scattering experiments) depends sensitively on this form factor.[27,10] However, the electromagnetic MEC form factor cannot be chosen arbitrarily because of the constraining continuity equation for the current. For example, in the seagull graph we could multiply the expression for the current, eqn (4.56), by a form factor, say $F_{\pi NN\gamma}(k^2)$. Likewise in the pion-current graph we could introduce a form factor at the $\pi\pi\gamma$ vertex of $F_{\pi\pi\gamma}(k^2)$ that would be included in the expression for the current, eqn (4.65). However, if the equation of continuity, eqn (4.44), is to remain satisfied, these form factors must be related to the isovector form factor in the one-body charge current

$$F_{\pi NN\gamma}(k^2) = F_{\pi\pi\gamma}(k^2) = F_1^{(1)}(k^2) \text{ or } G_E^{(1)}(k^2).$$

The choice between $F_1^{(1)}(k^2)$ and $G_E^{(1)}(k^2)$ depends on the choice between the Dirac and Sachs form for the one-body current.[28] In a non-relativistic theory the difference between F_1 and G_E should not matter as long as one stays within the non-relativistic domain. For example, the magnetic moment operator is evaluated at the low-energy limit, $k^2 \to 0$, where $G_E^{(1)}(k^2 \to 0) \to F_1^{(1)}(k^2 \to 0)$. But this is not the case for the Saclay data[29] on the electro-disintegration of the deuteron near threshold where the momentum transfer is in the range $6 \le k^2 \le 30\,\text{fm}^{-2}$. Fits to the data are much better using $F_1^{(1)}(k^2)$ rather than $G_E^{(1)}(k^2)$ as the electromagnetic form factor for the MEC terms. A very nice discussion of this problem has been given by Arenhövel.[30] We return to this point in Section 4.10.2.

Besides the electromagnetic vertices, we should also consider the meson–nucleon vertices. For example at the πNN vertex the coupling constant is not necessarily a simple constant but could be a function of a Lorentz scalar, say q^2, where q is the momentum transferred by the pion. It is quite common to introduce a form factor at meson–nucleon vertices in a phenomenological way in, for example, the construction of a nucleon–nucleon interaction as the sum of one-boson-exchange potentials.[25] A frequently used parametrization is

$$g_{\pi NN}(q^2) = g_{\pi NN}(0) \Gamma_\pi(q^2),$$

where $\Gamma_\pi(q^2)$ is of monopole form

$$\Gamma_\pi(q^2) = (\Lambda^2 - m_\pi^2)/(\Lambda^2 + q^2), \tag{4.81}$$

with $\Lambda \gg m_\pi$ and of order 1 GeV. If such form factors are to be introduced in the one-pion exchange potential, then again for consistency with the continuity equation they must be introduced in the construction of the MEC exchange current. This has to be done with some care as pointed out by Riska.[31]

The introduction of form factors and the selection of, say, monopole forms injects a certain model dependence into the exchange-current calculation. This is unavoidable. However, with Λ of the order of 1 GeV these modifications mainly affect the short-range behaviour of the deduced MEC operator. Also influencing the short-range behaviour will be exchange-current graphs involving heavy mesons such as ρ, ω, σ, A_1. Again it is quite straightforward to draw the appropriate pair and current Feynman diagrams and evaluate them once the appropriate Lagrangians have been specified. These Lagrangians will be model dependent, but with additional principles built in, such as chiral invariance and vector-meson dominance, they lead to a scheme that is convenient for practical calculation. This is discussed at some length in Adam, Truhlik, and Adamova,[10] and Towner[15] where expressions for the exchange currents for heavy mesons are given.

4.6 Construction of MEC magnetic moment operator

In the previous section we derived expressions for the exchange current j, corresponding to the Feynman graphs given in Figs. 4.2 and 4.3 to

4.6 MEC MAGNETIC MOMENT OPERATOR

leading order in their non-relativistic reductions. It remains to follow the procedure outlined in Section 4.2 of Fourier transformations and multipole decomposition to obtain the MEC magnetic moment operator. We will work this out in detail for the seagull current, eqn (4.56), to illustrate the method and just quote the results in other cases.

The first step is the Fourier transformation to coordinate space

$$j(x, r'_1, r'_2, r_1, r_2) = \frac{1}{(2\pi)^{12}} \int d^3k\, d^3p\, d^3q\, d^3P\, d^3Q$$
$$\times \exp[ip \cdot (r'_1 + r_1)/2]\exp(iq \cdot (r'_2 + r_2)/2]$$
$$\times \exp[iP \cdot (r'_1 - r_1)]\exp[iQ \cdot (r'_2 - r_2)]$$
$$\times \exp(-ik \cdot x)\delta(k - p - q)j(k, p, q, P, Q),$$

where relative and centre-of-mass coordinates are used: $p = p'_1 - p_1$, $q = p'_2 - p_2$, $2P = p'_1 + p_1$, $2Q = p'_2 + p_2$. Again note that if the current is independent of P, Q then it is said to be local, since the integrations over P and Q lead to delta functions $\delta(r'_1 - r_1)$ and $\delta(r'_2 - r_2)$. To handle non-local terms in P and Q we use the replacement $P\exp(iP \cdot r_1) = -i\nabla_1 \exp(iP \cdot r_1)$ and interchange the order of differentiation and integration in the Fourier transform. In this way the dependence on P and Q can be transformed away to become derivatives on the delta function. They produce in configuration space derivative operators and are simply obtained through the replacement rule: $P \to -i\nabla_1$ and $Q \to -i\nabla_2$. It is essential for the usefulness of this scheme that the dependence on P and Q be sufficiently weak; a quadratic dependence, for example, is barely tolerable. Here we will limit ourselves to non-local terms that are no higher than linear in derivative operators. The construction of the one-boson-exchange potential makes the same approximation. The spin–orbit potential is an example of a linear non-local term. Thus the current in configuration space is written

$$j(x, r_1, r_2) = \frac{1}{(2\pi)^6} \int d^3k\, d^3p\, d^3q \exp(ip \cdot r_1)\exp(iq \cdot r_2)$$
$$\times \exp(-ik \cdot x)\delta(k - p - q)j(k, p, q, -i\nabla_1, -i\nabla_2).$$

We choose to use the delta function to eliminate the integration over p

$$j(x, r_1, r_2) = \frac{1}{(2\pi)^6} \int d^3k\, d^3q \exp(-iq \cdot r)$$
$$\times \exp[ik \cdot (r_1 - x)]j(k, q, -i\nabla_1, -i\nabla_2), \quad (4.82)$$

where $r = r_1 - r_2$. The magnetic moment operator is given by eqn (4.24):

$$\mu(r_1, r_2) = \frac{1}{2}\int d^3x \times j(x, r_1, r_2),$$

and leads to a very simple calculation in the special case that j has no dependence on k. This, for example, is true for the pion seagull graph, eqn (4.56). Then the integration over k leads to a delta function that immediately removes the integration over x. In this special case

$$\mu(r_1, r_2) = \frac{1}{2}\frac{1}{(2\pi)^3}\int d^3q \exp(-i\boldsymbol{q}\cdot\boldsymbol{r})r_1 \times j(\boldsymbol{q}, -i\boldsymbol{\nabla}_1, -i\boldsymbol{\nabla}_2), \quad (4.83)$$

and the integration over q can be done analytically. Inserting eqn (4.56) for the seagull current we obtain

$$\mu(r_1, r_2) = -\tfrac{1}{2}ef_{\pi NN}^2(\boldsymbol{\tau}_1\times\boldsymbol{\tau}_2)_z r_1\times\boldsymbol{\sigma}_1\boldsymbol{\sigma}_2\cdot\hat{\boldsymbol{r}}Y_1(x_\pi)+(1\rightleftarrows 2), \quad (4.84)$$

where $Y_1(x)=(1+1/x)Y_0(x)$, $Y_0(x)=e^{-x}/x$, $x_\pi = m_\pi r$, and $f_{\pi NN}^2 = (4\pi)^{-1}[g_{\pi NN}m_\pi/(2M)]^2$. It is convenient to introduce relative and centre-of-mass coordinates, $r = r_1 - r_2$ and $2R = r_1 + r_2$, and obtain

$$\mu(r, R) = \tfrac{1}{2}ef_{\pi NN}^2(\boldsymbol{\tau}_1\times\boldsymbol{\tau}_2)_z[\tfrac{1}{2}(\tfrac{2}{3}\sigma^x - T^x)rY_1(x_\pi)$$
$$+(\boldsymbol{\sigma}_1\times\hat{\boldsymbol{R}}\boldsymbol{\sigma}_2\cdot\hat{\boldsymbol{r}}+\boldsymbol{\sigma}_2\times\hat{\boldsymbol{R}}\boldsymbol{\sigma}_1\cdot\hat{\boldsymbol{r}})RY_1(x_\pi)] \quad (4.85)$$

using the identity: $\boldsymbol{\sigma}_1\times\hat{\boldsymbol{r}}\boldsymbol{\sigma}_2\cdot\hat{\boldsymbol{r}} - \boldsymbol{\sigma}_1\cdot\hat{\boldsymbol{r}}\boldsymbol{\sigma}_2\times\hat{\boldsymbol{r}} = \tfrac{2}{3}\sigma^x - T^x$. Here σ^x is $\boldsymbol{\sigma}_1\times\boldsymbol{\sigma}_2$ and

$$T^x = [(\boldsymbol{\sigma}_1\times\boldsymbol{\sigma}_2)\cdot\hat{\boldsymbol{r}}\hat{\boldsymbol{r}} - \tfrac{1}{3}\boldsymbol{\sigma}_1\times\boldsymbol{\sigma}_2] = -\tfrac{1}{2}(8\pi)^{1/2}[Y_2(\hat{\boldsymbol{r}}), \boldsymbol{\sigma}_1\times\boldsymbol{\sigma}_2]^{(1)} \quad (4.86)$$

is a spherical tensor of rank 1 formed from coupling a spherical harmonic of rank 2 in orbital space, $Y_2(\hat{\boldsymbol{r}})$, with a spin tensor of rank 1, $\boldsymbol{\sigma}_1\times\boldsymbol{\sigma}_2$.

The calculation for the pion current term eqn (4.65) is quite a bit more complicated because of the double propagator. After the integration over the momentum p in the Fourier transform the pion current becomes

$$j(x, r_1, r_2) = -ie\frac{g_{\pi NN}^2}{(2M)^2}(\boldsymbol{\tau}_1\times\boldsymbol{\tau}_2)_z\frac{1}{(2\pi)^6}\int d^3k\, d^3q\, e^{-i\boldsymbol{q}\cdot\boldsymbol{r}}$$
$$\times\exp(i\boldsymbol{k}\cdot(r_1-x))\times\frac{\boldsymbol{\sigma}_1\cdot(\boldsymbol{k}-\boldsymbol{q})\boldsymbol{q}\boldsymbol{\sigma}_2\cdot\boldsymbol{q}}{[(\boldsymbol{k}-\boldsymbol{q})^2+m_\pi^2][q^2+m_\pi^2]}+(1\rightleftarrows 2). \quad (4.87)$$

To proceed, we use Feynman's integral relation[32]

$$\frac{1}{ab} = \int_0^1\frac{dz}{[az+b(1-z)]^2}$$

with $a = (\boldsymbol{k}-\boldsymbol{q})^2 + m_\pi^2$ and $b = q^2 + m_\pi^2$. Further, changing the integration from q to Q where $Q = q - kz$ gives

$$j(x, r_1, r_2) = -ie\frac{g_{\pi NN}^2}{(2M)^2}(\boldsymbol{\tau}_1\times\boldsymbol{\tau}_2)_z\frac{1}{(2\pi)^3}\int d^3k\int_0^1 dz\exp(i\boldsymbol{k}\cdot\boldsymbol{R})$$
$$\times\exp[i(1/2-z)\boldsymbol{k}\cdot\boldsymbol{r}]\exp(-i\boldsymbol{k}\cdot\boldsymbol{x})\frac{1}{(2\pi)^3}\int d^3Q\exp(-i\boldsymbol{Q}\cdot\boldsymbol{r})$$
$$\times\frac{[(1-z)\boldsymbol{\sigma}_1\cdot\boldsymbol{k}-\boldsymbol{\sigma}_1\cdot\boldsymbol{Q}][\boldsymbol{Q}+\boldsymbol{k}z][\boldsymbol{\sigma}_2\cdot\boldsymbol{Q}+z\boldsymbol{\sigma}_2\cdot\boldsymbol{k}]}{(Q^2+L^2)^2}+(1\rightleftarrows 2), \quad (4.88)$$

4.6 MEC MAGNETIC MOMENT OPERATOR

where $L^2 = k^2 z(1-z) + m_\pi^2$. Note that the term in the current proportional to the vector $\boldsymbol{k}$ gives no contribution to any observable. This is because the interaction Hamiltonian is proportional to $\boldsymbol{j} \cdot \boldsymbol{\varepsilon}$ where $\boldsymbol{\varepsilon}$ is the polarization vector associated with the photon. But because the photon is massless it only has two independent directions of polarization, both transverse to the photon momentum. Thus a subsidiary condition on the polarization vector is $\boldsymbol{k} \cdot \boldsymbol{\varepsilon} = 0$, and hence terms in the current proportional to $\boldsymbol{k}$ are redundant. The integrations over $\boldsymbol{Q}$ are elementary. However, because the current still has a functional dependence on $\boldsymbol{k}$, we cannot use the short cut of eqn (4.24) that we used for the seagull graph but must go laboriously through the construction of the multipole moment operator, eqn (4.20):

$$T^{\text{mag}}_{LM}(k, r_1, r_2) = -ie \frac{g^2_{\pi NN}}{(2M)^2} (\boldsymbol{\tau}_1 \times \boldsymbol{\tau}_2)_z \frac{1}{4\pi} i^{-L} \int_0^1 dz \int d^2\hat{k}$$
$$\times \exp(i\boldsymbol{k} \cdot \boldsymbol{R}) \exp(i(1/2 - z)\boldsymbol{k} \cdot \boldsymbol{r})[Y_L(\hat{k}), j]_{LM}$$

with

$$j = (1/8\pi)\{-i\boldsymbol{\sigma}_1 \cdot \hat{\boldsymbol{r}}\boldsymbol{r}\boldsymbol{\sigma}_2 \cdot \hat{\boldsymbol{r}} L^2 x_L Y_2(x_L)$$
$$+ i(\boldsymbol{\sigma}_1\boldsymbol{\sigma}_2 \cdot \hat{\boldsymbol{r}} + \boldsymbol{\sigma}_1 \cdot \boldsymbol{\sigma}_2 \hat{\boldsymbol{r}} + \boldsymbol{\sigma}_1 \cdot \hat{\boldsymbol{r}}\boldsymbol{\sigma}_2)L^2 Y_1(x_L)$$
$$- [(1-z)\boldsymbol{\sigma}_1 \cdot \boldsymbol{k}\hat{\boldsymbol{r}}\boldsymbol{\sigma}_2 \cdot \hat{\boldsymbol{r}} - \boldsymbol{\sigma}_1 \cdot \hat{\boldsymbol{r}}\boldsymbol{r}\boldsymbol{\sigma}_2 \cdot \boldsymbol{k}z](LY_1(x_L)/x_L)$$
$$- [(1-z)\boldsymbol{\sigma}_1 \cdot \boldsymbol{k}\boldsymbol{\sigma}_2 - z\boldsymbol{\sigma}_1\boldsymbol{\sigma}_2 \cdot \boldsymbol{k}] LY_0(x_L)$$
$$- i(1-z)z\boldsymbol{\sigma}_1 \cdot \boldsymbol{k}\boldsymbol{\sigma}_2 \cdot \boldsymbol{k}\hat{\boldsymbol{r}} x_L Y_0(x_L)\} + (1 \rightleftarrows 2). \tag{4.89}$$

Here $Y_2(x) = (1 + 3/x + 3/x^2)Y_0(x)$ and $x_L = Lr$. The plane-wave factors, $\exp(-i\boldsymbol{k} \cdot \boldsymbol{R})$ and $\exp[i(1/2 - z)\boldsymbol{k} \cdot \boldsymbol{r}]$, can be expanded in partial waves and the integration over the angles, $d^2\hat{k}$, becomes a simple integration over a product of spherical harmonics. The magnetic moment is then given by the low-energy limit, eqn (4.22), of $T^{\text{mag}}_{10}(k^2)/k$. Note that in the limit $k^2 \to 0$, the integral over z becomes elementary with the function L becoming independent of z, namely $L^2 \to m_\pi^2$, $x_L \to x_\pi$.

With that introduction to the method, we give a list of magnetic moment MEC operators corresponding to the graphs given in Figs. 4.2 and 4.3. It is convenient to separate the magnetic moment operator (as in eqn (4.47)) into

$$\boldsymbol{\mu}^{\text{ex}} = \boldsymbol{\mu}_r + \boldsymbol{\mu}_{\text{c.m.}}$$

where the first term depends only on the relative coordinate, $\boldsymbol{r} = \boldsymbol{r}_1 - \boldsymbol{r}_2$, and is translationally invariant while the second term depends on the centre-of-mass coordinate and is not translationally invariant. We will use the classification scheme of Chemtob and Rho[9] for $\boldsymbol{\mu}_r$:

$$\boldsymbol{\mu}_r = \frac{e}{2M} \tfrac{1}{2}[g_I \boldsymbol{\sigma}^x \boldsymbol{\tau}^x + g_{II} T^x \boldsymbol{\tau}^x + h_I \boldsymbol{\sigma}^- \boldsymbol{\tau}^- + h_{II} T^- \boldsymbol{\tau}^- + j_I \boldsymbol{\sigma}^+ \boldsymbol{\tau}^+ + j_{II} T^+ \boldsymbol{\tau}^+$$
$$+ l_I \boldsymbol{\sigma}^+ + l_{II} T^+ + m_I \boldsymbol{\sigma}^+ \boldsymbol{\tau}_1 \cdot \boldsymbol{\tau}_2 + m_{II} T^+ \boldsymbol{\tau}_1 \cdot \boldsymbol{\tau}_2] \tag{4.90}$$

where

$$\sigma^\odot = \sigma_1 \odot \sigma_2$$
$$T^\odot = [(\sigma_1 \odot \sigma_2) \cdot \hat{r}\hat{r} - \tfrac{1}{3}\sigma_1 \odot \sigma_2], \quad \odot = \pm, \times.$$

The functions $g_I, g_{II}, h_I, \ldots$ etc. are real scalar functions of r and are detailed below. Note the terms in l_I, l_{II}, m_I, and m_{II} are the isoscalar terms. The classification for the translationally non-invariant operators is taken from Hyuga et al.:[8]

$$\mu_{c.m.} = \frac{e}{2M} \{(\hat{r} \times \hat{R})\tau^x[F_I + F_{II}\sigma_1 \cdot \sigma_2 + F_{III}S_{12}]$$
$$+ \tau^x[\sigma_1 \cdot \hat{r}\sigma_2 \times \hat{R} + \sigma_1 \times \hat{R}\sigma_2 \cdot \hat{r}]G$$
$$+ \tau^-[\hat{R} \cdot \sigma^+\hat{r} - \hat{R} \cdot \hat{r}\sigma^+]J$$
$$+ [\tau^+ H + L + (\tau_1 \cdot \tau_2)M][\hat{R} \cdot \sigma^-\hat{r} - \hat{R} \cdot \hat{r}\sigma^-]\}, \quad (4.91)$$

where S_{12} is the tensor operator, $S_{12} = 3(\sigma_1 \cdot \hat{r})(\sigma_2 \cdot \hat{r}) - (\sigma_1 \cdot \sigma_2)$, $2\mathbf{R} = \mathbf{r}_1 + \mathbf{r}_2$ and $F, G, H, \ldots$ etc. are real scalar functions of r. Note that the terms in F_I, F_{II}, and F_{III} are precisely the ones that correspond to the Sachs moment, eqn (4.49).

Pion pair (seagull) current

$$g_I = \frac{2}{3}\frac{M}{m_\pi} f^2_{\pi NN} x^2_\pi V_{12}(x_\pi),$$
$$g_{II} = -\tfrac{3}{2}g_I, \quad (4.92)$$
$$G = \frac{M}{m_\pi} f^2_{\pi NN} X_\pi x_\pi V_{12}(x_\pi),$$

where $x_\pi = m_\pi r$ and $X_\pi = m_\pi R$.

Pion current

$$g_I = \frac{2}{3}\frac{M}{m_\pi} f^2_{\pi NN}(x^2_\pi V_{12}(x_\pi) - 3V_{00}(x_\pi)),$$
$$g_{II} = -\frac{M}{m_\pi} f^2_{\pi NN} x^2_\pi V_{12}(x_\pi),$$
$$G = -\frac{M}{m_\pi} f^2_{\pi NN} X_\pi x_\pi V_{12}(x_\pi), \quad (4.93)$$
$$F_{II} = \frac{1}{3}\frac{M}{m_\pi} f^2_{\pi NN} X_\pi x_\pi V_{02}(x_\pi),$$
$$F_{III} = \frac{1}{3}\frac{M}{m_\pi} f^2_{\pi NN} X_\pi x_\pi V_{22}(x_\pi).$$

4.6 MEC MAGNETIC MOMENT OPERATOR

Pion isobar current

$$g_{\rm I} = \tfrac{32}{75} G_M^{(1)} \frac{m_\pi}{\Delta E} f_{\pi NN}^2 V_{02}(x_\pi),$$

$$g_{\rm II} = -\tfrac{16}{25} G_M^{(1)} \frac{m_\pi}{\Delta E} f_{\pi NN}^2 V_{22}(x_\pi), \tag{4.94}$$

$$h_{\rm I} = j_{\rm I} = -g_{\rm I},$$

$$h_{\rm II} = j_{\rm II} = 2 g_{\rm II},$$

where $G_M^{(1)} = F_1^{(1)} + F_2^{(1)} = 4.7$ and $\Delta E = m_\Delta - M$ is the isobar–nucleon mass difference. We have used the quark model to relate isobar couplings to nucleon couplings, namely $G_{M,\gamma N\Delta}^{(1)}/G_{M,\gamma NN}^{(1)} = g_{\pi N\Delta}/g_{\pi NN} = 6\sqrt{2}/5$. If the Chew–Low result of $g_{\pi N\Delta}/g_{\pi NN} = 2$ is preferred, the above expressions should be multiplied by 100/72.

ρ–π graph

$$m_{\rm I} = -\frac{1}{3} \frac{g_{\rho NN} g_{\rho\pi\gamma} g_{\pi NN}}{2\pi} \frac{m_\pi^3}{m_\rho^3} Z_0(x_\pi, x_\rho),$$

$$m_{\rm II} = -\frac{g_{\rho NN} g_{\rho\pi\gamma} g_{\pi NN}}{2\pi} \frac{m_\pi^3}{m_\rho^3} Z_2(x_\pi, x_\rho). \tag{4.95}$$

ω–π graph

$$h_{\rm I} = j_{\rm I} = -\frac{1}{3} \frac{g_{\omega NN} g_{\omega\pi\gamma} g_{\pi NN}}{4\pi} \frac{m_\pi^3}{m_\omega^3} Z_0(x_\pi, x_\omega),$$

$$h_{\rm II} = j_{\rm II} = -\frac{g_{\omega NN} g_{\omega\pi\gamma} g_{\pi NN}}{4\pi} \frac{m_\pi^3}{m_\omega^3} Z_2(x_\pi, x_\omega). \tag{4.96}$$

The radial functions in eqns (4.92) to (4.96) are written out below and assume that monopole form factors, eqn (4.81), have been included at every meson–nucleon–nucleon vertex. In the limit that the range of the form factor tends to infinity, $\Lambda \to \infty$, the form factor $\Gamma(q^2) \to 1$ and the expressions reduce to that for point couplings. This limit is given by the first term in the expressions for V and the first two terms in the expressions for Z. The radial functions are

$$V_{00}(x) = Y_0(x) - \frac{\Lambda}{m} Y_0(x_\Lambda) - \frac{\Lambda^2 - m^2}{2\Lambda m} x_\Lambda Y_0(x_\Lambda),$$

$$V_{02}(x) = Y_0(x) - \frac{\Lambda^3}{m^3} Y_0(x_\Lambda) + \frac{\Lambda(\Lambda^2 - m^2)}{2m^3}(2 - x_\Lambda) Y_0(x_\Lambda),$$

$$V_{12}(x) = \frac{1}{x} Y_1(x) - \frac{\Lambda^3}{m^2} \frac{1}{x_\Lambda} Y_1(x_\Lambda) - \frac{\Lambda(\Lambda^2 - m^2)}{2m^3} Y_0(x_\Lambda), \tag{4.97}$$

$$V_{22}(x) = Y_2(x) - \frac{\Lambda^3}{m^3} Y_2(x_\Lambda) - \frac{\Lambda(\Lambda^2 - m^2)}{2m^3}(1 + x_\Lambda) Y_0(x_\Lambda),$$

where $Y_0(x) = e^{-x}/x$, $Y_1(x) = (1 + 1/x)Y_0(x)$, $Y_2(x) = (1 + 3/x + 3/x^2)Y_0(x)$, $x = mr$, and $x_\Lambda = \Lambda r$. In addition

$$Z_0(x_\pi, x_\rho) = a_1 Y_0(x_\pi) - a_2 Y_0(x_\rho) + a_3 Y_0(x_{\Lambda\pi}) - a_4 Y_0(x_{\Lambda\rho}),$$
$$Z_2(x_\pi, x_\rho) = a_1 Y_2(x_\pi) - a_2 Y_2(x_\rho) + a_3 Y_2(x_{\Lambda\pi}) - a_4 Y_2(x_{\Lambda\rho}),$$
(4.98)

with

$$a_1 = \frac{(\Lambda_\rho^2 - m_\rho^2)m_\rho^2}{(m_\rho^2 - m_\pi^2)(\Lambda_\rho^2 - m_\pi^2)}, \qquad a_2 = \left(\frac{m_\rho}{m_\pi}\right)^3 \frac{(\Lambda_\pi^2 - m_\pi^2)m_\rho^2}{(m_\rho^2 - m_\pi^2)(\Lambda_\pi^2 - m_\rho^2)},$$

$$a_3 = \left(\frac{\Lambda_\pi}{m_\pi}\right)^3 \frac{(\Lambda_\rho^2 - m_\rho^2)m_\rho^2}{(\Lambda_\rho^2 - \Lambda_\pi^2)(\Lambda_\pi^2 - m_\rho^2)}, \qquad a_4 = \left(\frac{\Lambda_\rho}{m_\pi}\right)^3 \frac{(\Lambda_\pi^2 - m_\pi^2)m_\rho^2}{(\Lambda_\rho^2 - m_\pi^2)(\Lambda_\rho^2 - \Lambda_\pi^2)}.$$

Similar expressions hold for $Z_0(x_\pi, x_\omega)$ and $Z_2(x_\pi, x_\omega)$.

Note that the translationally non-invariant term G cancels between the π-pair and π-current graphs. Thus in one-π range the only translationally non-invariant terms that remain are those from the π-current graph involving F_{II} and F_{III}. From these we have

$$\boldsymbol{\mu}_{\text{c.m.}} = \frac{e}{2M} \frac{1}{3} \frac{M}{m_\pi} f_{\pi NN}^2 X_\pi x_\pi [\boldsymbol{\sigma}_1 \cdot \boldsymbol{\sigma}_2 V_{02}(x_\pi) + S_{12} V_{22}(x_\pi)](\hat{\boldsymbol{r}} \times \hat{\boldsymbol{R}})\tau^x.$$

The one-π exchange potential $V_\pi(x_\pi)$ can be written as $v_{12}\boldsymbol{\tau}_1 \cdot \boldsymbol{\tau}_2$, where $v_{12} = \frac{1}{3}f_{\pi NN}^2 m_\pi[\boldsymbol{\sigma}_1 \cdot \boldsymbol{\sigma}_2 V_{02} + S_{12} V_{22}]$, so the translationally non-invariant magnetic moment operator reduces to

$$\boldsymbol{\mu}_{\text{c.m.}} = -\tfrac{1}{2}e(\boldsymbol{R} \times \boldsymbol{r})\tau^x v_{12},$$

which is precisely the form of the Sachs moment, eqn (4.49), deduced from the equation of continuity. Thus the diagram reduction method correctly gives the leading terms in the translationally non-invariant magnetic moment operator. For one-π exchange graphs the Sachs moment is the only contributor to $\boldsymbol{\mu}_{\text{c.m.}}$. For heavy-meson exchange the Sachs moment is also correctly produced, but additional terms are generated in $\boldsymbol{\mu}_{\text{c.m.}}$ that give no contribution to the continuity equation (since they come from currents that are transverse, $\boldsymbol{\nabla} \cdot \boldsymbol{j} = 0$) and whose precise form depends on the chosen form for the meson Lagrangians. We will include some numerical results of MEC of heavy-meson range in subsequent tables based on the Lagrangians and currents described in Towner.[15]

4.7 Calculation in closed-shell-plus-one nuclei

The development so far has led to the construction of a non-relativistic two-body magnetic moment operator representing the MEC processes depicted in Figs. 4.2 and 4.3. The last step is to evaluate matrix elements of these operators in a nuclear many-body system. We assume that the

4.7 CALCULATION IN CLOSED-SHELL-PLUS-ONE NUCLEI

shell model provides an adequate description of the many-body effects and that all the coupling constants introduced in the various Lagrangians maintain their known values from the free two-nucleon system when those two nucleons are embedded in a nuclear medium. This is exactly the same approximation as that used in the standard treatment with one-body operators. There the assumption is that the photon–nucleon coupling constant shall be the same in a nuclear medium as in a free nucleon. The assumption is called the *impulse approximation*. To facilitate the use of the shell model it is necessary to write the two-body operators, eqns (4.90) and (4.91), in terms of spherical tensors. The general structure of the operator is

$$T^{(\lambda,\lambda_T)} = f(r)g(R)[[C_{k_1}(\hat{r}), C_{k_2}(\hat{R})]^{(L)}, \Sigma_i^{(s)}]^{(\lambda)} \Sigma_j^{(\lambda_T)} \quad (4.99)$$

where square brackets denote vector coupling. The tensorial rank of the operator is λ ($\lambda = 1$ for magnetic moments) and λ_T in isospin space. The operator is decomposed into orbital and spin tensors. Here $f(r)$ and $g(R)$ are radial functions in the relative and centre-of-mass coordinates, $C_k(\hat{r})$ is an un-normalized spherical harmonic related to the standard spherical harmonic, $C_k = [4\pi/(2k+1)]^{1/2} Y_k$, and $\Sigma_i^{(S)}$ and $\Sigma_j^{(\lambda_T)}$ are two-body spin and isospin operators whose matrix elements are listed in Table 4.1. The decomposition of some of the operators in eqns (4.90) and (4.91) in terms of these tensors is given in Table 4.2. Because the two-body operator has been expressed in terms of relative and centre-of-mass coordinates, it would be convenient if the shell-model wavefunctions could likewise be expressed in terms of r and R. This is quite practicable if the one-body shell-model Hamiltonian is the harmonic oscillator. Then the coefficients of transformation from a single-particle basis to a relative and centre-of-mass basis are known[33] and are called Moshinsky brackets. There is one drawback to the harmonic oscillator Hamiltonian however. The eigenfunctions of the harmonic oscillator are not the eigenfunctions of the

Table 4.1 Two-body spin operators and their matrix elements $\langle(\frac{1}{2}\frac{1}{2})S_1\| \Sigma_i^{(S)} \|(\frac{1}{2}\frac{1}{2})S_2\rangle$ in the notation of Brink and Satchler.[21]
$\sigma^x = \sigma_1 \times \sigma_2 = -i\sqrt{2}\, T_1(\sigma_1, \sigma_2)$

			(S_1, S_2)			
i	S	$\Sigma_i^{(S)}$	(0, 0)	(1, 0)	(0, 1)	(1, 1)
1	0	$T_0(\sigma_1, \sigma_2)$	$\sqrt{3}$	0	0	$-1/\sqrt{3}$
2	1	$T_1(\sigma_1, \sigma_2)$	0	$\sqrt{2}$	$\sqrt{6}$	0
3	2	$T_2(\sigma_1, \sigma_2)$	0	0	0	$2\sqrt{5}/\sqrt{3}$
4	1	$\sigma_1 + \sigma_2$	0	0	0	$2\sqrt{2}$
5	1	$\sigma_1 - \sigma_2$	0	2	$-2\sqrt{3}$	0
6	0	1	1	0	0	1

Table 4.2 Representation of two-body magnetic moment operators in terms of spherical tensors

Operator	Coefficient	k_1	k_2	L	S	λ	λ_T	i	j
$\sigma^x \tau^x$	-2	0	0	0	1	1	1	2	2
$T^x \tau^x$	$2\sqrt{10}/3$	2	0	2	1	1	1	2	2
$\sigma^- \tau^-$	1	0	0	0	1	1	1	5	5
$T^- \tau^-$	$-\sqrt{10}/3$	2	0	2	1	1	1	5	5
$\sigma^+ \tau^+$	1	0	0	0	1	1	1	4	4
$T^+ \tau^+$	$-\sqrt{10}/3$	2	0	2	1	1	1	4	4
$\sigma^+ \tau_1 \cdot \tau_2$	$-\sqrt{3}$	0	0	0	1	1	0	4	1
$T^+ \tau_1 \cdot \tau_2$	$\sqrt{30}/3$	2	0	2	1	1	0	4	1
$(\hat{r} \times \hat{R})\sigma_1 \cdot \sigma_2 \tau^x$	$2\sqrt{3}$	1	1	1	0	1	1	1	2
$(\hat{r} \times \hat{R})S_{12}\tau^x$	$-2\sqrt{3}/\sqrt{5}$	1	1	1	2	1	1	3	2
	$-6/\sqrt{5}$	1	1	2	2	1	1	3	2
	$-2\sqrt{14}/\sqrt{5}$	3	1	2	2	1	1	3	2
	$-4\sqrt{7}/\sqrt{5}$	3	1	3	2	1	1	3	2
$[\sigma_1 \cdot \hat{r}\sigma_2 \times \hat{R} + \sigma_1 \times \hat{R}\sigma_2 \cdot \hat{r}]\tau^x$	$4/\sqrt{3}$	1	1	1	0	1	1	1	2
	$-2\sqrt{5}/\sqrt{3}$	1	1	1	2	1	1	3	2
	$-2\sqrt{5}$	1	1	2	2	1	1	3	2

nucleon–nucleon interaction. This is particularly important in the relative coordinate where the nucleon–nucleon interaction is known to have a strong short-range repulsion that makes the relative wavefunction go rapidly to zero, more rapidly than that given by uncorrelated oscillator functions. Thus to incorporate this piece of many-body physics in a simple way it is quite common to modify two-body operators by multiplying them by a short-range correlation function. Thus we write

$$\hat{\mu}^{ex}(r, R) = \mu^{ex}(r, R)\hat{g}(r), \qquad (4.100)$$

where $\hat{g}(r)$ is some function that tends to zero as $r \to 0$ and tends to one for large r. The precise choice of $\hat{g}(r)$ becomes part of the model dependence at short distances, along with the choice of vertex form factors at meson–nucleon couplings and the inclusion or not of specific heavy-meson exchange current processes. We take $\hat{g}(r)$ to be $\theta(r - d)$ with $d = 0.5\hbar/m_\pi c \approx 0.7$ fm. Here $\theta(x) = 0$ if $x < 0$ and $\theta(x) = 1$ if $x > 0$. Of course in very light systems, such as mass 2 and 3, where exact nuclear wavefunctions are calculable there is no need to introduce a correlation function and this source of uncertainty is avoided.

We will use a Roman letter to represent the quantum numbers of a single-particle state, e.g. $a \equiv (n_a, l_a, j_a)$; then we need to evaluate antisymmetrized two-body reduced matrix elements of the MEC opera-

4.7 CALCULATION IN CLOSED-SHELL-PLUS-ONE NUCLEI

tor, eqn (4.99). In jj coupling we have

$$\langle (ab)I_1 T_1 \| T^{(\lambda,\lambda_T)} \| (cd)I_2 T_2 \rangle$$

$$= \sum_{L_1 S_1 L_2 S_2} \begin{bmatrix} l_a & \frac{1}{2} & j_a \\ l_b & \frac{1}{2} & j_b \\ L_1 & S_1 & I_1 \end{bmatrix} \begin{bmatrix} l_c & \frac{1}{2} & j_c \\ l_d & \frac{1}{2} & j_d \\ L_2 & S_2 & I_2 \end{bmatrix} \begin{bmatrix} L_2 & L & L_1 \\ S_2 & S & S_1 \\ I_2 & \lambda & I_1 \end{bmatrix}$$

$$\times \langle (l_a l_b) L_1 S_1 T_1 \| T^{(L,S,\lambda_T)} \| (l_c l_d) L_2 S_2 T_2 \rangle, \quad (4.101)$$

where the array of nine elements is an angular momentum recoupling coefficient of four angular momenta (related to a $9j$ coefficient in eqn (1.18)) and where the reduced matrix element on the right is called the LS-coupled two-body reduced matrix element. It is given by

$$\langle (l_a l_b) L_1 S_1 T_1 \| T^{(L,S,\lambda_T)} \| (l_c l_d) L_2 S_2 T_2 \rangle$$

$$= \sum_{\substack{n_1 l_1 n_2 l_2 \\ \mathcal{N}_1 \mathcal{L}_1 \mathcal{N}_2 \mathcal{L}_2}} [1-(-)^{l_2+S_2+T_2}] \begin{bmatrix} l_2 & k_1 & l_1 \\ \mathcal{L}_2 & k_2 & \mathcal{L}_1 \\ L_2 & L & L_1 \end{bmatrix}$$

$$\times \langle n_1 l_1, \mathcal{N}_1 \mathcal{L}_1 : L_1 | n_a l_a, n_b l_b : L_1 \rangle \langle n_2 l_2, \mathcal{N}_2 \mathcal{L}_2 : L_2 | n_c l_c, n_d l_d : L_2 \rangle$$

$$\times \langle n_1 l_1 \| f(r) C_{k_1}(\hat{r}) \| n_2 l_2 \rangle \langle \mathcal{N}_1 \mathcal{L}_1 \| g(R) C_{k_2}(\hat{R}) \| \mathcal{N}_2 \mathcal{L}_2 \rangle$$

$$\times \langle (\tfrac{1}{2}\tfrac{1}{2}) S_1 \| \Sigma_i^{(S)} \| (\tfrac{1}{2}\tfrac{1}{2}) S_1 \rangle \langle (\tfrac{1}{2}\tfrac{1}{2}) T_1 \| \Sigma_j^{(\lambda_T)} \| (\tfrac{1}{2}\tfrac{1}{2}) T_2 \rangle, \quad (4.102)$$

where $\langle n_1 l_1 ; \mathcal{N}_1 \mathcal{L}_1 : L_1 | n_a l_a, n_b l_b : L_1 \rangle$ are Moshinsky brackets,[33] and

$$\langle n_1 l_1 \| f(r) C_{k_1}(\hat{r}) \| n_2 l_2 \rangle = (-)^{l_1} (2l_2+1)^{1/2} \begin{pmatrix} l_1 & k_1 & l_2 \\ 0 & 0 & 0 \end{pmatrix} I$$

$$\langle \mathcal{N}_1 \mathcal{L}_1 \| g(R) C_{k_2}(\hat{R}) \| \mathcal{N}_2 \mathcal{L}_2 \rangle = (-)^{\mathcal{L}_1} (2\mathcal{L}_2+1)^{1/2} \begin{pmatrix} \mathcal{L}_1 & k_2 & \mathcal{L}_2 \\ 0 & 0 & 0 \end{pmatrix} J$$

in terms of $3j$-symbols and radial integrals:

$$I = \int_{d/\sqrt{2}}^{\infty} R_{n_1 l_1}(x) f(\sqrt{2} x) R_{n_2 l_2}(x) x^2 \, dx,$$

$$J = \int_0^{\infty} R_{\mathcal{N}_1 \mathcal{L}_1}(X) g(X/\sqrt{2}) R_{\mathcal{N}_2 \mathcal{L}_2}(X) X^2 \, dX.$$

Here $x = (r_1 - r_2)/\sqrt{2}$, $X = (r_1 + r_2)/\sqrt{2}$, d is the cut-off parameter introduced in the correlation function, eqn (4.100), and $R(x)$ is the radial form of the harmonic oscillator wavefunction.

Our first application in finite nuclei is to calculate the expectation value of the two-body magnetic moment operator in a closed-shell-plus-one configuration. The computation boils down to an evaluation of two-body matrix elements between the valence nucleon and one of the nucleons in the closed-shell core summed over all the nucleons in the core. Let a and b be two quantum states of the valence nucleon, and c the quantum states of the occupied orbits in the core, then

$$\langle b\| T^{(\lambda,\lambda_T)} \|a\rangle = \sum_{\substack{c \\ I_1 I_2 T_1 T_2}} \frac{\hat{I}_1 \hat{I}_2 \hat{T}_1 \hat{T}_2}{\hat{j}_a \hat{j}_b \hat{\tfrac{1}{2}} \hat{\tfrac{1}{2}}} U(j_b j_c \lambda I_2; I_1 j_a)$$
$$\times U(\tfrac{1}{2}\tfrac{1}{2}\lambda_T T_2; T_1 \tfrac{1}{2}) \langle (bc) I_1 T_1 \| T^{(\lambda,\lambda_T)} \|(ac) I_2 T_2 \rangle \quad (4.103)$$

where $\hat{I} = (2I+1)^{1/2}$ and the U-coefficient is a recoupling coefficient of three angular momenta, eqn (1.17). If the closed shell occurs at LS closed shells it is computationally more efficient to express this result in terms of LS reduced matrix elements.

$$\langle b\| T^{(\lambda,\lambda_T)} \|a\rangle = \sum_{\substack{c \\ L_1 S_1 T_1 \\ L_2 S_2 T_2}} \begin{bmatrix} l_a & L & l_b \\ \tfrac{1}{2} & S & \tfrac{1}{2} \\ j_a & \lambda & j_b \end{bmatrix} \frac{\hat{L}_1 \hat{L}_2 \hat{S}_1 \hat{S}_2 \hat{T}_1 \hat{T}_2}{\hat{l}_a \hat{l}_b \hat{\tfrac{1}{2}} \hat{\tfrac{1}{2}} \hat{\tfrac{1}{2}} \hat{\tfrac{1}{2}}}$$
$$\times U(l_b l_c LL_2; L_1 l_a) U(\tfrac{1}{2}\tfrac{1}{2}SS_2; S_1 \tfrac{1}{2}) U(\tfrac{1}{2}\tfrac{1}{2}\lambda_T T_2; T_1 \tfrac{1}{2})$$
$$\times \langle (l_b l_c) L_1 S_1 T_1 \| T^{(L,S,\lambda_T)} \|(l_a l_c) L_2 S_2 T_2 \rangle. \quad (4.104)$$

This expression is more efficient than eqn (4.103) because the three 9j-coefficients that have to be computed in the jj matrix element, eqn (4.101), have been reduced to one 9j-coefficient in eqn (4.104). It will be useful to express the result of this computation in closed-shell-plus-one configuration in terms of an effective equivalent one-body operator, cf. eqn (1.48):

$$\boldsymbol{\mu}_{\text{eff}}^{\text{ex}} = \delta g_L \mathbf{L} + \delta g_S \mathbf{S} + \delta g_P [Y_2, \mathbf{S}]. \quad (4.105)$$

Then, for LS closed shells, the tensorial rank of the two-body exchange magnetic moment operator matches uniquely the tensorial rank of the equivalent one-body operator. For example, the terms in the exchange magnetic moment operator with tensorial rank $L = 1$ and $S = 0$ (the terms F_{II} and G in eqn (4.91), see Table 4.2) are the only ones that will contribute to δg_L. This is because the closed shells have $L = 0$ and $S = 0$. However, for closed shells with neutron excess orbits such as those occurring at ^{208}Pb, it is necessary to use jj coupling for the sums over the excess neutron orbits and the unique identification between terms in the two-body exchange operator and the equivalent effective one-body operator is lost.

We give some sample results in Table 4.3 for a 0p hole and 0d particle at an ^{16}O closed-shell core ($\hbar\omega = 13.3$ MeV), and for a 0$h_{9/2}$ proton and a 0$i_{13/2}$ neutron relative to a ^{208}Pb closed-shell core ($\hbar\omega = 7$ MeV). Let us look first at the δg_L value, which from pion graphs comes entirely from the Sachs moment. In the S-matrix approach, represented as a sum of low-order Feynman diagrams, the pion Born terms give a large contribution but there is a significant cancellation between the pair and current

4.7 CALCULATION IN CLOSED-SHELL-PLUS-ONE NUCLEI

Table 4.3 Meson-exchange current corrections to magnetic moments in closed-shell-plus-one configuration as a percentage of the single-particle Schmidt value. (From tables 17, 18 and 20 of Towner[15])

	$0p_{1/2}^{-1}$				$0d_{5/2}$			
	δg_L	δg_S	δg_P	$\delta\mu/\mu$ (%)	δg_L	δg_S	δg_P	$\delta\mu/\mu$ (%)
Isovector								
π pair	0.292	0.465	−0.300	−34.9	0.165	0.183	−0.369	12.0
π current	−0.215	−0.232	−0.300	14.4	−0.125	−0.134	−0.369	−10.1
π isobar[a]		−0.110	0.879	21.9		−0.118	1.031	−0.0
$\omega-\pi$		−0.014	0.224	6.1		−0.016	0.255	0.2
Heavy meson	0.098	0.045	−0.101	−15.9	0.062	−0.010	−0.112	3.2
Total	0.175	0.154	0.402	−8.4	0.102	−0.095	0.436	5.3
Isoscalar								
$\rho-\pi$		−0.002	0.025	−1.6		−0.002	0.026	0.0
Heavy meson	0.012	0.034	−0.008	1.8	0.009	0.022	−0.008	2.0
Total	0.012	0.032	0.017	0.2	0.009	0.020	0.018	2.0
	Proton $0h_{9/2}$				Neutron $0i_{13/2}$			
π pair	0.302	0.725	0.301	44.0	−0.168	−0.335	0.003	61.5
π current	−0.246	−0.382	−1.044	−35.8	0.136	0.214	1.160	−53.1
π isobar[a]	−0.003	−0.323	0.513	2.3	0.003	0.286	−0.752	−5.2
$\rho-\pi$	−0.000	−0.002	0.029	−0.1	−0.000	−0.003	0.026	0.0
$\omega-\pi$	−0.000	−0.052	0.158	0.1	0.000	0.049	−0.195	−0.6
Heavy meson	0.131	0.017	−0.100	24.7	−0.059	0.099	0.103	15.4
Total	0.184	−0.017	−0.143	35.2	−0.088	0.310	0.345	18.0

[a] Chew–Low value, $g_{\pi N\Delta}/g_{\pi NN} = 2$, is used for isobar couplings.

diagrams. (This is the cancellation of the term G in eqns (4.92) and (4.93)). The results in Table 4.3 include vertex form factors at the πNN vertices ($\Lambda_\pi = 1$ GeV) and we find, for example, for an $0h_{9/2}$ proton that $\delta g_L^{(\pi)} = 0.302 - 0.246 = 0.056$. Without the vertex form factors we would have had $\delta g_L^{(\pi)} = 0.325 - 0.230 = 0.095$. This latter value is very close to the estimate first obtained by Miyazawa[34] of $\delta g_L^{(\pi)} \simeq 0.1$ from pion-exchange currents. The non-Born terms of pion range are negligible, whereas heavy mesons make a significant contribution. The result in Table 4.3 is $\delta g_L^{(\pi)} = 0.053 + 0.131 = 0.184$ for an $h_{9/2}$ proton and $\delta g_L^{(v)} = -0.029 - 0.059 = -0.088$ for an $i_{13/2}$ neutron, where the first figure represents the contribution from pions (with vertex form factors) and the second figure that of heavy mesons. Arima and Hyuga[8] use the phenomenological Hamada–Johnston potential and the Sachs moment prescription to estimate the heavy-meson contribution. They obtain for an $h_{9/2}$ proton $\delta g_L^{(\pi)} = 0.093 + 0.062 = 0.155$ and for an $i_{13/2}$ neutron $\delta g_L^{(v)} = -0.052 - 0.034 = -0.086$. There are no vertex form factors in the pion contributions here. Note that the sum is quite similar irrespective of whether the Hamada–Johnston or the one-boson-exchange potential is used. This, of course, relates to the model independence of the δg_L calculation inasmuch as it is constrainted to the nucleon–nucleon potential through the Sachs moment.

The first experimental indication that the proton g_L value is enhanced in nuclei by about 10% over its free-nucleon value came in the measurement of Yamazaki et al.[35] of the magnetic moment of 11^- isomer in ^{210}Po. See the discussion in Section 1.9, where the isomer magnetic moment is interpreted as giving a correction to the proton g_L value of $\delta g_L^{(\pi)} = 0.16 \pm 0.01$. This is in excellent agreement with the MEC calculation. Recently a similar case involving a two-neutron 10^- isomer state in ^{190}Os has led to a direct determination[36] of the neutron δg_L value: $-0.12 \leq \delta g_L^{(\nu)} \leq -0.07$. Again the agreement with MEC calculations is good. A systematic analysis[37] of all the experimental magnetic moment data in the Pb region with an effective magnetic moment operator produces as best-fit values: $\delta g_L^{(\pi)} = 0.16 \pm 0.02$, $\delta g_L^{(\nu)} = -0.06 \pm 0.02$.

Turning to the spin operator, we again find cancellation among the two pion Born graphs and a significant contribution from the non-Born isobar graph. The net effect is quite a small contribution to δg_S. This is in contrast to core polarization, which we saw in Chapter 3 gives a strong quenching to g_S.

Finally, in Table 4.3, we see the contribution from MEC to isoscalar magnetic moments is quite small, a few per cent at most. This is because there is no contribution from pion Born graphs. The only pion-range component comes from the non-Born ρ–π graph, which is small, and heavy-meson pair graph contributions, which are cut down by short-range correlations.

4.8 MEC and core polarization

The core-polarization calculation, discussed in Chapter 3, corrects the matrix element of a one-body operator evaluated in the closed-shell-plus-one configuration for the presence of 2p–1h and 3p–2h admixtures in the single-particle wavefunction. The admixture amplitudes are estimated in perturbation theory. We found that the calculation must be taken at least to second order in the residual interaction (or to the fourth power in the meson–nucleon coupling constants). It is logical therefore that the matrix element of the two-body meson-exchange operators should likewise be corrected for 2p–1h and 3p–2h admixtures. Since the MEC operator itself involves the meson–nucleon coupling constants to the second power, it is sufficient to estimate this correction to first order in the residual interaction. The relevant Goldstone diagrams are shown in Fig.

4.8 MEC AND CORE POLARIZATION

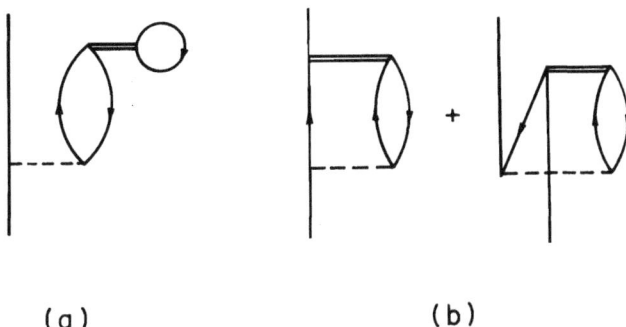

Fig. 4.4 Core-polarization corrections to a two-body MEC operator in a closed-shell-plus-one nucleus.

4.4, where the double horizontal line represents the two-body MEC operator.

Note that in graph (a), there is a limitation on the number of contributing particle–hole intermediate states coming from the restrictions imposed by the angular momentum algebra. The particle–hole state must be coupled to the same angular momentum, λ, as the multipolarity of the two-body MEC operator. We noted a similar limitation in the first-order core-polarization corrections for one-body operators discussed in Section 3.1. For this reason the contribution from Fig. 4.4(a) turns out to be small.

By contrast, the contributions from Fig. 4.4(b) are by no means negligible. Here there are no angular-momentum restrictions on the construction of intermediate states. As was the case of the second-order core polarization of one-body operators, discussed in Section 3.7, the tensor component of the residual interaction and the tensor component of the two-body MEC operator couple strongly to high-lying intermediate states. Thus the intermediate-state summation required in the evaluation of Fig. 4.4(b) must not be prematurely terminated. The importance of this graph was first pointed out by Arima et al.,[8] where is is known as the 'crossing term'. Its importance is that the graph gives a contribution to the renormalization of spin operators with the opposite sign to that from the second-order core-polarization graph involving one-body operators and cancels a large part of it. This has been demonstrated explicitly for triton β-decay in the work of Green and Schucan,[38] Ichimura et al.,[39] and Riska and Brown.[40] It is argued by Oset and Rho[41] that this cancellation should persist in heavier nuclei. In the calculations reported by Towner,[15] such a cancellation is clearly present but in detail it varies from case to case and the quoted numbers are model dependent and parameter dependent. Nevertheless the indications are clear. It would be a mistake to omit this MEC contribution to the tensor correlations.

Finally we are ready to put together the corrections from the modifications of the wavefunction (core polarization) and from the modifications of the operator (meson-exchange currents) for the case of closed-shell-plus-one nuclei and compare the computed magnetic moments with experiment. The results taken from Towner,[15] are given in Table 4.4.

First in light nuclei, the core-polarization correction comes entirely from the second-order graphs discussed in Section 3.7. While for the operator corrections we put together the lowest-order MEC terms listed in Table 4.3 with the core-polarization correction to the MEC (to be denoted MEC-CP) and the relativistic correction discussed in Section 4.3. Note in particular the large impact the MEC-CP terms have by comparing the entries in Tables 4.3 and 4.4. The isovector calculation is characterized by the conflicting interplay of many competing processes. For example, core polarization and MEC both give large corrections but they are of the opposite sign. Thus the final result is a delicate balance between these contributing pieces. Indeed it is something of an achievement to get the right sign for the correction to the Schmidt isovector magnetic moment. The entries in the table show the calculation underestimates the correction in the $0p_{1/2}^{-1}$ moment in $A=15$ and the $0f_{7/2}$ moment in $A=41$, but is in reasonable agreement in other cases. The isoscalar calculation, on the other hand, is dominated by the second-order core polarization. The MEC corrections are predominantly isovector, the only isoscalar contribution of pion range coming from the $\rho-\pi$ graph and this is small. The results are generally in good agreement with experiment although the correction for the $0f_{7/2}$ moment in $A=41$ is overestimated by a factor of two.

In heavy nuclei with shell closures separating the single-particle spin–orbit partner states there is a large correction to the magnetic moment coming from first-order core polarization, as discussed in Section 3.3, and as extended to higher orders in the RPA in Section 3.4. However, under the heading of core polarization we must also consider other second-order graphs that are not part of the RPA series. These graphs present a computationally difficult task, but estimates have been given by Shimizu[45] using a closure approximation. In addition it must be remembered that single-particle states in the Pb region, especially the high-spin states, are not pure single-particle configurations but have some core admixtures. For example, the lowest $13/2^+$ state in ^{209}Bi is not simply a proton $i_{13/2}$ state but has a sizeable admixture of $h_{9/2} \times 3^-$. Hamamoto[42] has estimated the impact of core excitations on the magnetic moment expectation values in a particle-vibration model, and we include her results with the core polarization values in Table 4.4. Hamamoto's corrections are small except for $i_{13/2}$ states.

The results show for the orbital δg_L a considerable cancellation

4.8 MEC AND CORE POLARIZATION

Table 4.4 Corrections to single-particle magnetic moments (in nuclear magnetons) for closed-shell-plus-one configurations from core polarization and meson-exchange currents

	Core polarization[a]				MEC[b]				$\delta\mu$	
	δg_L	δg_S	δg_P	$\delta\mu$	δg_L	δg_S	δg_P	$\delta\mu$	Sum	Expt.[c]
Isovector										
$A=15$ $0p_{1/2}^{-1}$	−0.183	−0.699	0.177	−0.029	0.287	0.396	0.760	0.024	−0.005	−0.050
$A=17$ $0d_{5/2}$	−0.101	−0.508	0.146	−0.448	0.169	−0.033	0.958	0.376	−0.072	−0.044
$A=39$ $0d_{3/2}^{-1}$	−0.198	−0.845	0.286	−0.137	0.278	0.211	0.616	0.363	0.226	0.197
$A=41$ $0f_{7/2}$	−0.127	−0.646	0.225	−0.689	0.193	0.022	0.731	0.639	−0.050	−0.343
Isoscalar										
$A=15$ $0p_{1/2}^{-1}$	0.013	−0.138	−0.022	0.035	0.009	0.004	−0.010	0.007	0.041	0.031
$A=17$ $0d_{5/2}$	0.010	−0.097	−0.012	−0.029	0.000	0.000	−0.018	−0.001	−0.030	−0.026
$A=39$ $0d_{3/2}^{-1}$	0.010	−0.158	−0.007	0.066	0.004	0.015	−0.010	0.004	0.070	0.071
$A=41$ $0f_{7/2}$	0.006	−0.121	0.002	−0.042	−0.001	0.000	−0.018	−0.004	−0.047	−0.019
Proton										
$A=209$ $0h_{9/2}$	−0.175	−2.535	0.481	0.13	0.282	0.183	−0.184	1.33	1.45	1.49
$A=209$ $0i_{13/2}$	−0.157	−2.475	0.596	−2.13	0.222	−0.004	−0.482	1.29	−0.83	−0.78
$A=207$ $2s_{1/2}^{-1}$		−2.086		−1.04		0.253		0.13	−0.92	−0.92
$A=207$ $1d_{3/2}^{-1}$	−0.164	−2.140	0.790	0.25	0.338	0.283	0.310	0.49	0.74	0.64
Neutron										
$A=209$ $1g_{9/2}$	0.079	1.158	−0.344	0.86	−0.143	0.317	−0.244	−0.43	0.44	0.58
$A=207$ $2p_{1/2}^{-1}$	0.133	1.241	−0.279	−0.08	−0.183	0.260	−0.021	−0.16	−0.24	−0.05
$A=207$ $1f_{5/2}^{-1}$	0.095	1.149	−0.292	−0.11	−0.190	0.213	0.212	−0.64	−0.74	−0.58
$A=207$ $0i_{13/2}^{-1}$	0.085	1.502	0.102	1.27	−0.150	0.241	0.345	−0.75	0.52	0.90

[a] In light nuclei: from table 15 of Towner.[15] In heavy nuclei: sum of CP(RPA) + CP(2nd) + Vib from tables 29, 30 of Towner[15]
[b] Sum of MEC + Δ + (MEC-CP) + Rel from tables 25, 26, 29, 30 of Towner.[15]
[c] Experimental data from references quoted in Towner.[15]

between meson-exchange currents and core polarization as was the case in light nuclei. The average values in the Pb region of $\delta g_L^{(\pi)} = 0.13 \pm 0.02$ and $\delta g_L^{(v)} = -0.07 \pm 0.01$ (see Table 1.1) are not too far from the empirical values determined by Yamazaki[37] in a best fit analysis to magnetic moment data of 0.16 ± 0.02 and -0.06 ± 0.02 respectively. The spin δg_S values are dominated by core polarization, all other effects more or less cancelling each other. The average results of $\delta g_S/g_S = -(40 \pm 3)\%$ show a large quenching in the g_S value, and justifies the empirical relation often used: $g_{S,\text{eff}} = 0.6 g_S$. Lastly for the magnetic moments themselves, the agreement between theory and experiment is within $0.2\mu_N$ in all cases except one, the neutron $i_{13/2}$ state in ^{207}Pb, where the strong cancellation between core polarization and MEC seems to be too severe.

4.9 M1 and GT giant resonances in the Pb region

We return to the problem of quenching in the M1 and Gamow–Teller giant resonances in ^{208}Pb introduced in Section 3.5. The effects of meson-exchange currents and isobar currents in particular can now be introduced into the previous results by simply replacing the one-body operator by an effective one, eqn (4.105), where the corrections δg_L, δg_S, etc. have been computed as recorded in Table 4.4. The wavefunctions are still given by a nucleons-only RPA calculation but the transition operator connecting the closed-shell and 1p–1h configurations has become modified. This procedure is correct to first order in perturbation theory. The results of this calculation are given in Table 4.5. The first line reproduces the RPA results of Table 3.3 and gives the percentage of the sum rule strength in the strongest state for the same two choices of effective interaction used before.

We have decided to separate the isobar contributions from other MEC corrections. We do this because the remaining MEC terms have a rather small effect on the transition strength, while the isobar terms are frequently commented on in the literature. It is curious that the isobar gives only a small contribution to the correction to diagonal matrix elements such as magnetic moments but has a significant impact on off-diagonal $l + \frac{1}{2} \to l - \frac{1}{2}$ matrix elements. This can be understood from the results in Table 4.3 where it is noted that the calculated δg_S and δg_P values have opposite signs. Consider the case of a nucleon in a $j = l + \frac{1}{2}$ orbit. In the diagonal matrix element, these two components of the effective one-body operator come together in the combination: $\delta g'_S = \delta g_S + [2l/(2l+3)](8\pi)^{-1/2} \delta g_P$. Hence there is a cancellation between the central and tensor pieces with the result that the corrections to magnetic moments are rather small. By contrast, M1 spin–flip matrix elements mainly responsible for transition strength in the giant resonance bring the

Table 4.5 Percentage of sum-rule strength in the strongest state in RPA calculations in ^{208}Pb for M1 and GT operators with harmonic oscillator wavefunctions and experimental single-particle energies including MEC and isobar corrections for two choices of residual interaction

	Zero-range, $g' = 0.6$				OBEP			
	$B(M1;\uparrow)$	% Sum	$B(GT;\uparrow)$	% Sum	$B(M1;\uparrow)$	% Sum	$B(GT;\uparrow)$	% Sum
RPA	37	75	134	64	44	88	156	75
RPA + MEC	41	82	133	64	48	96	155	74
RPA + MEC + Δ	22	45	100	48	38	76	125	60
RPA + MEC + Δ + CP[a]					31	62	89	43

[a] Second-order core polarization (non-RPA) and core-polarization correction to the MEC operator estimated from effective operator of Arima and Hyuga.[8]

two components together in the combination: $\delta g'_S = \delta g_S - \frac{1}{2}(8\pi)^{-1/2}\delta g_P$. Thus for off-diagonal matrix elements isobar currents are clearly inducing quenching.

Returning then to Table 4.5 the meson-exchange current corrections (excluding isobars) are seen to have a negligible influence on the Gamow–Teller resonance, while for the M1 resonance they actually enhance the strength in the strongest state by about $4\mu_N^2$ in the $B(M1; 0^+ \to 1^+)$. This is understood as coming principally from a modification to the g_L value from the Sachs moment.

The isobar currents, on the other hand, noticeably reduce the strength in the strongest state but the degree of quenching depends very much on the choice of the isobar–hole interaction. For a zero-range effective interaction, with $g' = 0.6$, isobar currents reduce the resonance strength by roughly 50%. Typically, a $B(M1; 0^+ \to 1^+) \simeq 20\mu_N^2$ is obtained, close to the current experimental value.[43] Similarly for the GT resonance, the zero-range interaction gives roughly a 70% reduction in strength in the strongest state resulting in a $B(GT; 0^+ \to 1^+)$ that is about 50% of $3(N-Z)g_A^2$ sum rule and comparable to what has been experimentally observed.[44]

The one-boson-exchange interaction on the other hand gives noticeably less quenching from isobar–hole states to the transition strength as can be seen in Table 4.5. Unlike the zero-range effective interaction, which is constructed for use in RPA calculations alone, the OBEP interaction is a more fundamental one and can not be singled out for use in just a certain subset of possible graphs such as the RPA. All second-order and possibly higher-order processes must be considered on an equal footing.

For nuclei in the Pb region, a calculation of all second-order processes that contribute to the giant resonance excitation energy and transition strength is a formidable task. It is likely, however, that the most important graphs beyond the RPA are the second-order core polarization graphs discussed in Section 3.7. Shimizu[45] has estimated the impact of these graphs on such properties as the magnetic moments of single-particle states using a closure approximation. His results combined with estimates of the core-polarization correction to MEC from Hyuga et al.[8] are listed in line 4 of Table 4.5. It is seen that the combination of isobar currents and these other second-order effects reduce the calculated $B(M1)$ strength by roughly a factor of two, much the same as the reduction from isobar currents alone with the zero-range effective interaction. Thus the interpretation of the parametrized zero-range interaction is clear: the parameter g' is compensating for the other second-order processes not included in the RPA. Cha and Speth[46] give some quantitative estimates of this.

4.10 MEC in few-nucleon systems

As has been evident in the last few sections unambiguous information on meson-exchange currents from finite nuclei is difficult to obtain because the impreciseness of the nuclear structure calculations confuses the signal. The core-polarization corrections are comparable in magnitude to the MEC corrections and often of opposite sign. In very light systems, $A = 2$ and $A = 3$, there is little or no uncertainty in the nuclear structure calculation. The wavefunction can be calculated 'exactly' for a given nucleon–nucleon potential. We will briefly comment here on three examples of isovector M1 transitions that show without doubt the need for MEC corrections.

4.10.1 Radiative thermal neutron capture

Consider the capture of slow neutrons by protons

$$n + p \rightarrow d + \gamma(M1)$$

with the emission of a photon of energy equal to the deuteron binding energy. The neutrons being of very low energy, the np-scattering state can be considered simply as an s wave ($^1S_0$) and from symmetry arguments has isospin $T = 1$. The deuteron ground state on the other hand has spin 1^+, isospin $T = 0$, and is a mixture of S and D states ($^3S_1 + {}^3D_1$). The radiative decay is an isovector M1 transition. The best calculation with nucleons only[47] gives a capture cross-section of $\sigma_{\text{theor.}} \simeq 302.5 \pm 4$ mb and is 10% short of the experimental value $\sigma_{\text{expt.}} = 334.2 \pm 0.5$ mb.[46] No reasonable variation of the nucleons-only calculation can shift the theoretical value outside its stated error limit. This suggests that a correction from MEC to the matrix element of

$$\delta = \delta_{SS} + \delta_{SD} = \frac{\langle {}^1S_0| \text{ MEC } |{}^3S_1 + {}^3D_1\rangle}{\langle {}^1S_0| \text{ IA } |{}^3S_1\rangle} = 5.2 \pm 0.7\%$$

is required to reconcile theory and experiment. The denominator is the standard one-body matrix element with free-nucleon coupling constants (impulse approximation). With the MEC operator given in eqn (4.90), we have

$$\delta_{SS} = I_{SS}/I_0,$$
$$\delta_{SD} = -\tfrac{1}{3}\sqrt{2}\, I_{SD}/I_0,$$

with

$$I_{SS} = \int_0^\infty u_0(r)u(r)(g_I(r) + h_I(r))\, dr,$$

$$I_{SD} = \int_0^\infty u_0(r)w(r)(g_{II}(r) + h_{II}(r))\, dr,$$

$$I_0 = (g_S^{(\pi)} - g_S^{(\nu)}) \int_0^\infty u_0(r)u(r)\, dr,$$

Table 4.6 Calculated MEC corrections to thermal neutron radiative capture by proton in the one-pion-exchange approximation (no vertex form factors). From Riska and Brown.[49]

	δ_{SS} (%)	δ_{SD} (%)
π pair	3.29	0.69
π current	−1.39	0.69
π isobar	0.00	1.45
Total	1.90	2.83

where $u_0(r)$ is the s-wave scattering wavefunction $\sin(kr + \delta)/k \cos \delta$ of phase shift δ and wave number k, and $u(r)/r$ and $w(r)/r$ are the S- and D-wave bound state wavefunctions of the deuteron. Riska and Brown[49] were the first to include the deuteron D state in the MEC calculation and as a consequence resolve the discrepancy. Their results are given in Table 4.6 for pion-exchange processes and have been confirmed by many other calculations.[50] Mathiot,[51] in particular, has examined the question of vertex form factors at the πNN-vertex and the role of heavy-meson exchange (particularly ρ-meson), and finds each come in at the $\delta \simeq 0.5\%$ level, but with opposite sign. This example of thermal np-radiative capture is often cited as the best illustration for the presence of mesonic effects in nuclei.

4.10.2 Electrodisintegration of the deuteron at threshold

A more demanding test of the theory is supplied by the electrodisintegration of the deuteron. There, the scattered electron acts as the source of the electromagnetic field or virtual photon. The important difference with respect to radiative capture is that the momentum and energy transferred to the deuteron by the virtual photon can be varied independently. Indeed the one-body currents (impulse approximation) and two-body currents (MEC) contributing to the cross-section have strong destructive interferences that occur successively at different momentum transfers. Cross-sections for this reaction have been measured up to momentum transfers of 28 fm^{-2}. Experimental data,[29] averaged over the energy of the n-p system near threshold ($E_{np} \leqslant 3$ MeV) are shown in Fig. 4.5 together with theoretical predictions[52] using the Paris potential for deuteron wavefunctions. Earlier calculations[53] based on the Reid hard-core and soft-core potentials give similar results. One notes that the impulse approximation alone leads to a deep minimum around momentum transfers of 12 fm^{-2} resulting from the destructive interference between the $^3S_1$–$^1S_0$ and $^3D_1$–$^1S_0$ amplitudes. Meson-exchange currents are essential for the interpretation of the data; indeed

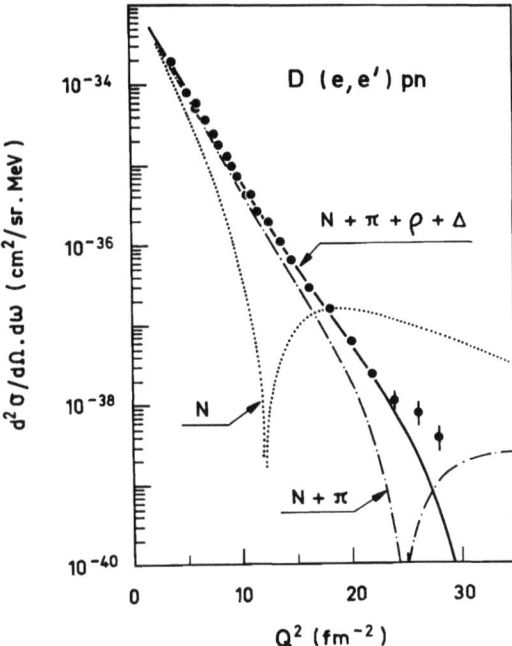

Fig. 4.5 Cross-section for threshold electro-disintegration of deuterium from impulse approximation (dotted curve) with successive additions of pion-exchange Born graphs (dot-dash curve) and isobar and heavy-meson contributions (solid curve). (From Frois and Papanicolas.[54])

between 10 and 15 fm^{-2} they provide essentially the bulk of the experimental cross-section.

Up to about 15 fm^{-2} there is little sensitivity in the calculation to details of the meson-exchange currents. The Born terms (pion-pair and pion-current graphs) shown by the dash-dot line in Fig. 4.5 are essentially model independent. The solid curve in Fig. 4.5 includes as well the non-Born terms (isobar, heavy-meson exchange, and vertex form factors) and gives a satisfactory description of the data. The theoretical predictions at the higher momentum transfers depend on the detailed structure of the currents and wavefunctions. Again we mention that the calculations of Mathiot[52] use the Dirac form, $F_1^{(1)}(Q^2)$, rather than the Sachs form, $G_E^{(1)}(Q^2)$, for the electromagnetic form factor. With all other inputs left unchanged a switch from F_1 to G_E produces a very bad fit to the data. This is discussed in some detail by Arenhövel.[30] His main point is that the choice of nucleon–nucleon potential introduces another source of uncertainty. While a variation in the potential leaves the MEC calculation (up to 15 fm^{-2}) unchanged, there can be considerable impact on the one-body impulse approximation. For example the minimum in the IA calculation is shifted from $Q^2 = 12.5\,\text{fm}^{-2}$ with the Paris potential to $Q^2 = 14\,\text{fm}^{-2}$ for the Bonn potential. The reason for this is the notably smaller D-state strength in the deuteron for the Bonn potential thus weakening the destructive interference between S and D states. Thus Arenhövel can get an equally good fit to the data using $G_E(Q^2)$ and the Bonn potential as Mathiot obtained with $F_1(Q^2)$ and the Paris potential. This merely emphasizes the range of model dependence present in non-relativistic theories.

4.10.3 Elastic magnetic electron scattering from ^{3}He and ^{3}H

Elastic electron scattering cross-section measurements from ^{3}He and ^{3}H have recently been able to separate the charge from the magnetic scattering form factors up to a momentum transfer of $Q^2 \simeq 25\,\text{fm}^{-2}$.[55] From Fig. 4.6 it is quite clear that the impulse approximation alone cannot explain the experimental data. In the region of $Q^2 \simeq 8\,\text{fm}^{-2}$ for ^{3}He and $Q^2 \simeq 12\,\text{fm}^{-2}$ for ^{3}H the cross-section is entirely due to meson-exchange processes because of destructive interference among the nucleonic amplitudes. The theoretical curve in Fig. 4.6 is taken from Hajduk et al.[56] using the Paris potential and the Dirac form factor $F_1(Q^2)$. Similar results have been given by others.[57] The computational

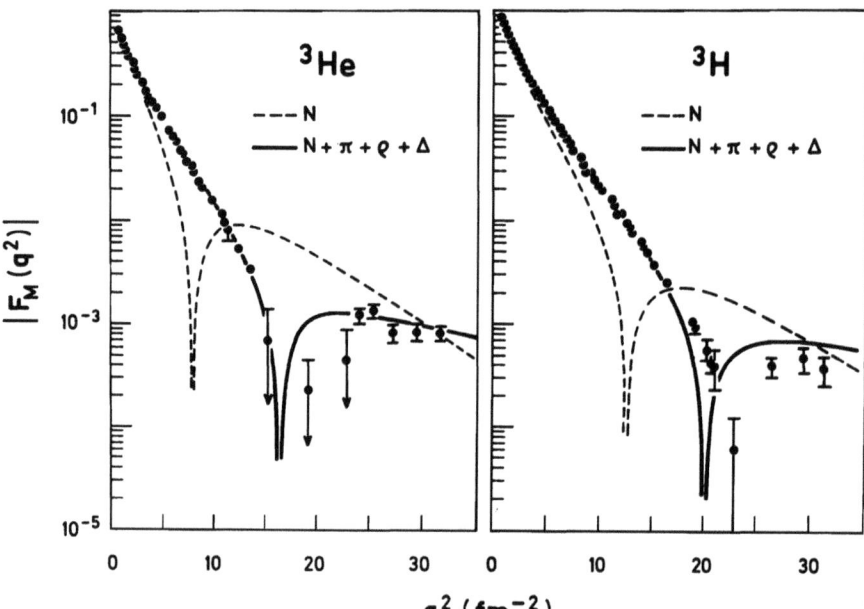

Fig. 4.6 Magnetic elastic electron scattering form factors from impulse approximation (dashed line) plus MEC contributions (solid curve) for ^{3}He and ^{3}H. (From Frois and Papanicolas.[54])

results agree with the data up the first diffraction minimum. Again there is considerable model dependence (F_1 versus G_E, choice of nucleon–nucleon potential) at the higher momentum transfers.[14,58]

In summary, the presence of MEC currents are well established by these examples. The remaining uncertainties arise from relativistic corrections and the short-range part of the nucleon–nucleon force.

4.11 Summary

We began this chapter by introducing a quantum field theory for nucleons and mesons. We then showed, using the standard procedures of minimal substitution, how the electromagnetic current of the nucleon is derived.

The same procedure was then used in a more complex situation where two nucleons are interacting through the exchange of a meson. We have concentrated our efforts on π-meson processes because this, being the lightest mass meson, leads to the longest-range operators that are the more important in nuclear physics.

The two-nucleon electromagnetic current coming from the pion-exchange processes derived here through minimal substitution are essentially model independent and unambiguous. Alternative derivations through soft-pion theorems and current algebra are not discussed here but lead to essentially similar results.

We have also shown how the theory remains gauge invariant and that current conservation is guaranteed.

Finally, we have briefly described some higher-order processes such as the excitation of the nucleon to the delta resonance as well as the inclusion of heavier mesons. These meson-exchange processes lead to two-body operators which are evaluated for finite nuclei. Typically, they lead to 10% effects for isovector magnetic properties and are of the order of 1% for isoscalar ones. Their effects on quadrupole moments are negligible.

We concluded this chapter by showing that the importance of the meson-exchange corrections exhibits itself dramatically through the magnetic electron scattering results of the deuteron and ^{3}He.

4.12 References

1. de Wit, B. and Smith, J. (1986). *Field theory in particle physics*, Vol. 1 North-Holland, Amsterdam.
2. Scheck, F. (1983). *Leptons, hadrons and nuclei*. North-Holland, Amsterdam.
3. Bernstein, J. (1968). *Elementary particles and their currents*. Freeman, San Francisco.
4. de Forest, T. and Walecka, J. D. (1966). *Adv. Phys.*, **15**, 1; Überall, H. (1971). *Electron scattering from complex nuclei*. Academic, New York;

Donnelly, T. W. and Walecka, J. D. (1975). *Ann. Rev. Nucl. Sci.*, **25**, 329; Donnelly, T. W. and Sick, I. (1984). *Rev. Mod. Phys.*, **56**, 461.
5. Bell, J. S. and Blin-Stoyle, R. J. (1957). *Nucl. Phys.*, **6**, 87; Ohtsubo, H., Sano, M., and Morita, M. (1973). *Prog. Theor. Phys.*, **49**, 877; Mukhopodhyay, N. C. and Miller, L. D. (1973). *Phys. Lett.*, **B47**, 415.
6. Friar, J. L. (1973). *Ann. Phys. (N.Y.)*, **81**, 332.
7. Dalitz, R. H. (1954). *Phys. Rev.*, **95**, 799.
8. Arima, A. and Hyuga, H. (1979). In *Mesons in nuclei* (ed. D. H. Wilkinson and M. Rho), p. 685. North-Holland, Amsterdam; Hyuga, H., Arima, A., and Shimizu, K. (1980). *Nucl. Phys.*, **A336**, 363.
9. Chemtob, M. and Rho, M. (1971). *Nucl. Phys.*, **A163**, 1; Chemtob, M. and Rho, M. (1969). *Phys. Lett.*, **B29**, 540.
10. Adam, J. and Truhlik, E. (1984). *Czech. J. Phys.*, **B34**, 1157; Adam, J., Truhlik, E., and Adamova, D. (1989). *Nucl. Phys.*, **A492**, 556.
11. Kroll, N. M. and Ruderman, M. A. (1954). *Phys. Rev.*, **93**, 233.
12. de Alfaro, V., Fubini, S., Furlan, G., and Rossetti, C. (1973). *Currents in hadron physics*. North Holland, Amsterdam.
13. Beyer, M., Drechsel, D., and Giannini, M. M. (1983). *Phys. Lett.*, **122B**, 1.
14. Lina, J. M. and Goulard, B. (1986). *Phys. Rev.*, **C34**, 714.
15. Towner, I. S. (1987). *Phys. Rep.*, **155**, 263.
16. Rarita, W. and Schwinger, J. (1941). *Phys. Rev.*, **60**, 61.
17. Sugawara, H. and von Hippel, F. (1968). *Phys. Rev.*, **172**, 1764.
18. Chew, G. F. and Low, F. E. (1956). *Phys. Rev.*, **191**, 1570.
19. Oset, E., Toki, H., and Weise, W. (1982). *Phys. Rep.*, **83**, 281.
20. Nath, L. M., Etemadi, B., and Kimel, J. D. (1971). *Phys. Rev.*, **D3**, 2153.
21. Brink, D. M. and Satchler, G. R. (1968). *Angular momentum*. Clarendon, Oxford.
22. Dumbrajs, O., et al. (1983). *Nucl. Phys.*, **B216**, 277.
23. Jensen, T., et al. (1983). *Phys. Rev.*, **D27**, 26.
24. Höhler, G. and Pietarinen, E. (1975). *Nucl. Phys.*, **B95**, 210.
25. Machleidt, R., Holinde, K., and Elster, Ch. (1987). *Phys. Rep.*, **149**, 1.
26. Particle Data Group (1986). *Phys. Lett.*, **B170**, 1.
27. Sauer, P. U. (1986). *Prog. Part. Nucl. Phys.*, **16**, 35; Delorme, J. (1985). *Nucl. Phys.*, **A446**, 65c; Hadjimichael, E. (1986). *Phys. Lett.*, **B172**, 156.
28. Fabian, W. and Arenhövel, H. (1976). *Nucl. Phys.*, **A258**, 461; Friar, J. L. and Fallieros, S. (1976). *Phys. Rev.*, **C13**, 2571.
29. Auffret, S., et al. (1985). *Phys. Rev. Lett.*, **55**, 1362; Bernheim, M., et al. (1981). *Phys. Rev. Lett.*, **46**, 402; Simon, G., et al. (1979). *Nucl. Phys.*, **A234**, 277.
30. Arenhövel, H. (1987). Subnuclear degrees of freedom in electromagnetic interactions. *Lecture at the Yukawa int. seminar on mesons and quarks in nuclei, Kyoto, April 1987*; Leidemann, W. and Arenhövel, H. (1987). *Z. Phys.*, **326**, 333; Singh, S. K., Leidemann, W., and Arenhövel, H. (1988). *Z. Phys.*, **A331**, 509.
31. Riska, D. O. (1984). *Prog. Part. Nucl. Phys.*, **11**, 199; Mathiot, J. F. (1984). *Nucl. Phys.*, **A142**, 201.
32. Bjorken, J. D. and Drell, S. D. (1964). *Relativistic quantum mechanics*. McGraw-Hill, New York.

4.12 REFERENCES

33. Brody, T. A. and Moshinsky, M. (1960). *Tables of transformation brackets*. Instituto de fisica, Mexico.
34. Miyazawa, H. (1951). *Prog. Theor. Phys.*, **6**, 801.
35. Yamazaki, T., Nomura, T., Nagamiya, S., and Katou, T. (1970). *Phys. Rev. Lett.*, **25**, 547.
36. Beck, R., Eder, R., Hagn, E., and Zech, E. (1987). *Phys. Rev. Lett.*, **59**, 2923.
37. Yamazaki, T. (1979). In *Mesons in nuclei* (ed. D. H. Wilkinson and M. Rho), p. 651. North-Holland, Amsterdam.
38. Green, A. M. and Schucan, T. H. (1972). *Nucl. Phys.*, **A188**, 289.
39. Ichimura, J., Hyuga, H., and Brown, G. E. (1972). *Nucl. Phys.*, **A196**, 17.
40. Riska, D. O. and Brown, G. E. (1971). *Phys. Lett.*, **B32**, 662.
41. Oset, E. and Rho, M. (1979). *Phys. Rev. Lett.*, **42**, 47.
42. Hamamoto, I. (1976). *Phys. Lett.*, **B61**, 343.
43. Laszewski, R. M., Alarcon, R., Dale, D. S., and Hoblit, S. D. (1988). *Phys. Rev. Lett.*, **61**, 1710.
44. Bainum, D. E., *et al.* (1980). *Phys. Rev. Lett.*, **44**, 1751; Horen, D. J., *et al.* (1980). *Phys. Lett.*, **B95**, 27; Rapaport, J. (1983). *American Institute of Physics Conference Proceedings*, No. 42 (ed. M. O. Meyer), p. 365. American Institute of Physics, New York.
45. Shimizu, K. (1976). *Z. Phys.*, **A278**, 201.
46. Cha, D. and Speth, J. (1984). *Phys. Lett.*, **B143**, 297.
47. Noyes, H. P. (1965). *Nucl. Phys.*, **74**, 508.
48. Cox, A., Wynchank, S., and Collie, C. (1965). *Nucl. Phys.*, **74**, 481.
49. Riska, D. O. and Brown, G. E. (1972). *Phys. Lett.*, **38B**, 193.
50. Kaschluhn, F. and Lewin, K. (1972). *Nucl. Phys.*, **B49**, 525; Gari, M. and Huffman, A. H. (1973). *Phys. Rev.*, **C7**, 994; Thakur, J. and Foldy, L. L. (1973). *Phys. Rev.*, **C8**, 1957; Colocci, M., Masconi, B., and Ricci, P. (1973). *Phys. Lett.*, **45B**, 224; Arenhövel, H., Fabian, W., and Miller, H. G. (1974). *Phys. Lett.*, **52B**, 303; Craver, B. A., Tubis, A., and Kim, Y. E. (1978). *Phys. Rev.*, **C18**, 1559.
51. Mathiot, J. F. (1982). *Phys. Lett.*, **115B**, 174.
52. Mathiot, J. F. (1984). *Nucl. Phys.*, **A412**, 201.
53. Hockert, J., Riska, D. O., Gari, M., and Huffman, A. (1973). *Nucl. Phys.*, **A217**, 14.
54. Frois, B. and Papanicolas, C. N. (1987). *Ann. Rev. Nucl. Part. Sci.*, **37**, 133.
55. Juster, E. P., *et al.* (1985). *Phys. Rev. Lett.*, **55**, 2261.
56. Hajduk, Ch., Sauer, P. U., and Strueve, W. (1983). *Nucl. Phys.*, **A405**, 581.
57. Hadjimichael, E., Goulard, B., and Bornais, R. (1983). *Phys. Rev.*, **C27**, 831; Maize, M. A. and Kim, Y. E. (1984). *Nucl. Phys.*, **A420**, 365; Riska, D. O. (1980). *Nucl. Phys.*, **A350**, 227.
58. Strueve, W., Hajduk, Ch., and Sauer, P. U. (1983). *Nucl. Phys.*, **A405**, 620.

5
COLLECTIVE MODELS

While the static moments of nuclei located close to the major shell closures are well described by the independent particle model (with core polarization and meson-exchange current corrections) nuclei further away from shell closures are characterized by non-spherical shapes and large quadrupole moments. To give a quantitative discussion of such nuclei will require a more phenomenological approach. It will be necessary to introduce some collective variables to describe cooperative modes of motion and a few intrinsic variables to describe the few extra-core nucleons not participating in the collective motion. Any coupling between the collective and intrinsic variables is taken as weak and studied through perturbation theory. Collective variables are only useful if the energy Hamiltonian can be separated into terms that depend on the collective variables and terms that do not. Then the eigenstates are of product form

$$\Psi_{\alpha,I} = \Phi_\alpha(q)\phi_{\alpha,I}(\omega), \qquad (5.1)$$

where α specifies the intrinsic state (with variables q), ω specifies the collective variables, and I is the total angular momentum.

Rotational motion in three dimensions involves three angular variables, such as the Euler angles $\omega = \phi, \theta, \psi$, and requires three quantum numbers to specify the state of motion.[1] The total angular momentum I and its component $M = I_z$ on a space-fixed axis provide two of these quantum numbers; the third may be obtained by considering the components of I with respect to an intrinsic (or body-fixed) coordinate system. Let I_x, I_y, I_z be the projections of I on the space-fixed axes and I_1, I_2, I_3 be the projections on the body-fixed axes then $I_{1,2,3}$ will commute with $I_{x,y,z}$ respectively. Thus as a commuting set of angular momentum variables we choose I^2, I_z, and I_3. The eigenvalues of I_3 are denoted by K, $-I \leq K \leq +I$. For specified values of I, K, and M the rotational wavefunction is

$$\phi_{KIM}(\omega) = \left(\frac{2I+1}{8\pi^2}\right)^{1/2} \mathscr{D}^I_{MK}(\omega) \qquad (5.2)$$

where the functions $\mathscr{D}^I_{MK}(\omega)$ are the rotation matrices.[2] While I^2 and I_z are constants of motion for any rotationally invariant Hamiltonian, the commutator of I_3 with the Hamiltonian depends on intrinsic properties of the system. In general, stationary states involve a superposition of states, eqn (5.2), with different values of K.

5.1 Axially symmetric nuclei

If the system possesses axial symmetry, then the projection I_3 on the symmetry axis is a constant of motion, and K is a good quantum number. If, further, the intrinsic Hamiltonian is invariant with respect to a rotation of π about an axis perpendicular to the symmetry axis, say the two-axis, then there is a further reduction in the rotational degrees of freedom. Intrinsic states with $K = 0$ can be labelled by the eigenvalue, r, of $\mathcal{R}_2(\pi)$

$$\mathcal{R}_2(\pi)\Phi_{r,K=0}(q) = r\Phi_{r,K=0}(q),$$

with $r = \pm 1$, because $\mathcal{R}_2^2(\pi) = \mathcal{R}_2(2\pi) = +1$ for a system with integer angular momentum. Thus

$$\Psi_{r,K=0,IM} = \left(\frac{2I+1}{8\pi^2}\right)^{1/2} \Phi_{r,K=0}(q)\mathcal{D}^I_{MK=0}(\omega) \tag{5.3}$$

with $I = 0, 2, 4 \ldots$ when $r = +1$ and $I = 1, 3, 5 \ldots$ when $r = -1$. The intrinsic states with $K \neq 0$ are twofold degenerate as a consequence of the $\mathcal{R}_2(\pi)$ invariance, and so the wavefunction takes the form

$$\Psi_{KIM} = \left(\frac{2I+1}{16\pi^2}\right)^{1/2} \{\Phi_K(q)\mathcal{D}^I_{MK}(\omega) + (-)^{I+K}\Phi_{\bar{K}}(q)\mathcal{D}^I_{M,-K}(\omega)\} \tag{5.4}$$

where $\Phi_{\bar{K}}(q) = \mathcal{R}_2^{-1}(\pi)\Phi_K(q)$. If the intrinsic states are expanded in components of total angular momentum, J, then

$$\Phi_K(q) = \sum_J C_J \Phi_{JK}(q),$$

$$\Phi_{\bar{K}}(q) = e^{i\pi J_2}\Phi_K(q) = \sum_J (-)^{J+K} C_J \Phi_{J,-K}(q).$$

Note the presence of a phase factor $\sigma = (-)^{I+K}$ in the second term in eqn (5.4). This is referred to as the *signature*. The contribution from this second term to the matrix element of an operator will alternate in sign for successive values of I. The signature-dependent term implies that the rotational bands with $K \neq 0$ in a system with axial symmetry and $\mathcal{R}_2(\pi)$ invariance tend to separate into two families distinguished from each other by the quantum number σ.[1]

For a slowly rotating system, the energy is given to a first approximation by the intrinsic energy E_K associated with the state $\Phi_K(q)$ and is therefore the same for all members of the band. The superimposed rotational motion gives an additional energy depending on the rotational angular momentum, $I_{1,2,3}$. For $K = 0$ bands, the rotational Hamiltonian is diagonal with respect to I_3 and can only depend on the combination $I_1^2 + I_2^2$:

$$(H_{\rm rot})_{K=0} = h_0(q)(I_1^2 + I_2^2)$$

$$E_{\rm rot} = \frac{\hbar^2}{2\mathscr{I}} I(I+1) \tag{5.5}$$

where the effective moment of inertia is $\hbar^2/2\mathscr{I} = \langle K| h_0 |K\rangle$ with h_0 a function of the intrinsic variables. This form is the familiar expression obtained by quantizing the classical Hamiltonian for a symmetric top. More generally the rotational energy is a function of $I(I+1)$. Many low-lying spectra of even–even nuclei show a sequence of states $I = 0, 2, 4, \ldots$ with spacings given by the $I(I+1)$ rule. This is the characteristic signal of axial symmetry. Furthermore the absence of odd values of I reveals the $\mathscr{R}_2(\pi)$ symmetry in the nuclear shape.

For $K \neq 0$, the nuclear states involve a combination of intrinsic wavefunctions with $I_3 = +K$ and $I_3 = -K$. Thus the rotational energy consists partly of terms with $\Delta K = 0$, which are of the same form as for $K = 0$ bands, and partly of terms with $\Delta K = \pm 2K$ having the form

$$(H_{\rm rot})_{\Delta K = \pm 2K} = h_{2K}(q)(I_-)^{2K} + h_{2\bar K}(q)(I_+)^{2K}$$

with $I_\pm = I_1 \pm iI_2$. The second term is the conjugate of the first under the symmetry $\mathscr{R}_2(\pi)$. The expectation value of the additional rotation energy is[1]

$$(\Delta E)_{\rm rot} = (-)^{I+K} A_{2K} \frac{(I+K)!}{(I-K)!}$$

with $A_{2K} = \langle K| h_{2K} |\bar K\rangle$. In particular, for $K = \frac{1}{2}$ bands, $E_{\rm rot}$ becomes

$$E_{\rm rot} = \frac{\hbar^2}{2\mathscr{I}} [I(I+1) + a(-)^{I+1/2}(I+\tfrac{1}{2}) \delta_{K,1/2}], \tag{5.6}$$

where $A_1 = a\hbar^2/2\mathscr{I}$. The quantity a is referred to as the decoupling parameter. Note that, apart from a constant, the rotational energy goes as $[I + (1+a\sigma)/2]^2$, where σ is the signature. Thus for $I = \frac{3}{2}, \frac{7}{2}, \ldots$ $E_{\rm rot}$ is proportional to $[I + (1+a)/2]^2$ while for $I = \frac{1}{2}, \frac{5}{2}, \ldots$ there is a displacement with $E_{\rm rot}$ proportional to $[I + (1-a)/2]^2$. Indeed with $a > 1$ the displacements lead to an inversion of the normal spin sequence.

An axially symmetric deformation can be characterized by an intrinsic electric quadrupole moment defined as

$$Q_0 \equiv \left\langle K \left| \int \rho(r')(3z'^2 - r'^2) \, d^3r' \right| K \right\rangle$$

$$= 2\langle K| T_{2, \nu=0}(E) |K\rangle. \tag{5.7}$$

5.1 AXIALLY SYMMETRIC NUCLEI

The primed coordinates refer to the intrinsic (body-fixed) system, and $T_{2,v=0}(E)$, introduced in eqn (1.37), denotes the components of the electric quadrupole tensor relative to the intrinsic system. The E2 moments referred to the space-fixed system are obtained from the intrinsic moments by a rotation of the tensor operator:[1]

$$T_{2\mu}(E;r) = \sum_v T_{2v}(E,r')\mathscr{D}^2_{\mu v}(\omega) = \tfrac{1}{2}Q_0\mathscr{D}^2_{\mu v=0}(\omega) \tag{5.8}$$

where $\omega = \phi, \theta, \psi$ are the Euler angles describing the orientation of the body-fixed system. This operator, eqn (5.8), describes only the collective part of the E2 operator associated with the average deformation of the intrinsic state.

In general, we need to calculate matrix elements of the tensor operators, $T_{\lambda\mu}(\xi)$ with $\xi = E$ for electric operators and $\xi = M$ for magnetic operators as introduced in eqn (1.36), using the collective wavefunction eqn (5.4) for states in specific rotation bands K_1 and K_2. The matrix element involves an integration over the Euler angles ω, and an integration over the intrinsic variables q. The former integral is elementary, following the properties of rotation matrices:[2]

$$\langle \mathscr{D}^{I_1}_{M_1K_1} | \mathscr{D}^{\lambda}_{\mu v} | \mathscr{D}^{I_2}_{M_2K_2} \rangle = \frac{8\pi^2}{2I_1+1} \langle I_2M_2\lambda\mu \mid I_1M_1 \rangle \langle I_2K_2\lambda v \mid I_1K_1 \rangle. \tag{5.9}$$

The general result is

$$\langle K_1I_1M_1 | T_{\lambda\mu} | K_2I_2M_2 \rangle = \hat{I}_2\hat{I}_1^{-1} \langle I_2M_2\lambda\mu \mid I_1M_1 \rangle$$
$$\times \{\langle I_2K_2\lambda K_1 - K_2 \mid I_1K_1 \rangle \langle K_1 | T_{\lambda, K_1-K_2} | K_2 \rangle$$
$$+ (-)^{I_2+K_2} \langle I_2 - K_2\lambda K_1 + K_2 \mid I_1K_1 \rangle \langle K_1 | T_{\lambda, K_1+K_2} | \bar{K}_2 \rangle \} \tag{5.10}$$

where $\hat{I} = (2I+1)^{1/2}$. Note that we have assumed that the matrix elements are real, which follows if the tensor operators have specific transformation properties under time reversal, $\mathscr{T}$

$$\mathscr{T} T_{\lambda\mu} \mathscr{T}^1 = c(-)^{\lambda+\mu} T_{\lambda,-\mu}$$

with $c = +1$. (If the electromagnetic operators introduced in eqn (1.36) are multiplied by i^λ for electric operators and by $i^{\lambda-1}$ for magnetic operators, then $c = +1$ as required.) There is an intimate connection between invariance under time reversal and invariance under rotation $\mathscr{R}_2(\pi)$. Indeed for a $\mathscr{T}$- and $\mathscr{R}_2(\pi)$-invariant intrinsic Hamiltonian, the phases of the intrinsic states can be chosen [1] such that $\mathscr{R}_2(\pi)\mathscr{T} = 1$. Then it follows that

$$\mathscr{T}\Phi_K(q) = \Phi_{\bar{K}}(q); \qquad \mathscr{T}\Phi_{\bar{K}}(q) = (-)^{2K}\Phi_K(q)$$

and hence

$$\langle \bar{K}_1| T_{\lambda v} |K_2\rangle = (-)^{\lambda+v+2K_1}\langle K_1| T_{\lambda,-v} |\bar{K}_2\rangle$$
$$\langle \bar{K}_1| T_{\lambda\mu} |\bar{K}_2\rangle = (-)^{\lambda+v+2K_1+2K_2}\langle K_1| T_{\lambda,-v} |K\rangle. \quad (5.11)$$

Using these relations, the four terms that appear when matrix elements, eqn (5.10), are evaluated with wavefunctions, eqn (5.4), are reduced to two. The second of these terms carries the signature phase, $\sigma = (-)^{I_2+K_2}$, and is known as the signature-dependent term. Its contribution is usually small when compared with the first term. Naturally it only contributes if $K_1 + K_2 \leq \lambda$.

In the case that $K_1 = K_2 \neq 0$, a consideration of the combined symmetries, Hermitian conjugation, and time reversal, leads to a restriction on the signature-dependent matrix element. Suppose under Hermitian conjugation the tensor operator transforms as $T^\dagger_{\lambda v} = (-)^{k+v} T_{\lambda,-v}$. For example, the electromagnetic tensor operators, eqn (1.36), (multiplied by i^λ if electric and $i^{\lambda-1}$ if magnetic to satisfy reality conditions), have a phase $(-)^k$ equal to $(-)^\lambda$ if the operator is $T_{\lambda v}(E)$ and equal to $(-)^{\lambda-1}$ if the operator is $T_{\lambda v}(M)$. Hermitian conjugation relates the matrix elements of T to the matrix elements with the initial and final states transposed: $\langle K_1| T^\dagger |K_2\rangle = \langle K_2| T |K_1\rangle^*$. Assuming the matrix elements are real, then the signature-dependent matrix element has the property

$$\langle K| T_{\lambda,2K} |\bar{K}\rangle = \langle \bar{K}| T^\dagger_{\lambda,2K} |K\rangle$$
$$= (-)^{2K}\langle K| \mathcal{T} T^\dagger_{\lambda,2K} \mathcal{T}^{-1} |\bar{K}\rangle$$
$$= (-)^{2K}(-)^{\lambda+k}\langle K| T_{\lambda,2K} |\bar{K}\rangle.$$

Hence $(-)^{2K+\lambda+k}$ must equal $+1$. For electric operators with $k = \lambda$ this leads to a requirement that the signature-dependent matrix element must vanish if K is a half integer. Conversely for magnetic operators with $k = \lambda - 1$ the signature-dependent matrix element must vanish if K is an integer. Thus for E2 transition rates the signature-dependent term only contributes in $K = 1$ bands while for M1 transition rates it only contributes in $K = \frac{1}{2}$ bands.

The signature-dependent term is also absent in the special case $K_1 = K_2 = K = 0$. This is because eqn (5.3) has to be used instead of eqn (5.4), to obtain

$$\langle K = 0, I_1 M_1| T_{\lambda\mu} |K = 0, I_2 M_2\rangle = \hat{I}_2 \hat{I}_1^{-1} \langle I_2 M_2 \lambda\mu | I_1 M_1\rangle \langle I_2 0 \lambda 0 | I_1 0\rangle$$
$$\times \langle r, K = 0| T_{\lambda,0} |r, K = 0\rangle \delta_{\lambda,\text{even}}. \quad (5.12)$$

Note the additional requirement that λ has to be even. This follows from the property of the intrinsic matrix element under time reversal

$$\langle r, K = 0| T_{\lambda,0} |r, K = 0\rangle = \langle \mathcal{T}r, K = 0| \mathcal{T} T_{\lambda,0} \mathcal{T}^{-1} |\mathcal{T}r, K = 0\rangle$$
$$= (-)^\lambda \langle r, K = 0| T_{\lambda,0} |r, K = 0\rangle.$$

Thus, for example, there are no non-zero magnetic moments or M1 matrix elements in $K = 0$ rotation bands in this model. We return to this shortly.

In what follows we will drop the signature-dependent term in the general expression, eqn (5.10), and consider the reduced matrix element of the electric quadrupole tensor, eqn (5.8), within a given rotation band $K_1 = K_2 = K$:

$$\langle KI_1\| T_2(E) \|KI_2\rangle = \hat{I}_2\hat{I}_1^{-1}\langle I_2 K 20 | I_1 K\rangle \tfrac{1}{2}Q_0. \tag{5.13}$$

Our definition of reduced matrix element is given in eqn (1.4). The reduced E2 transition probabilities, eqn (1.35), between members of the band and the diagonal quadrupole moments of states in the band therefore only depend on one parameter Q_0:

$$B(E2; I \to I - 2) = \frac{5}{16\pi}\frac{2I-3}{2I+1}\langle I - 2K20 | IK\rangle^2 Q_0^2$$

$$= \frac{5}{16\pi}\frac{6(I^2-K^2)[(I-1)^2-K^2]}{(2I+1)(2I)(2I-1)(2I-2)}Q_0^2 \tag{5.14a}$$

$$Q = \langle II20 | II\rangle\langle IK20 | IK\rangle Q_0$$

$$= \frac{3K^2 - I(I+1)}{(I+1)(2I+3)}Q_0. \tag{5.14b}$$

Thus a number of simple relationships between $B(E2)$ rates and quadrupole moments are trivially established. For example, many even–even nuclei in the rare-earth region exhibit a ground-state rotation band, $I = 0, 2, 4\ldots$ with level spacings obeying the $I(I+1)$ rule. The ratio of $B(E2)$ rates, $B(E2; 4\to 2)/B(E2; 2\to 0)$, should therefore be the same in all these nuclei and equal to 10/7 for $K = 0$. This result is well supported by many data.

However, we will not belabour the relations obtained from eqn (5.14) as they only represent leading terms. There will be corrections, for example, whenever there is a term in the Hamiltonian coupling the intrinsic and rotational variables. A Coriolis term can do this. We will discuss an example of this in the next section with the particle–rotor model. There will also be corrections from non-axially symmetric components in the nuclear charge distribution. A systematic analysis may be obtained by expanding the intrinsic E2 operator $T_{2\nu}(E)$ in eqn (5.8) in powers of the rotational angular momentum I, in which the leading I-independent term is the term that leads to relations such as eqn (5.14). Bohr and Mottelson[1] discuss some examples of corrections such as these.

We have already demonstrated for an M1 operator (whose form depends only on the intrinsic variables and is independent of the rotational variables) that its matrix element is zero in $K=0$ bands from time-reversal invariance. However, a term could be introduced that depends on the rotational angular momentum. The obvious choice is

$$T_{10}(M) = g_R I \mu_N, \tag{5.15}$$

where the parameter g_R is the effective g-factor for rotational motion and μ_N is the unit of nuclear magneton. Higher-order terms can be added by treating g_R as a function of $I(I+1)$. The rotational angular momentum is primarily associated with orbital motion of the nucleons. Magnetic effects arise from the motion of charged particles ($g_L = 1$ for protons and $g_L = 0$ for neutrons, see eqn (1.2)), thus for a uniform charge distribution we would anticipate $g_R = Z/A$. The g_R factors for even–even nuclei can be deduced from the measured magnetic moment of the first 2+ state. The measured values of g_R are generally comparable with, although in most cases somewhat smaller than, the value Z/A. We will return to this in Section 5.10. For $K=0$ there are no M1 transitions within the band since successive states have $\Delta I = 2$.

In bands with $K \neq 0$, the intrinsic nucleonic motion generates a magnetic moment in addition to that produced by the rotational motion and the M1 operator takes the form:[1]

$$T_{1\mu}(M) = \sum_\nu T_{1\nu}(M) \mathcal{D}^1_{\mu\nu} + g_R(I_\mu - I_3 \mathcal{D}^1_{\mu 0}) \mu_N. \tag{5.16}$$

The first term is the intrinsic moment and is I-independent. The second term represents the rotational motion and is proportional to the angular momentum perpendicular to the symmetry axis. The reduced matrix element of eqn (5.16) is

$$\langle KI_1 \| T_1(M) \| KI_2 \rangle = \hat{I}_2 \hat{I}_1^{-1} \{ (g_K - g_R) K \langle I_2 K 10 | I_1 K \rangle$$
$$+ g_R [I_1(I_1+1)]^{1/2} \delta_{I_1, I_2}$$
$$- \frac{1}{\sqrt{2}} (-)^{I_2+1/2} (g_K - g_R) b \langle I_2 - \tfrac{1}{2} 11 | I_1 \tfrac{1}{2} \rangle \delta_{K, 1/2} \} \mu_N, \tag{5.17}$$

where the following notation has been introduced for the intrinsic matrix elements:

$$g_K K \mu_N \equiv \langle K | T_{10}(M) | K \rangle,$$
$$(g_K - g_R) b \mu_N \equiv -\sqrt{2} \langle K | T_{11}(M) | \bar{K} \rangle \delta_{K, 1/2}. \tag{5.18}$$

The signature-dependent term involving the parameter b only contributes for $K = \tfrac{1}{2}$ bands. The quantity b is referred to as the decoupling parameter. The expressions for the static moments and $B(M1)$ transition rates within a band are given by

$$B(M1; I \to I-1) = \frac{3}{4\pi}(g_K - g_R)^2 \frac{(I-K)(I+K)}{I(2I+1)}$$
$$\times (K + \tfrac{1}{2}b(-)^{I+1/2}\delta_{K,1/2})^2, \quad (5.19a)$$

$$\mu = g_R I + (g_K - g_R)\frac{K^2}{I+1}[1 + (2I+1)b(-)^{I+1/2}\delta_{K,1/2}], \quad (5.19b)$$

in nuclear magneton, μ_N, units. Again these expressions give leading-order terms for the $B(M1)$ transition rates and magnetic moments in a given band in terms of a few parameters and simple geometric factors. The parameters, in fact, are defined in terms of intrinsic matrix elements and can in principle be calculated given a model for the intrinsic state. Alternatively they can be taken as parameters and determined in fits to experimental data. Often the derived parameters in the latter case are found to vary from nucleus to nucleus. For example the observed g_R values for odd-A nuclei deviate from those of the neighbouring even–even nuclei. Such a difference arises from the renormalization of g_R associated with a particle–core coupling through the Coriolis force. We turn then to a very simple model for which such effects can be evaluated.

5.2 Particle–rotor model

The model of a particle coupled to a rotor provides an approximate description of many of the properties of low-lying bands in odd-A nuclei. The extension to several valence particles interacting through pairing forces and coupled to a rotor can be easily included within the framework of the model. Review articles on the particle–rotor model can be found in Bunker and Reich, Ogle *et al.*, and Engeland.[3,4]

The basic assumption is that the Hamiltonian can be separated into a rotational part and an intrinsic part:

$$H = T_{\text{rot}} + H_{\text{intr}}. \quad (5.20)$$

We take the rotor to have axial symmetry and $\mathcal{R}_2(\pi)$ invariance with quantum numbers $R_3 = 0$, $r = +1$ as for the ground states of even–even nuclei. The angular momentum of the rotor is denoted by $\boldsymbol{R}$, with component R_3 with respect to the symmetry axis and the spectrum consists of the sequence $R = 0, 2, 4 \ldots$. The kinetic energy of rotational motion then takes the form

$$T_{\text{rot}} = \frac{\hbar^2}{2\mathcal{I}}(R_1^2 + R_2^2) = \frac{\hbar^2}{2\mathcal{I}}\boldsymbol{R}^2, \quad (5.21)$$

where $\mathscr{I}$ is the moment of inertia. For the intrinsic Hamiltonian we write

$$H_{\text{intr}} = H_0 + H_{\text{pair}} = \sum_{i=1}^{n} h_0(i) + \sum_{i \neq j}^{n} v_{\text{pair}}(i,j), \quad (5.22)$$

where H_0 contains one-body kinetic and potential energy terms representative of a deformed quadrupole field, typically given in terms of a Nilsson potential.[5] The H_{pair} accounts for the two-body interactions of the extra-core nucleons. In the particle–rotor model the total angular momentum I is composed of two terms, the collective rotation of the core R and the angular momentum due to the valence nucleons, j. Thus we have $I = R + j$. Since I^2 is a conserved quantity, we replace R in eqn (5.21) by $I - j$ to obtain the following well known expression for the particle–rotor model Hamiltonian

$$H = H_I + H_{\text{intr}} + H_C, \quad (5.23)$$

where

$$H_I = \frac{\hbar^2}{2\mathscr{I}}(I^2 - I_3^2),$$

$$H_{\text{intr}} = H_0 + H_{\text{pair}} + \frac{\hbar^2}{2\mathscr{I}}(j^2 - j_3^2), \quad (5.24)$$

$$H_C = -\frac{\hbar^2}{2\mathscr{I}}(I_+ j_- + I_- j_+),$$

with $I_\pm = I_1 \pm iI_2$ and $j_\pm = j_1 \pm ij_2$. The Hamiltonian therefore is written as three terms. The rotational term H_I represents the kinetic energy of rotational motion and produces energy differences between states in a rotational band. The intrinsic term is as written in eqn (5.22) but with a recoil term added. For just one particle outside a core this term can simply be absorbed in the single-particle potential field in H_0. But for several valence particles, the recoil term contains two-body interactions that must be included with H_{pair}. The Coriolis term, H_C, represents a coupling between rotational and intrinsic motion and distorts the spectrum of rotational bands. It is a general experience[4] that the effect of this term is far too strong compared with experiment and some artificial way of reducing its effect is often introduced. Solutions in a particle–rotor model calculation are obtained in two steps. First, the intrinsic eigenvalue problem is solved

$$H_{\text{intr}} \Phi_K(q) = \varepsilon_K \Phi_K(q) \quad (5.25)$$

to give a set of intrinsic states and intrinsic energies ε_K. From these, rotational wavefunctions of the form eqn (5.4) are constructed. These form the basis for the calculation. Then in the second step the matrix elements of the Coriolis term are evaluated in the basis states

$$\langle K_1 IM| H_C |K_2 IM \rangle = -\frac{\hbar^2}{2\mathscr{I}} \{(-)^{I+1/2}(I+\tfrac{1}{2})\langle K_1| j_+ |\bar{K}_2 \rangle \delta_{K_1,1/2} \delta_{K_2,1/2}$$
$$+ [(I \mp K_2)(I \pm K_2 + 1)]^{1/2} \langle K_1 | j_\pm |K_2\rangle \delta_{K_1, K_2 \pm 1}\}. \quad (5.26)$$

The off-diagonal matrix elements of H_C are non-zero only when $|K_1 - K_2| = 1$ or when $K_1 = K_2 = \tfrac{1}{2}$. Thus the Coriolis interaction introduces band mixing. The energy eigenvalues and eigenvectors are obtained from the diagonalization of the matrix $\langle H_C \rangle$ and have the form

$$|iIM\rangle = \sum_K A^i_{IK} |KIM\rangle \quad (5.27)$$

where the A^i_{IK} are the calculated mixing coefficients, the eigenvectors of the matrix diagonalization.

In the extreme case of one particle outside a strongly deformed core we can use the Nilsson model[5] to evaluate the intrinsic matrix elements. The Hamiltonian adopted in the original version of the model can be written as

$$h_0 = (p^2/2M) + \tfrac{1}{2}M[\omega_\perp^2(x_1^2 + x_2^2) + \omega_3^2 x_3^2] + \kappa\hbar\omega_{00}(2\mathbf{l}\cdot\mathbf{s} + \mu l^2). \quad (5.28)$$

The first two terms correspond to the deformed harmonic oscillator potential where p and M are the momentum and mass of the odd nucleon and (x_1, x_2, x_3) are its position coordinates in a coordinate system fixed in the nucleus. The $\mathbf{l} \cdot \mathbf{s}$ term is a spin–orbit term, and the l^2 term is introduced to reproduce the fact that the high angular momentum states occur lower in energy than predicted in the simple harmonic-oscillator potential. The frequencies $\omega_\perp$ and ω_3 are defined by

$$\omega_\perp^2 = \omega_0(\delta)^2(1 + 2\delta/3),$$
$$\omega_3^2 = \omega_0(\delta)^2(1 - 4\delta/3),$$
$$\omega_0(\delta) = \omega_{00}(1 - 4\delta^2/3 - 16\delta^3/27)^{-1/6}, \quad (5.29a)$$
$$\hbar\omega_{00} = \hbar\omega_0(0) \approx 41 A^{-1/3} \text{ MeV},$$

and δ is the deformation parameter. Its definition is

$$\delta = 3(\omega_\perp - \omega_3)/(2\omega_\perp + \omega_3) = 3(c - a)/(2c + a), \quad (5.29b)$$

where c is the length of the semi-major axis (along the symmetry axis) and a the length of the semi-minor axis. For $\delta > 0$, the nucleus has a prolate shape, and for $\delta < 0$ it has an oblate shape. The parameter δ is related to the intrinsic quadrupole moment, eqn (5.7),

$$Q_0 = \tfrac{2}{5}Ze(c^2 - a^2) = \tfrac{4}{5}ZeR_0^2\beta(1 + \tfrac{1}{6}\beta), \quad (5.30)$$

where ΔR is the difference between the major and minor axis, $\Delta R = c - a$, R_0 the mean radius $R_0 = (c + 2a)/3$, and $\beta = \Delta R/R_0$.

In solving for the eigenstates of h_0, a spherical harmonic-oscillator representation is employed, and the values of κ and μ are chosen in such a way that for $\delta = 0$ the well established sequence of levels of the spherical shell model is reproduced. Because of the non-spherical symmetry of h_0 the total angular momentum, j, of the odd particle is not conserved for a non-zero deformation. Thus the eigenfunctions of h_0 are expressed as a linear combination of the appropriate spherical harmonic-oscillator functions, namely

$$\Phi_K = \sum_{jl} c_{ljK} |NljK\rangle \tag{5.31}$$

or, equivalently,

$$\Phi_K = \sum_{l\Lambda\Sigma} a_{l\Lambda\Sigma K} |Nl\Lambda\Sigma\rangle,$$

where $a_{l\Lambda\Sigma K}$ and c_{ljK} are related by the Clebsch–Gordan transformation

$$c_{ljk} = \sum_{\Lambda\Sigma} a_{l\Lambda\Sigma K} \langle l\Lambda\tfrac{1}{2}\Sigma | jK \rangle.$$

Here K is the magnitude of the eigenvalue of j_3, the projection of j along the nuclear symmetry axis, Λ and Σ are the projections of l and s on the symmetry axis ($K = \Lambda + \Sigma$), and N is the total number of oscillator quanta, assumed to be a good quantum number.

With Nilsson wavefunctions, the intrinsic matrix elements of $j_\pm$ in the matrix elements of $\langle H_C \rangle$ are trivially evaluated

$$\langle K_1 | j_\pm | K_2 \rangle = \sum_j c_{ljK_1} c_{ljK_2} [(j \mp K_2)(j \pm K_2 + 1)]^{1/2} \delta_{K_1, K_2 \pm 1}, \tag{5.32a}$$

$$a = \langle K_1 | j_+ | \tilde{K}_2 \rangle = \sum_j |c_{lj1/2}|^2 (-)^{j+1/2} (j + \tfrac{1}{2}) \delta_{K_1, 1/2} \delta_{K_2, 1/2}. \tag{5.32b}$$

The latter matrix element is the quantity a introduced earlier, eqn (5.6), as the decoupling parameter in the expression for the energy of the states I in a rotation band with $K = \tfrac{1}{2}$.

Likewise for the magnetic moment and $B(M1)$ transition rates, the intrinsic matrix elements can be evaluated in the particle–rotor model with Nilsson wavefunctions. From the definitions in eqn (5.18) we have

$$g_K K = \langle K | g_L L + g_S S | K \rangle$$

$$= \sum_j |c_{ljK}|^2 K[g_L \pm (g_S - g_L)/(2l+1)], \tag{5.33a}$$

$$(g_K - g_R)b = \sum_j |c_{ljk}|^2 (-)^{j+1/2} (j + \tfrac{1}{2}) [g_L \pm (g_S - g_L)/(2l+1)], \tag{5.33b}$$

for $j = l \pm \tfrac{1}{2}$. Here g_L and g_S are the orbital and spin g-factors for the last odd nucleon. The gyromagnetic factors g_K and g_R for a rotational band with $K \neq \frac{1}{2}$ can be obtained from experimental data by combining a measured magnetic moment with an M1 transition matrix element. For $K = \frac{1}{2}$ bands, the M1-matrix elements involve the additional parameter b and the analysis requires the determination of an additional moment or transition probability. An analysis of M1 data in odd mass nuclei in the rare-earth region is given in Bohr and Mottelson[1] and by Prior et al.[6] The most striking feature of the determined g_R values is the systematic tendency for the odd-proton values to exceed those of the odd-neutron nuclei, while the values from the neighbouring even–even nuclei tend to be somewhere in between. The deduced g_K and b values can be compared with calculations based on Nilsson wavefunctions, eqn (5.33). The empirical g_K values deviate systematically from these calculations in a manner that can be approximately accounted for by replacing the free-nucleon g_S value by an effective one with $g_{S,\text{eff}}/g_S$ in the range 0.6 to 0.7. This is of the same magnitude as core-polarization quenching discussed in Chapter 3 involving a single particle outside closed-shells core. Note that for the decoupling parameter b, the intrinsic matrix element being calculated is $\langle K| S_+ |\bar{K}\rangle$ with $K = \frac{1}{2}$, compared with $\langle K| S_3 |K\rangle$ for the parameter g_K. Bochnacki and Ogaza[7] have argued that the core-polarization correction for S_+ would be larger than for S_3 and empirical fits to data seem to support this.

The model of an odd-mass nucleus of a single nucleon moving in the field of a deformed core is still rather crude. Improvements are needed. One common procedure is to include some many-particle correlations by converting the odd particle to a quasiparticle through a standard pairing model calculation. Briefly, in a second quantization notation, let the operator $a^\dagger_{\alpha K}$ ($a_{\alpha K}$) create (destroy) a particle in a Nilsson orbital, α. The residual interaction between the valence particles is given by the pairing term, H_{pair}, which in second quantization is written

$$H_{\text{pair}} = G \sum_{\alpha \alpha'} a^\dagger_\alpha a^\dagger_{\bar{\alpha}} a_{\bar{\alpha}'} a_{\alpha'},$$

where $\bar{\alpha}$ stands for the time-reversed Nilsson orbital. An approximate solution to this problem is given by the BCS method[8] in which a trial wavefunction is written

$$|\alpha\rangle = a^\dagger_\alpha |\bar{0}\rangle_\alpha$$
$$= a^\dagger_\alpha \prod_{\beta \neq \alpha} (U_\alpha + V_\alpha a^\dagger_\alpha a^\dagger_{\bar{\alpha}}) |0\rangle,$$

where $|\bar{0}\rangle_\alpha$ is the BCS vacuum excluding the pair orbitals ($\alpha\bar{\alpha}$) and $|0\rangle$ is the particle vacuum. The odd particle occupies the Nilsson orbital α, the remaining particles are distributed in time-reversed pairs over the available Nilsson orbitals. The input to the calculation is the strength of the pairing force, G, and the single-particle energies ε_α of the m Nilsson orbitals retained in the calculation. The output is the occupancy of these m orbitals,

$$V_\beta^2 = \tfrac{1}{2}(1 - \eta_\beta/E_\beta), \qquad U_\beta^2 = 1 - V_\beta^2,$$

where $\eta_\beta = \varepsilon_\beta - \lambda$ and $E_\beta = (\eta_\beta^2 + \Delta^2)^{1/2}$. The gap energy Δ and the Lagrange multiplier λ are obtained from a minimization of the energy to variations in the occupancy V_β^2 subject to the constraint that the number of particles be n. This minimization leads to the gap equations

$$1 = \tfrac{1}{2}G \sum_\beta 1/E_\beta,$$

$$n = \sum_\beta (1 - \eta_\beta/E_\beta),$$

to be solved for λ and Δ.

The expressions given earlier, eqns (5.32) and (5.33), for single-particle intrinsic matrix elements in a Nilsson basis can be approximately corrected for the transformation from an uncorrelated (single-particle) to a correlated (quasiparticle) basis by multiplying them by pairing factors. For the matrix elements of $\langle H_C \rangle$ the factor is $P_{K_1,K_2} \simeq (U_{\alpha_1} U_{\alpha_2} + V_{\alpha_1} V_{\alpha_2})$ where α_1 is the Nilsson orbital with quantum number K_1. For matrix elements of the electromagnetic transition operator the pairing factor is $P_{K_1,K_2} \simeq (U_{\alpha_1} U_{\alpha_2} \pm V_{\alpha_1} V_{\alpha_2})$, where the plus sign applies for magnetic transitions and the minus sign for electric transitions. In particular for matrix elements diagonal in K, such as the gyromagnetic factor g_K and decoupling parameter b of eqn (5.33) for magnetic moments and M1 transitions, the pairing factor $P_{K,K} \simeq U_\alpha^2 + V_\alpha^2 = 1$. Thus pairing corrections leave magnetic moment and in-band M1 transitions essentially unaffected. However, E2 transition rates between different one-quasiparticle states near the Fermi surface may be reduced two or three orders of magnitude below the Nilsson-model estimates.

5.3 Collective states at high spin

Deformed nuclei characterized by a non-spherical spatial distribution of nuclear density are known to exhibit rotational bands in their spectra. For example consider the case of a prolate nucleus with axial symmetry (taken as the three-axis in the body-fixed frame) rotating around an axis (say the one-axis) perpendicular to the nuclear symmetry axis. The relation between the excitation energy E and the spin I is often a smooth one and for spins not too high can be approximated by $E = AI(I+1)$. Such a series of states is called a rotation band. The lowest energy state of the band is known as a bandhead. It is the state for which $I = K$. Many states of different intrinsic structure can in principle become bandheads and have rotational bands built on them. The band built on the ground state of the nucleus is referred to as the ground-state band; all other

5.3 COLLECTIVE STATES AT HIGH SPIN

bands are called excited (or side) bands. The lowest energy state of a given angular momentum in a nucleus is known as the yrast state (from the Swedish: *yr* means dizzy, *yrast* dizziest). The sequence of all yrast states, represented in a plot of E versus I, is called the yrast line.

On examining different rotation bands of a given nucleus on an E versus I plot, it is evident that different bands can cross in the (E, I) plane. Near the crossing point between the two bands several physical quantities related to yrast line (e.g. energy, angular momentum, moment of inertia) represented as functions of the rotational frequency ω ($\mathcal{I}\omega = \hbar I$) often show a characteristic multivalued behaviour in the form of an S shape. This fact is referred to as a *backbending* effect. Typical examples of nuclei exhibiting such a collective rotation with crossing bands are provided in the regions of mass number $150 \leq A \leq 190$ (rare-earths) and $A \geq 220$ (actinides).

The first such backbend was identified by Johnson et al.[9] in 1972. Since then nuclear spectroscopy at high spin has become a growth industry, occasioned by the advent of heavy-ion accelerators to populate the states and the development of new techniques to detect the de-exciting γ rays. We will not endeavour to survey all this activity but refer the reader to recent reviews.[10] Instead we will concentrate on just one topic: what happens to the magnetic moment of the yrast states in the regime of the band crossing.

Consider a deformed even–even nucleus in the rare-earth region. Its ground-state rotation band has an $I = 0, 2, 4 \ldots$ sequence characteristic of axial symmetry with $K = 0$. The intrinsic structure is that of a rigid rotor, or more generally that of a rigid core with the extra-core nucleons in a zero quasiparticle solution of the pairing Hamiltonian. An excited intrinsic state then is a two-quasiparticle solution, namely one in which two like nucleons (either two protons or two neutrons) are unpaired. Such a two-quasiparticle state lies about 2 MeV above the ground state, i.e. $E \simeq 2\Delta$, where Δ is the pairing gap, $\Delta \simeq 1$ MeV. A rotation band can be built on this two-quasiparticle intrinsic state. If this excited-state rotation band is to intersect the ground-state band in an E versus I plot it must have a larger moment of inertia, i.e. be more deformed. This can be achieved if the unpaired nucleons are in high-spin orbitals and if these spins are aligned along the direction of rotation (the one-axis) perpendicular to the symmetry axis (three-axis). In the rare-earth nuclei the Fermi surface between occupied and unoccupied orbitals lies at the $i_{13/2}$ orbital for neutrons. Thus the interpretation of the backbending is a crossing between the ground-state rotational band (g-band) and a band in which a pair of $i_{13/2}$ quasineutrons are aligned along the one-axis (s band).

The mechanism for the rotational alignment is the Coriolis interaction, introduced in eqn (5.24) in the particle–rotor model. The matrix element of H_C given there will evidently be a maximum when the quasiparticles' angular momentum and the rotor's angular momentum vectors are parallel. Then we have,[11]

$$|\langle H_C \rangle|_{max} \simeq 2 \frac{\hbar^2}{2\mathscr{I}} Ij.$$

In rare-earth nuclei, a typical value of the moment of the inertia is $\hbar^2/2\mathscr{I} = 0.01$ MeV. Thus when the aligned quasiparticles' angular momentum is of order $j \simeq 10$, $|\langle H_C \rangle|_{max}$ will be around 2 MeV when the total spin I is of order 10. That is with $I \simeq 10$ the Coriolis interaction overcomes the pairing gap. In these situations, the Coriolis force is strong and a use of perturbation theory (as outlined in the last section for the particle–rotor model) may be inappropriate. Calculations in these circumstances are usually done in a cranking model.[10] However, for our present purpose, it is sufficient to argue semiclassically, drawing vector diagrams of the relevant angular momentum vectors and finding the projection of the magnetic moment vector along the direction of the total angular momentum, I, (i.e. the $M = I$ substate) to obtain an estimate of the static magnetic moment.

Let the angular momentum in the ground-state band be I_g. Evidently $I_g = R$ where R is the rotational angular momentum directed along the one-axis. In the s band, the angular momentum I_s is composed of a rotational angular momentum, R, and the aligned angular momentum of the two-quasiparticles, i_2, also directed along the one-axis. (In general, the two quasiparticles would have angular momentum j_2 lying in the 1–2 plane. The projection on the three-axis is zero, so the s band has $K = 0$. However, the projection of j_2 on the one-axis, i_2, is not necessarily quantized and i_2 does not have to be an integer. The assumption here is that j_2 is very close to lying along the one-axis and that i_2 is very close to j_2.) Thus $I_s = R + i_2$. Since the aligned angular momentum, i_2, of the two quasiparticles has the same direction as the collective angular momentum, R, the magnetic moment has the direction of the rotational axis and an absolute value

$$\mu = g_R R + g_2 i_2$$

in units of the nuclear magneton. The g-factor, $g = \mu/I$, for the yrast states is then[12]

$$g = g_R + (g_2 - g_R) i_2/I \tag{5.34}$$

where i_2 is zero below the backend and equal to its aligned value above. Here g_R is the gyromagnetic ratio in the g-band (obtained perhaps from the magnetic moment of the first 2^+ state). Alternatively the collective value $g_R \simeq Z/A$ can be taken as a first estimate. For g_2, shell-model values are used, assuming that the two aligned quasiparticles belong to a

5.3 COLLECTIVE STATES AT HIGH SPIN

pure j-shell with $j = l + 1/2$. Then g_2 is given by

$$g_2 = g_L + (g_S - g_L)/(2l + 1),$$

with g_L and g_S the free-nucleon gyromagnetic ratios but with g_S attenuated by a factor 0.7 to accommodate core-polarization corrections. Note that g_2 is quite different for protons and neutrons having a value for a high-spin orbital of $g_2 \simeq 1.3$ for a proton and $g_2 \simeq -0.2$ for a neutron. Thus a measurement of the g-factors of yrast states around the backbending region provides a sensitive test of the rotation alignment picture as this quantity is a clear signature of the quasiparticles involved. For example, if the backbending is caused by a complete alignment of two $i_{13/2}$ neutrons, a drastic reduction of the g-factors in the backbending region is expected, $g \simeq 0$ compared to the collective value $g_R \simeq 0.35$. In contrast, the g-factor is expected to be $g \simeq 1.0$ in the case where the backbending is caused by the alignment of two $h_{11/2}$ protons.

To evaluate eqn (5.34) it remains to specify the aligned quasiparticle angular momentum, i_2, which must be deduced from the experimental spectrum. In the case where there is a sharp crossing between the g and s bands, Frauendorf[12] recommends the following procedure:

1. Determine the rotational frequency, ω, as a function of the angular momentum I for the yrast states below and above the backbending region. This can be deduced from the energy of E2 gamma rays in the $I \rightarrow I - 2$ transitions in the yrast states

$$\hbar\omega(I) = dE/dI \simeq \tfrac{1}{2}[E(I+1) - E(I-1)] = \tfrac{1}{2}E_\gamma(I+1 \rightarrow I-1). \quad (5.35)$$

2. Determine $I_g(\omega)$, the rotational angular momentum due to collective motion for the ground-state band at the rotational frequencies ω determined in eqn (5.35). Again

$$\hbar\omega = dE/dI \simeq \frac{1}{2}\frac{\hbar^2}{2\mathscr{I}}[(I+1)(I+2) - (I-1)I]$$

$$\simeq \frac{\hbar^2}{\mathscr{I}}(I + \tfrac{1}{2}),$$

in the collective model. Rewriting this for the g band, we get

$$I_g(\omega) = \frac{\mathscr{I}}{\hbar^2}(\hbar\omega) - \tfrac{1}{2}.$$

A more accurate parametrization is given by the Harris polynomial

$$I_g(\omega) = \frac{\mathscr{I}^{(0)}}{\hbar^2}(\hbar\omega) + \frac{\mathscr{I}^{(1)}}{\hbar^2}(\hbar\omega)^3 - \tfrac{1}{2}. \quad (5.36)$$

For yrast states below the backbend, fit the Harris polynomial to the experimental data at the frequencies determined in eqn (5.35) to obtain the moments of inertia $\mathcal{J}^{(0)}$ and $\mathcal{J}^{(1)}$.

3. Determine $i_2(\omega)$, the aligned angular momentum of the two quasiparticles, by extrapolating the values of $I_g(\omega)$ found in eqn (5.36) to states above the backbend and subtract from the observed $I(\omega)$, namely

$$i_2(\omega) = I(\omega) - I_g(\omega). \qquad (5.37)$$

4. Determine $g(I)$, the g-factor for the yrast state of angular momentum, I, using eqn (5.34) with i_2 as given by eqn (5.37). For the case where the band crossings are not sharp Chen and Frauendorf[13] recommend a modified procedure, which we will not elaborate here.

In Fig. 5.1 we show examples of this procedure for two nuclei ^{156}Dy and ^{168}W. Below the backbend region the angular momentum of

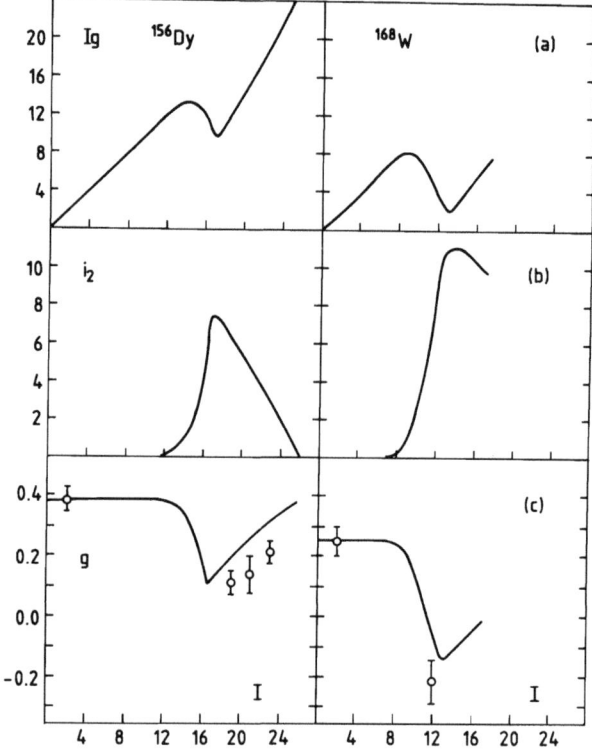

Fig. 5.1 Rotational alignment (backbending) in ^{156}Dy and ^{168}W. In panel (a) the component of the angular momentum of the g-band, I_g, as a function of the yrast angular momentum, I; (b) the aligned angular momentum of the two quasiparticles, i_2, as a function of I; and (c) the g-factor of the yrast state as a function of I.

5.3 COLLECTIVE STATES AT HIGH SPIN

collective rotation, I_g, equals the observed angular momentum of the yrast states, I. In the backbend region there is mixing between the g-band and s-band states of the same angular momentum and the curve I_g versus I shows a pronounced kink as seen in panel (a). Above the backbend, I_g continues to increase uniformly. The aligned momentum of the two quasiparticles, i_2, in the yrast state sharply rises to a maximum in the backbend region and falls off for higher values of I. This behaviour is immediately reflected in the calculated g-factor, which for an alignment of two $i_{13/2}$ neutrons illustrated in panel (c), indicates a sharp decrease in the region of the backbend. Experimental data on ^{156}Dy,[14] ^{168}W,[15] and ^{158}Dy[16] (not illustrated here) confirm this trend. Had the backbending been due to the alignment of protons rather than neutrons then there would have been an increase rather than a decrease in g in the yrast states. The magnetic moment data give a strong confirmation to the picture of the rotational alignment of neutrons in the backbending region.

Backbending also occurs in odd-mass nuclei. Essentially the same picture of rotational alignment applies here too. Let j_1 be the angular momentum of the odd particle. Its projection on the three-axis is K. Thus below the backbend we have the standard picture of particle–rotor coupling. Let i_1 be the projection of j_1 on the one-axis. Again this is not quantized, but under the assumption of rotational alignment we assume the vector j_1 lies in the 1–3 plane, and hence $i_1^2 = j_1^2 - K^2$. As before the rotational angular momentum R is directed along the one-axis and the total angular momentum is $I = R + j_1$. If we write the g-factor of the odd nucleon as g_1 then the magnetic moment operator is $\mu = g_R R + g_1 j_1 = g_R I + (g_1 - g_R) j_1$ and its expectation value in the $M = I$ substate is given (semiclassically) by the projection of μ on I. Thus we get

$$\mu = g_R I + (g_1 - g_R)[K^2 + i_1(I^2 - K^2)^{1/2}]/I,$$

where the projection of j_1 on I can be deduced from the geometry of the vectors, and the g-factor is

$$g = g_R + (g_1 - g_R)(1/I^2)[K^2 + i_1(I^2 - K^2)^{1/2}]. \quad (5.38)$$

Note the similarity of this equation (with $I^2 \to I(I+1)$ as a quantal correction) to eqn (5.19b) for the magnetic moment of an odd-A nucleus in the particle–rotor model. The leading terms agree in the limit that i_1 is small. Equation (5.38) applies for states in the ground-state rotation band below the backbending region. The s band has a similar particle–core structure except that in addition there are two quasiparticles with angular momentum i_2 aligned along the rotation axis. The magnetic momentum operator is $\mu = g_R I + (g_1 - g_R) j_1 + (g_2 - g_R) i_2$ and its expectation value given by its projection on I. Thus[17] we have

$$g = g_R + (g_1 - g_R)(1/I^2)[K^2 + i_1(I^2 - K^2)^{1/2}]$$
$$+ (g_2 - g_R)(i_2/I^2)(I^2 - K^2)^{1/2}. \tag{5.39}$$

Again i_1 and i_2 need to be deduced from the systematics of the energy spectrum in the yrast states.

In odd-mass nuclei we can discuss the $I \to I - 1$ M1 transitions in the band. The $B(M1)$ value is

$$B(M1; I \to I - 1) = \frac{3}{4\pi} |\langle II| T_{11}(M) |I - 1\, I - 1\rangle|^2.$$

In the semiclassical approach the matrix element of $T_{11}(M)$ in the $M = I$ substate is interpreted as the projection of the magnetic moment operator perpendicular to the spin, I. (There is an extra factor of $\sqrt{2}$ related to the construction of spherical tensors, compared to the geometric approach, $\langle T_{11}\rangle = \mu_\perp/\sqrt{2}$, see[17].) Thus

$$B(M1; I \to I - 1) = \frac{3}{8\pi} \frac{K^2}{I^2} ((g_1 - g_R)\{(I^2 - K^2)^{1/2} - i_1\} + (g_2 - g_R)i_2)^2. \tag{5.40}$$

Similar expressions can be given for the E2/M1 mixing ratio, δ, defined as

$$\delta = \frac{\langle II| T_{21}(E) |I - 1\, I - 1\rangle}{\langle II| T_{11}(M) |I - 1\, I - 1\rangle}$$
$$= \sqrt{3}\, Q_0 \left(1 - \frac{K^2}{I^2}\right)^{1/2} ((g_1 - g_R)\{(I^2 - K^2)^{1/2} - i_1\} + (g_2 - g_R)i_2)^{-1}, \tag{5.41}$$

and the branching ratio

$$\frac{B(M1; I \to I - 1)}{B(E2; I \to I - 2)} = \frac{3/4\pi}{5/16\pi} \frac{|\langle II| T_{11}(M) |I - 1\, I - 1\rangle|^2}{|\langle II| T_{22}(E) |I - 2\, I - 2\rangle|^2}$$
$$= \tfrac{16}{5} Q_0^{-2} \left(1 - \frac{K^2}{I^2}\right)^{-2} \frac{K^2}{I^2} ((g_1 - g_R)\{(I^2 - K^2)^{1/2} - i_1\} + (g_2 - g_R)i_2)^2, \tag{5.42}$$

to leading order in $1/I$. These expressions, which are sensitive to changes in i_2 in the backbending region, provide further tests of the rotational alignment picture.

5.4 Superdeformation at high spin

For spherical nuclei the most stable configurations are associated with shell closures. With the harmonic oscillator Hamiltonian, the single-particle orbitals are degenerate in energy for a given major shell $N = 2n + l$, where n is the principal quantum number (the number of radial nodes) and l the orbital angular momentum quantum number. The energy spacing between major shells is $\hbar\omega$, where ω is the oscillator frequency—the same in each of the 1, 2, and 3 directions. Major-shell closures occur for nucleon numbers 2, 8, 20, 40, 70, 112, 168 .... The addition of a strong spin–orbit term lowers the energy of a high-l orbital with $j = l + \frac{1}{2}$ and the shell closures are shifted to ... 40, 82, 126, 184 .... These are known as the magic numbers.

For deformed nuclei with axial symmetry described by an anisotropic harmonic oscillator potential, major degeneracies, in the single-particle spectra occur when the oscillator frequencies are in the ratio of (small) integers. For example, if the frequencies perpendicular and parallel to the symmetry axis are in the ratio $\omega_\perp : \omega_3 = 2:1$, corresponding to a deformation $\delta = 0.6$, see eqn (5.29b), the closed-shell numbers in a pure harmonic potential are 2, 4, 10, 16, 28, 40, 60, 80, 110, 140, ... Again a strong spin–orbit force will shift the high-j orbitals one shell downwards. Bohr and Mottelson[1] show that for a 2:1 axis ratio the deformed magic numbers are ... 66, 86, 116, 148 .... The nucleus $Z = 66$, $N = 86$ (^{152}Dy) is anticipated to have a stable highly deformed configuration. Recently, high-spin γ-ray spectroscopy studies at the Daresbury Laboratory[18] discovered the 'superdeformed' band in ^{152}Dy. A hint of possible shell structure at an axis ratio of 2:1 has earlier been revealed by the discovery of fission isomers,[19] although this evidence is less direct.

At least three different structures have been identified in the spectroscopy of ^{152}Dy. The yrast states at low spin are believed to be slightly oblate, having two equal major axes. At about 0.5 to 1.5 MeV above the yrast states lies a low-deformation ($\delta \approx 0.15$) prolate rotational band extending up to $I = 40$. At higher energies and spins lies the superdeformed band. In Fig. 5.2 the energies of the oblate, prolate, and superdeformed states are plotted as a function of spin with the assumption that the superformed band becomes yrast between $50\hbar$ and $60\hbar$. Some 19 discrete γ-ray lines were observed in the Daresbury experiment corresponding to E2 stretched decays de-exciting the nucleus from spin $60\hbar$ to spin $22\hbar$. At lower spins the transition probability out of the superdeformed band rapidly increases and the states de-excite to the yrast line.

From the discrete γ-ray energies, the moment of inertia for the superdeformed band can be inferred. There are two expressions for this. The first,

$$\mathcal{J}^{(1)} = \tfrac{1}{2}\hbar^2\left(\frac{dE(I)}{d(I^2)}\right)^{-1} \simeq \frac{2I-1}{E_\gamma(I \to I-2)}\hbar^2, \qquad (5.34a)$$

is referred to as the 'kinematical' moment of inertia and depends on the

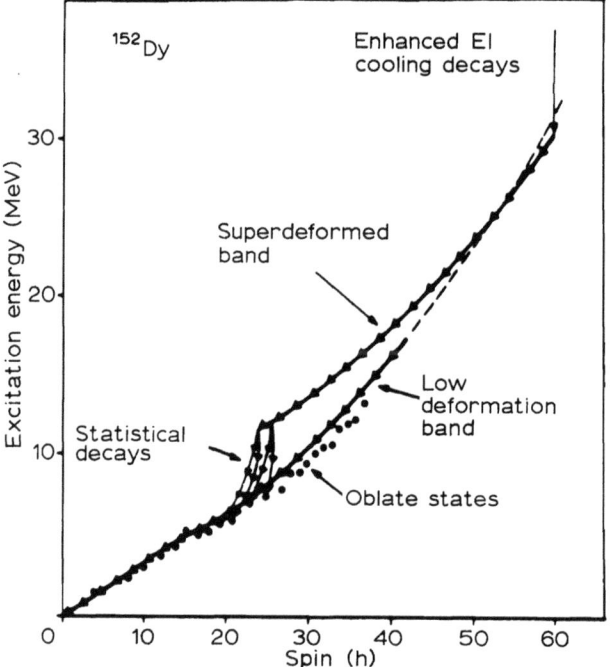

Fig. 5.2 A schematic picture of the three structures, oblate states, low-deformation band, and superdeformed band, in the spectroscopy of ^{152}Dy. It is assumed the superdeformed band became yrast at spin $55\hbar$. (From Twin.[22])

first difference of adjacent energy levels in a rotation band. The second,

$$\mathscr{I}^{(2)} = \hbar \left(\frac{d^2 E(I)}{dI^2}\right)^{-1} \simeq \frac{4}{[E_\gamma(I+2 \to I) - E_\gamma(I \to I-2)]} \hbar^2, \quad (5.43b)$$

is referred to as the 'dynamical' moment of inertia and depends on the second difference or the curvature of the energy $E(I)$ versus spin I line. Both definitions coincide for rigid-body rotation. For the observed γ-rays in the superdeformed bands both definitions agree at the highest spins, $I \simeq 60\hbar$, to give $\mathscr{I}_{\text{expt}} \simeq 83\hbar^2 \text{MeV}^{-1}$, but $\mathscr{I}^{(1)}$ falls slightly (by 10%) and $\mathscr{I}^{(2)}$ increases slightly (by 10%) at lower spins $I \simeq 24\hbar$. Theoretically the moment of inertia of a rigid prolate deformed body rotating about an axis perpendicular to the symmetry axis is (de Voigt et al.[10])

$$\mathscr{I} = \mathscr{I}_0 (1 + \beta/3),$$

where $\beta = \Delta R/R_0$, $\Delta R = c - a$, the difference in length of the semi-major and semi-minor axes, and R_0 the mean radius, $R_0 = (c + 2a)/3$. Also $\mathscr{I}_0$ is the moment of inertia of a uniform rigid sphere of radius R_0, $\mathscr{I}_0 = 2AMR_0^2/5 \simeq 60\hbar^2 \text{MeV}^{-1}$, with M the mass of a nucleon, A the atomic mass number, and $R_0 \simeq 1.2 A^{1/3} \simeq 6.4$ fm for ^{152}Dy. For a 2:1 ratio of axes, the rigid-body moment of inertia is $\mathscr{I} \simeq 75\hbar^2$ MeV^{-1} quite close to the experimental value. More complete calculations by Ragnarsson and Åberg[20] obtain $\mathscr{I}^{(2)}$ in very good agreement with experiment.

Energy differences alone do not provide a foolproof method of determining nuclear shapes because a rotational coupling between the core and the extra-core nucleons can cause a large apparent increase in the moment of inertia. A more definitive test is a measurement of the B(E2) transition rates for the stretched transitions in the band, which from eqn (5.14a) leads to a determination of the intrinsic quadrupole moment Q_0. For an ellipsoid of uniform charge Ze, the intrinsic quadrupole moment is, eqn (5.30),

$$Q_0 = \tfrac{4}{5} Z e R_0^2 \beta (1 + \tfrac{1}{6}\beta),$$

giving $Q_0 \simeq 18\,e$ b for a 2:1 axis ratio for which $\beta = \Delta R/R_0 = 3/4$. Contrast this with the low-deformation band in ^{152}Dy with $\delta = 0.15$ corresponding to $\beta = 0.17$ and giving $Q_0 = 4\,e$ b typical of the smaller deformations common in rare-earth nuclei. A measurement of the lifetimes within the superdeformed band in ^{152}Dy[21] by the Doppler-shift attenuation method yielded $Q_0 = 19 \pm 3\,e$ b. This direct measurement of enhanced collectivity is conclusive evidence that the collectivity is associated with the superdeformed shape.

A second example of a superdeformed band has been found by the Chalk River group[23] in ^{149}Gd and again it has a large moment of inertia, $\mathscr{I}^{(2)} \simeq 77\hbar^2$ MeV^{-1} and large intrinsic quadrupole moment $Q_0 \simeq 17 \pm 2\,e$ b. Despite extensive searches it was two years before this second example was discovered fuelling for a while the idea that ^{152}Dy was a unique case. Now several examples have been found. In ^{152}Dy the dynamic moment of inertia $\mathscr{I}^{(2)}$ was nearly constant, with $\mathscr{I}^{(1)} < \mathscr{I}^{(2)}$ over the whole spin range. In ^{149}Gd the trend is different: it is $\mathscr{I}^{(1)}$ that is nearly constant, with $\mathscr{I}^{(2)}$ decreasing with decreasing spin and with $\mathscr{I}^{(1)} > \mathscr{I}^{(2)}$. These trends can generally be understood[23] although not in detail.

5.5 Vibrational spectra

So far we have only considered the spectra of nuclei taken as rigid rotors with axially symmetric non-spherical density distributions. For a great variety of nuclei, however, it is possible to identify in the excitation spectra another form of collective motion associated with elementary modes of vibration. These vibrations represent fluctuations in the density about some equilibrium shape. It is convenient to expand this density fluctuation in spherical harmonics

$$\delta\rho(r) = f_\lambda(r) \sum_\mu Y^*_{\lambda\mu}(\theta, \phi) \alpha_{\lambda\mu}, \tag{5.44}$$

where the coefficient $\alpha_{\lambda\mu}$ is the amplitude describing the displacement from the equilibrium shape. The multipole quantum number λ represents the angular momentum of the vibrational mode. Each mode is $(2\lambda + 1)$-fold degenerate corresponding to the different values of the component μ. For small values of $\alpha_{\lambda\mu}$, the vibrational energy may be expressed in powers of $\alpha_{\lambda\mu}$ and its time derivative $\dot{\alpha}_{\lambda\mu}$. To leading order it is written as

$$H = T + V = \tfrac{1}{2} D_\lambda \sum_\mu \dot{\alpha}_{\lambda\mu}^2 + \tfrac{1}{2} C_\lambda \sum_\mu \alpha_{\lambda\mu}^2. \tag{5.45}$$

The first term is the kinetic energy and the coefficient D_λ is referred to as the mass parameter. The second term is the potential energy with C_λ the restoring force parameter. Introducing a conjugate variable

$$\pi_{\lambda\mu} = \frac{\partial}{\partial \dot{\alpha}_{\lambda\mu}} (T - V) = D_\lambda \dot{\alpha}_{\lambda\mu} \tag{5.46}$$

and requiring amplitudes $\alpha_{\lambda\mu}$ and $\pi_{\lambda\mu}$ satisfy the canonical commutation relation $[\pi_{\lambda\mu}, \alpha_{\lambda\mu'}] = -i\hbar \delta_{\mu,\mu'}$. leads to the familiar relation for the energy spectrum of a harmonic oscillator

$$H = \hbar \omega_\lambda \sum_\mu (n_{\lambda\mu} + \tfrac{1}{2}) \tag{5.47}$$

with frequency $\omega_\lambda = (C_\lambda/D_\lambda)^{1/2}$. The excitation energy is

$$E(n_\lambda) - E(n_\lambda = 0) = n_\lambda \hbar \omega_\lambda \tag{5.48}$$

where n_λ is the number of quanta $n_\lambda = \sum n_{\lambda\mu}$ summed over the degeneracy, μ. States with $n_\lambda = 1$ are referred to as the one-phonon states, $n_\lambda = 2$ the two-phonon states, and so on. The spectrum is that of equally spaced phonon excitations. For given n_λ, the values of the vibrational angular momentum I can be obtained by coupling the angular momenta λ of the individual quanta and imposing the restrictions of Bose statistics. (For example with $n_\lambda = 2$ the state must be symmetric.) Thus for quadrupole fluctuations the one-phonon state has angular momentum $I = \lambda = 2$, while the two-phonon state is a triplet with angular momenta $I = 0, 2, 4$ occurring at an excitation energy twice that of the one-phonon state. The explicit construction of many-phonon states is discussed in Bohr and Mottelson.[1]

So far, this discussion is based on a spherical equilibrium shape. In many instances, however, the equilibrium shape is deformed and there can be a rotational band of states built on each phonon. Then the wavefunction for the system will be a product function, eqn (5.1), with the intrinsic state a harmonic oscillator function in the amplitude $\alpha_{\lambda\mu}$ and the collective state a function of the Euler angles describing the orientation of the deformed equilibrium shape. For an axially symmetric equilibrium shape, the normal modes can be characterized by the

5.5 VIBRATIONAL SPECTRA

quantum number v representing the component of vibrational angular momentum along the symmetry axis. Moreover, a shape with axial symmetry is invariant with respect to a rotation, $\mathcal{R}_2(\pi)$, about an axis perpendicular to the symmetry axis. This implies a degeneracy of modes with $\pm v$. If the intrinsic angular momentum in the absence of a vibration has the value $K_0 = 0$ as for the ground-state configuration of even–even nuclei, the excitation of a quantum $v = 0$ gives rise to a $K = 0$ rotation band. This is known as a β vibration. For $v \neq 0$, the conjugate modes together generate a single band with $K = |v|$ and $I = K, K+1, \ldots$, called a γ vibration. For example, with quadrupole vibrations, $\lambda = 2$, the spectrum of a deformed even–even nucleus will have a ground-state $K = 0$ rotation band, $I = 0, 2, 4, \ldots$, an excited one-phonon excitation with $v = 0$ giving another $K = 0$ band with $I = 0, 2, 4, \ldots$ (the β band), and an excited one-phonon excitation with $v = 2$ giving a $K = 2$ band with $I = 2, 3, 4, \ldots$ (the γ band). Many rare-earth nuclei display these features in their spectra. A schematic representation is given in Fig. 5.3.

A useful model that builds in both rotations and vibrations for quadrupole deformations is the five-dimensional quadrupole oscillator considered by Bohr[24] and analysed in detail by Kumar and Baranger.[25] Here the deformation is defined by five variables. Two of the variables β

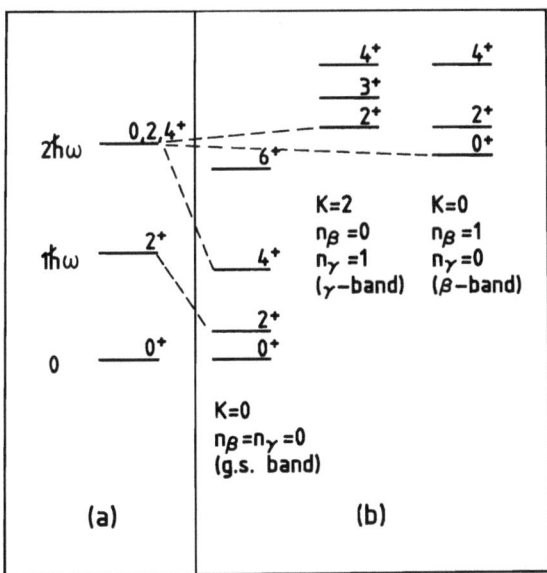

Fig. 5.3 Schematic level diagram relating the spectra of quadrupole vibrations in a spherical nucleus (a) with those in a deformed nucleus (b). The quantum numbers n_β and n_γ denote the number of phonons of excitation in β and γ vibrational modes respectively.

and γ define the quadrupole deformation in the body-fixed frame of references. The orientation of this body-fixed frame in space is defined by the Euler angles which are adopted as the other three variables. A second-order differential operator in these variables is written down (Bohr Hamiltonian) whose eigenfunctions should give a good description of collective states in nuclei. Once a potential energy is defined as a function of β and γ in the Bohr Hamiltonian it is possible to solve the differential equation. For certain potentials, approximate solutions were obtained that are particularly simple. These correspond to vibrational, rotational, or γ-unstable nuclei. In general, the solution has to be obtained by rather involved numerical integrations.

We will not pursue this topic here, preferring to postpone a discussion of electromagnetic moments and transition rates in vibrational nuclei in favour of introducing an alternative model, the interacting boson model, which has built into it rotational and vibrational Hamiltonians as special limits of a more general Hamiltonian.

5.6 Interacting boson model-1 (IBM-1)

The interacting boson model, state Iachello and Arima in their recent book,[26] orginated from early ideas of Feshbach and Iachello,[27] who in 1969 described some properties of light nuclei in terms of interacting bosons (each boson in principle being thought of as a pair of nucleons), and from the work of Janssen et al.[28] who suggested a description of collective quadrupole states in nuclei in terms of an SU(6) group. The latter description was subsequently cast into a different mathematical form by Arima and Iachello[29] with the introduction of an s-boson which made the SU(6), or rather U(6), structure more apparent. The success of this phenomenological approach to the structure of nuclei has led to major developments in the understanding of nuclear structure. Recent reviews can be found in Iachello and Talmi, Casten and Warner, and Arima and Iachello.[30]

The model, however, generated some controversy among nuclear physicists because it is equivalent to other models of collective nuclear structure, most notably the geometrical model of Bohr and Mottelson.[1] The difference is that in the five-dimensional quadrupole oscillator model Hamiltonian of Bohr[24] the kinetic energy is a differential operator in five variables. Differential equations have to be solved. In the interacting boson model the group structures of the boson algebras are exploited and all manipulations are algebraic.

Low-lying collective states in even–even medium-mass nuclei are dominated by pairing and quadrupole correlations. Hence their algebraic description relies on the introduction of six boson degrees of freedom, divided into a scalar, called s, with angular momentum and parity $I^\pi = 0^+$

5.6 INTERACTING BOSON MODEL-1 (IBM-1)

and a quadrupole, called d, with $I^\pi = 2^+$. Since the six boson operators b_α ($\alpha = 1 \ldots 6$) span a six-dimensional space, the structure of the algebraic model describing the low-lying collective states is that of the group of unitary transformations in six dimensions, U(6). The model consists of writing the Hamiltonian, H, and other operators (such as the electromagnetic transition operators) in terms of the boson operators b_α, or conversely, in terms of the 36 generators of the group U(6), $G_{\alpha\alpha'} = b_\alpha^\dagger b_{\alpha'}$. For example, the Hamiltonian H when limited to no more than two-body boson interactions can be written as

$$H = H_0 + \sum_{\alpha\alpha'} \varepsilon_{\alpha\alpha'} G_{\alpha\alpha'} + \tfrac{1}{2} \sum_{\alpha\alpha'\beta\beta'} u_{\alpha\alpha'\beta\beta'} G_{\alpha\alpha'} G_{\beta\beta'}. \quad (5.49)$$

All the model information is contained in the parameters $\varepsilon_{\alpha\alpha'}$ and $u_{\alpha\alpha'\beta\beta'}$ describing the single-boson energies and boson–boson interactions. Note that the number of bosons, N, is conserved by this Hamiltonian. With just s and d bosons, the Hamiltonian contains nine parameters. For a given nucleus, with N fixed, the number of parameters reduces to six when only excitation energies are considered. There are two other Hamiltonian forms commonly in use with six terms. The first is the multipole form

$$H = E_0 + \varepsilon n_d + a_0(P^\dagger \cdot P) + a_1(L \cdot L) + a_2(Q \cdot Q)$$
$$+ a_3(U \cdot U) + a_4(V \cdot V), \quad (5.50a)$$

and the second is in terms of Casimir operators associated with the group decomposition of U(6)

$$H = e_0 + \varepsilon C_1(U5) + \alpha C_2(U5) + \beta C_2(O5) + \gamma C_2(O3)$$
$$+ \delta C_2(SU3) + \eta C_2(O6). \quad (5.50b)$$

The operators are defined in Table 5.1 in terms of the basic boson operators: $s^\dagger$, $d_\mu^\dagger$ ($\mu = 0, \pm 1, \pm 2$). Note that $\tilde{s} = s$ and $\tilde{d} = (-)^\mu d_{-\mu}$.

Analytic solutions and closed expressions for the energy eigenvalues can be found in certain limiting cases of these Hamiltonians. The

Table 5.1 Boson operators introduced in the multipole and Casimir form of the model Hamiltonian. Square brackets denote vector coupling, namely

$$[d^\dagger \times \tilde{d}]_M^{(L)} = \sum_{\mu_1\mu_2} \langle 2\mu_1 2\mu_2 | LM \rangle d_{\mu_1}^\dagger \tilde{d}_{\mu_2}$$

$n_d = (d^\dagger \cdot \tilde{d})$	$C_1(U5) = n_d$
$P = \tfrac{1}{2}(\tilde{d} \cdot \tilde{d}) - \tfrac{1}{2}(\tilde{s} \cdot \tilde{s})$	$C_2(U5) = n_d(n_d + 4)$
$L = \sqrt{10}\,[d^\dagger \times \tilde{d}]^{(1)}$	$C_2(O5) = 4[\tfrac{1}{10} L \cdot L + U \cdot U]$
$Q = [d^\dagger \times \tilde{s} + s^\dagger \times \tilde{d}]^{(2)} - \tfrac{1}{2}\sqrt{7}\,[d^\dagger \times \tilde{d}]^{(2)}$	$C_2(O3) = 2(L \cdot L)$
$U = [d^\dagger \times \tilde{d}]^{(3)}$	$C_2(SU3) = \tfrac{2}{3}[2Q \cdot Q + \tfrac{3}{4} L \cdot L]$
$V = [d^\dagger \times \tilde{d}]^{(4)}$	$C_2(O6) = 2[N(N+4) - 4P^\dagger \cdot P]$

technique is to look for subalgebras of U(6). There are three and only three chains of subalgebras if one wants to include the rotation group, O(3), as a subalgebra. The three chains are:[26]

$$\begin{array}{l} \nearrow U(5) \supset O(5) \supset O(3) \supset O(2) \\ U(6) \rightarrow SU(3) \supset O(3) \supset O(2) \\ \searrow O(6) \supset O(5) \supset O(3) \supset O(2) \end{array} \quad (5.51)$$

Using the Casimir form of the Hamiltonian in the limit $\delta = \eta = 0$ leads to an algebraic solution for the energy eigenvalue for the first chain (to be known as the U(5) chain):

$$U(5): E = \varepsilon n_d + \alpha n_d(n_d + 4) + 2\beta v(v + 3) + 2\gamma L(L + 1). \quad (5.52a)$$

The limit $\varepsilon = \alpha = \beta = \eta = 0$ gives the SU(3) chain

$$SU(3): E = \tfrac{2}{3}\delta[\lambda^2 + \mu^2 + \lambda\mu + 3(\lambda + \mu)] + 2\gamma L(L + 1), \quad (5.52b)$$

and the limit $\varepsilon = \alpha = \delta = 0$ the O(6) chain

$$O(6): E = 2\beta\tau(\tau + 3) + 2\gamma L(L + 1) + 2\eta\sigma(\sigma + 4). \quad (5.52c)$$

The quantum numbers n_d, v, L, λ, μ, τ, and σ are taken[26] from the labels needed to classify the states in the three group chains. The values of these quantum numbers occuring for a given number of bosons, N, is specified by the algebras. For example in the U(5) chain, the values of n_d characterizing the representations of U(5) are

$$n_d = N, N-1, \ldots 1, 0$$

and the values of v for the representations of O(5) contained in one particular representation of U(5) is

$$v = n_d, n_d - 2, \ldots 1 \text{ or } 0.$$

The quantum number L is the angular momentum and in the simplest cases (where quantum numbers n_d, v, and L completely specify the basis states) has the values

$$L = v, v+1, \ldots 2v-2, 2v.$$

Note that $2v - 1$ is missing.

In the SU(3) chain the quantum numbers (λ, μ) are the quantum numbers introduced by Elliott.[31] The lowest energy states have $(\lambda, \mu) = (2N, 0), (2N-4, 2), (2N-8, 4), \ldots$.

Finally in the O(6) chain, the values of the quantum number σ contained in a given representation N of U(6) are

$$\sigma = N, N-2 \ldots 1 \text{ or } 0$$

and the values of the quantum number τ contained in a given representation σ of O(6) are

$$\tau = \sigma, \sigma - 1, \ldots 1, 0.$$

The step from O(5) to O(3) is not fully reducible and an additional quantum number may be needed. In the simplest cases, the angular momentum values are

$$L = \tau, \tau + 1, \ldots 2\tau - 2, 2\tau.$$

Again note that $2\tau - 1$ is missing.

The first few collective states for each of these three chains are shown in Fig. 5.4. The energy spectra of the U(5), SU(3), and O(6) limits correspond closely to the geometrical cases of a spherical anharmonic quadrupole vibrator, a deformed axial rotor with β and γ vibrations and a deformed γ-unstable rotor, respectively.

5.7 Transition rates in IBM-1

Operators representing static moments and electromagnetic transition rates are also written in this model in terms of the boson operators. The most general form for the M1 and E2 operators in lowest order is

$$T_{1\mu}(M) = \beta_1 [d^\dagger \times \tilde{d}]^{(1)}_\mu,$$
$$T_{2\mu}(E) = \alpha_2 [d^\dagger \times \tilde{s} + s^\dagger \times \tilde{d}]^{(2)}_\mu + \beta_2 [d^\dagger \times \tilde{d}]^{(2)}_\mu,$$
(5.53)

where the coefficients β_1, α_2, and β_2 are treated as parameters to be determined in a fit to data. Just as the decomposition of the U(6) group structure could be used to identify simple cases of the Hamiltonian, so

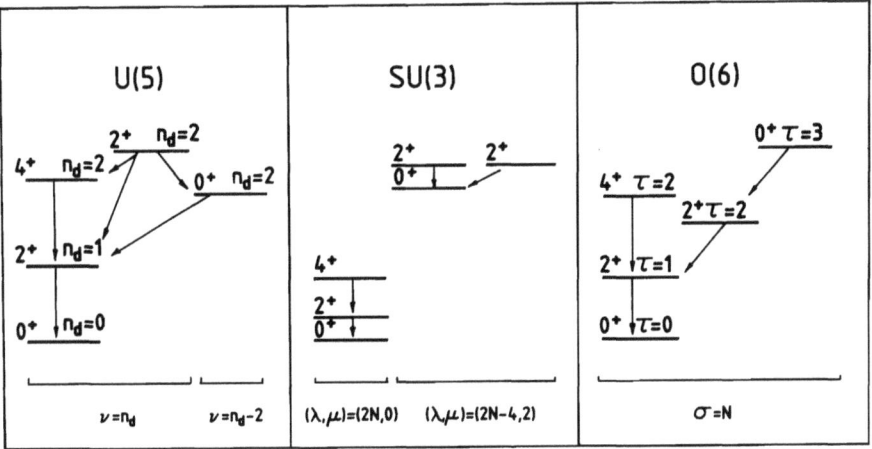

Fig. 5.4 The first few collective states for the U(5), SU(3), and O(6) limits of IBM-1. The allowed E2 transitions are illustrated.

too it can be used to identify simple cases of the operators. In these situations matrix elements of the operators can be calculated explicitly and certain selection rules become evident.

Consider first the M1 operator. It can clearly be written in terms of the angular momentum operator, $\boldsymbol{L}$:

$$T_{1\mu}(M) = (3/4\pi)^{1/2} g_B L_\mu \tag{5.54}$$

where we introduce a new constant g_B (to replace β_1) to be known as the effective boson g-factor. Since $\boldsymbol{L}$ is a generator operator of the group O(3) that occurs in all subgroup decompositions of U(6) and since $\boldsymbol{L}$ is therefore diagonal in any basis, no M1 transitions can occur in this approximation. The diagonal matrix elements depend only on L and are the same for all three (U(5), SU(3), and O(5)) cases. The magnetic moment is

$$\mu = (4\pi/3)^{1/2} \langle L, M_L = L | T_{10}(M) | L, M_L = L \rangle,$$
$$= g_B L. \tag{5.55}$$

Thus the g-factors, $g = \mu/L$, are all equal in this approximation and given by $g = g_B$. Only by introducing higher-order terms (involving, say, two creation and two annihilation operators) can non-zero M1 transition rates be obtained in this model. This avenue is not generally followed, rather a second version of the IBM is used, which we discuss in the next section.

For the E2 operator there are a number of limiting cases of interest:

1. If $\beta_2/\alpha_2 = -\sqrt{\frac{7}{2}}$, then $T_{2\mu}(E)$ is just the quadrupole operator, Q_μ, that is a generator of the SU(3) subgroup. In this case the operator will not be able to produce transitions between different irreducible representations of the subgroup. This leads to selection rules for E2 transitions in the SU(3) limit, namely only in-band $\Delta\lambda = 0$, $\Delta\mu = 0$ transitions can occur, where (λ, μ) are the SU(3) quantum numbers labelling a rotation band.

2. If $\beta_2 = 0$, then $T_{2\mu}(E)$ is a generator of the O(6) subgroup. This produces the selection rules $\Delta\tau = \pm 1$, $\Delta\sigma = 0$. Note in particular that the $\Delta\tau = \pm 1$ selection rule leads to the result that the quadrupole moment of any state vanishes in the O(6) limit.

3. In the U(5) limit, for all values of α_2 and β_2, the E2 selection rule $\Delta n_d = 0, \pm 1$ holds: the term α_2 controls $\Delta n_d = \pm 1$ transitions and the term β_2 controls $\Delta n_d = 0$ transitions.

Allowed E2 transitions for the U(5), SU(3), and O(6) limiting cases are shown in Fig. 5.4. In Table 5.2 we give the algebraic result for the quadrupole moment

$$Q = \left(\frac{16\pi}{5}\right)^{1/2} \langle L, M_L = L | T_{20}(E) | L, M_L = L \rangle, \tag{5.56}$$

5.8 INTERACTING BOSON MODEL 2 (IBM-2)

Table 5.2 Quadrupole moments and $B(E2)$ transition rates in IBM-1 for the three group chains introduced in eqn (5.51). $R_1 = B(E2; 4_1^+ \to 2_1^+)/B(E2; 2_1^+ \to 0_1^+)$, $R_2 = B(E2; 2_2^+ \to 2_1^+)/B(E2; 2_1^+ \to 0_1^+)$, $R_3 = B(E2; 0_2^+ \to 2_1^+)/B(E2; 2_1^+ \to 0_1^+)$

	U(5)	SU(3)	O(6)
$Q(2_1^+)$	$\beta_2 \left(\dfrac{32\pi}{35}\right)^{1/2}$	$-\alpha_2 \left(\dfrac{8\pi}{245}\right)^{1/2} (4N+3)$	0
$Q(4_1^+)/Q(2_1^+)$	2	14/11	0
$Q(2_2^+)/Q(2_1^+)$	$-3/7$	1	0
$B(E2; 2_1^+ \to 0^+)$	$\alpha_2^2 N$	$\alpha_2^2 N(2N+3)/5$	$\alpha_2^2 N(N+4)/5$
R_1	$2(N-1)/N$	$\dfrac{10(N-1)(2N+5)}{7N(2N+3)}$	$\dfrac{10(N-1)(N+5)}{7N(N+4)}$
R_2	$2(N-1)/N$	0	$\dfrac{10(N-1)(N+5)}{7N(N+4)}$
R_3	$2(N-1)/N$	0	0

and the $B(E2)$ transition rates for a number of low-lying states in each of the three group chains in terms of the number of bosons, N, active in the particular nucleus. Typically the number of bosons is the number of pairs of nucleons (or pairs of nucleon holes) outside major closed shells.

Note that in the limit $N \to \infty$, the ratio of transition rates in the U(5) limit all tend to $R_1 = R_2 = R_3 = 2$, which is exactly the vibrational model limit in which the $B(E2)$ rate is proportional to the phonon number. The 0^+, 2^+, 4^+ excited state triplet are two-phonon states, the lower 2^+ state a one-phonon state in a quadrupole vibrational model. The nucleus ^{110}Cd is often cited as a good example of a vibrational nucleus. The experimental ratio of rates is $R_1 = 1.53 \pm 0.19$ and $R_2 = 1.08 \pm 0.29$,[26] less than the canonical value of 2. The interacting boson model with the number of bosons $N = 7$ predicts $R_1 = R_2 = R_3 = 1.7$ for ^{110}Cd.

In the SU(3) chain the limit $N \to \infty$ leads to the axially symmetric rigid rotor result discussed in eqn (5.14). The characteristic signal of nuclei of this type are strong in-band transitions and weak cross-band transitions. In the O(6) limit, by contrast, strong cross-band transitions, $R_2 = R_1 = 10/7$, are allowed. Casten and von Brentano[32] find Xe and Ba nuclei near mass $A \sim 130$ exhibit features characteristic of O(6) symmetry.

5.8 Interacting boson model 2 (IBM-2)

An extension of the interacting boson model is to consider a system of interacting bosons of two types: proton bosons and neutron bosons. The proton and neutron bosons with spin $I^\pi = 0^+$ are denoted s_π and s_ν

respectively, and with spin 2^+ by d_π and d_ν. The number of proton bosons, N_π, and neutron bosons, N_ν, is counted from the nearest closed shell; that is, if more than half of the shell is full N_π or N_ν is taken as the number of hole pairs. The total number of bosons, N, is fixed to be $N_\pi + N_\nu$. This extension was introduced by Otsuka et al.[33] in an attempt to give a physical interpretation of the bosons as correlated pairs of fermions coupled to spin 0^+ or 2^+. A separate treatment of proton and neutron degrees of freedom comes naturally in this interpretation. Thus there are 12 creation and 12 annihilation boson operators in the model denoted generally as $b_\alpha^\dagger$ and b_α. The label α distinguishes proton π and neutron ν, the orbital angular momentum $l = 0$ or 2, and the magnetic projection m, $-l \leq m \leq +l$. Instead of writing π or ν explicitly, proton and neutron bosons can be considered as having an intrinsic quantity, called F-spin, of value $F = \frac{1}{2}$.[33] Proton bosons have z-projection, $m_f = +\frac{1}{2}$, while neutron bosons have $m_f = -\frac{1}{2}$. The general boson operator, b_α, then has a label $\alpha = 1/2, m_f, l, m$, where the first two indices represent F-spin quantum numbers, and the last two ordinary spin quantum numbers. To form operators that transform appropriately under rotations both in ordinary space and in F-spin space, coupled operators, $b_\alpha^\dagger \times \tilde{b}_{\alpha'}$, are required where

$$\tilde{b}_{1/2, m_f, l, m} = (-)^{1/2 + m_f + l + m} b_{1/2, -m_f, l, -m}.$$

The Hamiltonian operator can be written as a sum of three pieces

$$H = H_\pi + H_\nu + V_{\pi\nu}, \tag{5.57}$$

where H_π and H_ν are both of the general structure used in IBM-1 eqn (5.52), and $V_{\pi\nu}$ represents a proton–neutron boson interaction. If the number of proton and neutron bosons are separately conserved and H limited to no more than two-body boson interactions then the number of parameters (single-boson energies and boson–boson interactions) in the Hamiltonian is 30. For fixed N_π and N_ν not all terms are independent. Furthermore, considering just excitation energies reduces the number of parameters to 21. This is still too many terms for a direct phenomenological study. However, the microscopic structure of the model suggests that only a few terms are important. This is because the residual nucleon–nucleon interaction in the spherical shell model is dominated by a pairing term between identical nucleons. This produces in the boson Hamiltonian a term of the type $\varepsilon_\pi n_{d\pi} + \varepsilon_\nu n_{d\nu}$. In addition there is a quadrupole–quadrupole interaction between non-identical nucleons producing a term of type $\kappa Q_\pi^\chi \cdot Q_\nu^\chi$, where

$$Q_\rho^\chi = [d_\rho^\dagger \times \tilde{s}_\rho + s_\rho^\dagger \times \tilde{d}_\rho]^{(2)} + \chi_\rho [d_\rho^\dagger \times \tilde{d}_\rho]^{(2)}, \qquad \rho = \pi \text{ or } \nu. \tag{5.58}$$

Finally there is a symmetry energy favouring states in which protons and neutrons move in phase. This symmetry energy is introduced through the

5.8 INTERACTING BOSON MODEL 2 (IBM-2)

Majorana operator

$$M_{\pi\nu} = [s_\nu^\dagger \times d_\pi^\dagger - s_\pi^\dagger \times d_\nu^\dagger]^{(2)} \cdot [\tilde{s}_\nu \times \tilde{d}_\pi - \tilde{s}_\pi \times \tilde{d}_\nu]^{(2)}$$
$$- 2 \sum_{k=1,3} [d_\nu^\dagger \times d_\pi^\dagger]^{(k)} \cdot [\tilde{d}_\nu \times \tilde{d}_\pi]^{(k)}. \quad (5.59)$$

Thus the adopted Hamiltonian is

$$H = E_0 + \varepsilon_\pi n_{d\pi} + \varepsilon_\nu n_{d\nu} + 2\kappa Q_\pi^\chi \cdot Q_\nu^\chi + \lambda' M_{\pi\nu}, \quad (5.60)$$

where quite often one puts $\varepsilon_\pi = \varepsilon_\nu = \varepsilon$. This Hamiltonian is specified by ε, κ, χ_π, χ_ν in addition to E_0 and λ'. The fact that only four parameters determine the excitation energies allows for straightforward phenomenological studies. This Hamiltonian was suggested by Talmi[33] and has become known as the Talmi Hamiltonian. In some applications quadrupole-quadrupole interactions between identical nucleons are added as well, giving two additional terms, $\kappa_\pi Q_\pi^\chi \cdot Q_\pi^\chi + \kappa_\nu Q_\nu^\chi \cdot Q_\nu^\chi$ and two more parameters κ_π and κ_ν.

The algebraic structure of model IBM-2 is much richer than that of the simpler IBM-1. This is because there are now two sets of 36 operators: $G_{\alpha\beta}^{(\pi)} = b_{\pi,\alpha}^\dagger b_{\pi,\beta}$ and $G_{\alpha\beta}^{(\nu)} = b_{\nu,\alpha}^\dagger b_{\nu,\beta}$. Taken together, the 72 operators are the generators of the group formed by the direct product, $U_\pi(6) \otimes U_\nu(6)$. The question, as before, is to find a chain of subgroups that include the rotation group, O(3), which is required since nuclear states are characterized by a good value of the angular momentum. There are many possibilities as discussed by Iachello and Arima.[26] The chains that have the closest similarity to IBM-1 are

$$\begin{array}{c} U_\pi(6) \searrow \\ \\ U_\nu(6) \nearrow \end{array} U_\pi(6) \otimes U_\nu(6) \rightarrow \begin{array}{c} \nearrow U_{\pi+\nu}(5) \supset O_{\pi+\nu}(5) \searrow \\ SU_{\pi+\nu}(3) \\ \searrow O_{\pi+\nu}(6) \supset O_{\pi+\nu}(5) \nearrow \end{array} \rightarrow O_{\pi+\nu}(3) \supset O_{\pi+\nu}(2).$$

(5.61)

Each of the chains starts by joining the algebras of $U_\pi(6)$ and $U_\nu(6)$ at the first step. This is relatively simple since both representations, $U_\pi(6)$ and $U_\nu(6)$, are totally symmetric describing N_π proton boson states and N_ν neutron boson states. On joining, it follows that one can obtain not only states that are symmetric in the proton and neutron degrees of freedom but also states with partial symmetry. A convenient way to display the symmetry character of the resulting wavefunctions is through the use of Young tableaux.[34] In this case, the resulting Young tableaux obtained by multiplying the appropriate Young tableaux for totally symmetric proton and neutron boson states are at most two-rowed. For example the wavefunction corresponding to $N_\pi = 1$ and $N_\nu = 1$ can be obtained from the product

$$[1] \otimes [1] = [2] + [11].$$

The wavefunctions [2] are totally symmetric while those with label [11] are antisymmetric. In general, the direct product is

$$[N_\pi] \otimes [N_\nu] = \sum_{k=0}^{\min(N_\pi, N_\nu)} [N_\pi + N_\nu - k, k], \quad (5.62)$$

and the wavefunction carries a symmetry label, $[N_1, N_2]$, represented by a Young tableau with N_1 boxes in the first row and N_2 boxes in the second row. (With reference to eqn (5.62), $N_1 = N_\pi + N_\nu - k$ and $N_2 = k$). Instead of the two numbers, N_1 and N_2 one can use the quantities

$$\begin{aligned} N &= N_1 + N_2 = N_\pi + N_\nu, \\ F &= \tfrac{1}{2}(N_1 - N_2), \end{aligned} \quad (5.63)$$

to characterize the state, i.e. the total boson number $N = N_\pi + N_\nu$ and the value of the F-spin. A fully symmetric product state is one in which the Young tableau has no boxes in the second row, $N_2 = 0$. Then F takes its maximum value $F_{\max} = N/2$. However, one can also generate states where the F-spin is less than its maximum value. The occurrence of such states, which are not totally symmetric, is the new aspect brought in by the coupling of protons and neutrons.

It is clear that F-spin is an exact quantum number if the IBM-2 Hamiltonian is fully symmetric between protons and neutrons; that is, if the Hamiltonian is a scalar in F-spin or contains at most explicit dependence on F^2 and F_z, which are diagonal in F-spin. In actual cases, the Hamiltonian will not have this property. For example, the Talmi Hamiltonian, eqn (5.60), with $\varepsilon_\pi = \varepsilon_\nu = \varepsilon$ has only quadrupole–quadrupole interactions between proton and neutron bosons. Since no $\kappa_\pi Q_\pi \cdot Q_\pi + \kappa_\nu Q_\nu \cdot Q_\nu$ interaction is present, the Talmi Hamiltonian is not fully symmetric. To restore the symmetry requires adding these terms and setting the coefficients all equal, $\kappa = \kappa_\pi = \kappa_\nu$. Furthermore the χ parameter in the definition of Q, eqn (5.58), must be the same for protons and neutrons, $\chi_\pi = \chi_\nu$. The Majorana operator in the Talmi Hamiltonian is diagonal in F-spin and has the eigenvalue

$$\langle M_{\pi\nu} \rangle = (\tfrac{1}{2}N - F)(\tfrac{1}{2}N + F + 1). \quad (5.64)$$

Note the eigenvalue is zero for totally symmetric states $F = F_{\max} = N/2$. If the coefficient of $M_{\pi\nu}$ in the Talmi Hamiltonian, λ', is set positive then the totally symmetric states with $F = F_{\max}$ will lie lowest in the spectrum and partially symmetric states $F < F_{\max}$ are pushed upwards. The prediction and resulting discovery[35] of partially symmetric states is one of the main achievements of IBM-2.

5.9 Electric quadrupole transitions in IBM-2

The electric quadrupole operator is written in IBM-2 as

$$T_2(E) = e_\pi Q_\pi^\chi + e_\nu Q_\nu^\chi, \tag{5.65}$$

where e_π and e_ν are the boson effective charges to be determined in fits to data. Algebraic expressions for the matrix elements of this operator can be obtained in certain limiting cases represented by the group chains given in eqn (5.61). We consider a Hamiltonian that is fully symmetric between protons and neutrons, such that F-spin is a good quantum number for the states. In particular the ground-state band has $F = F_{\max} = (N_\pi + N_\nu)/2$. We write the 0^+ and 2^+ states of this band as 0_s^+ and 2_s^+, s standing for symmetric. We also consider a 2^+ state of $F = F_{\max} - 1$, denoting this as 2_a^+. In a typical nucleus this will be the fourth or fifth 2^+ state lying at an excitation energy of order 3 MeV. In Table 5.3 we give algebraic expressions[26] for the $B(E2; 2_s^+ \to 0_s^+)$ and the $B(E2; 2_a^+ \to 0_s^+)$ transition rates for the three group chains. Note that in the E2 operator the same Q^χ operators appear as in the Hamiltonian with $\chi_\pi = \chi_\nu$. This requirement is known as the consistent-Q formalism. Table 5.3 indicates that the in-band transitions are proportional to $(e_\pi N_\pi + e_\nu N_\nu)^2$ and the cross-band transitions between states of $F = F_{\max}$ and $F = F_{\max} - 1$ are proportional to $(e_\pi - e_\nu)^2$. In the limit that $e_\pi = e_\nu$ these cross-band transition rates go to zero while the in-band rates reduce to those given in IBM-1, see Table 5.2, with the parameter e_π identified as α_2. It is obvious that in the limit $e_\pi = e_\nu$, the operator $T_2(E)$ is a generator of the direct product group $U_\pi(6) \otimes U_\nu(6)$ and cannot induce transitions between states of different representations of the group, (i.e. states of differing F-spin).

Applications of IBM-2 in phenomenological studies indicate that for most observables, such as energy levels and E2 matrix elements, the results obtained for low-lying states are very similar to the ones for the simpler IBM-1. The reasons for this are, first, that the low-lying states in the IBM-2 approach are predominantly symmetric in the neutron–proton degree of freedom and, second, that in the E2 operator the boson effective charges e_π and e_ν are empirically found to be about equal and therefore its matrix elements, as we have just discussed, depend essentially only on the total boson number $N = N_\pi + N_\nu$ and not on N_π and N_ν separately. This is in sharp contrast to the M1 operator where the neutron–proton degree of freedom appears essential for the description of magnetic dipole properties. The next two sections discuss this.

Table 5.3 $B(E2)$ transition rates in the consistent-Q formalism between states of F-spin, $F = F_{max}$ (denoted by subscript s) and states of $F = F_{max} - 1$ (denoted by subscript a) for the three group chains, eqn (5.61).

$$N = N_\pi + N_\nu$$

	U(5)	SU(3)	O(6)
$B(E2; 2_s^+ \to 0_s^+)$	$(e_\pi N_\pi + e_\nu N_\nu)^2/N$	$(e_\pi N_\pi + e_\nu N_\nu)^2 \dfrac{2N+3}{5N}$	$(e_\pi N_\pi + e_\nu N_\nu)^2 \dfrac{N+4}{5N}$
$B(E2; 2_a^+ \to 0_s^+)$	$N_\pi N_\nu (e_\pi - e_\nu)^2/N$	$N_\pi N_\nu (e_\pi - e_\nu)^2 \dfrac{3(N-1)}{5N(2N-1)}$	$N_\pi N_\nu (e_\pi - e_\nu)^2 \dfrac{2(N+2)}{5N(N+1)}$

5.10 g-factors of 2^+ states

The magnetic dipole transition operator in IBM-2 is written as

$$T_1(M) = (3/4\pi)^{1/2}(g_\pi L_\pi + g_\nu L_\nu) \qquad (5.66)$$

and the magnetic moment is the diagonal matrix element for a state of spin I evaluated in the $M = I$ magnetic substate

$$\mu = \langle I, M = I | g_\pi L_\pi + g_\nu L_\nu | I, M = I \rangle. \qquad (5.67)$$

Here g_π and g_ν are the proton and neutron boson g-factors to be determined empirically and L_ρ is the orbital angular operator given in terms of boson operators as $L_\rho = \sqrt{10}\,[d_\rho^\dagger \times \tilde{d}_\rho]^{(1)}$ with $\rho = \pi$ or ν. In this section we will be concerned with the g-factor of the first excited 2^+ state in even–even nuclei. Phenomenological IBM-2 investigations indicate that these states can to a good approximation be characterized by a maximum F-spin. Then a simple analytic expression is obtained for the g-factor, $(g = \mu/I)$, namely

$$g(2_s^+) = (g_\pi N_\pi + g_\nu N_\nu)/N, \qquad (5.68)$$

with $N = N_\pi + N_\nu$. Microscopically the boson operators are constructed from pairs of fermions. In even–even nuclei the net contribution from the spin part of the magnetic moment operator in fermionic calculations is very small, thus one expects the boson g-factors to follow from the orbital part of the fermionic operator, namely $g_\pi = 1$ and $g_\nu = 0$. Then eqn (5.68) reduces to the canonical collective model result, $g(2^+) = Z/A$, if Z and A are regarded as the effective number of valence nucleons. In the IBM-2 model boson numbers N_π and N_ν represent the number of pairs of valence nucleons (particles or holes) relative to the nearest major closed shell. Thus the main trend of the g-factor (assuming $g_\nu = 0$) should follow from the very simple expression $g \sim N_\pi/(N_\pi + N_\nu)$. That is, in the lower half of a major shell, where neutron and proton bosons are treated as particles, g should be a decreasing function of neutron number for fixed N_π, whereas in the upper half of a major shell, where the bosons are built from holes, the opposite behaviour is predicted. In Fig. 5.5 we plot the g-factors for the 2^+ states in Ba isotopes ($N_\pi = 3$), as an example of a case in the lower half of a major shell, and in Os isotopes ($N_\pi = 3$) as an example from the upper half of a shell. Clearly the expected trends are evident and significantly different from the collective model Z/A value. Sambataro et al.[36] have looked at all the data on $g(2^+)$ for nuclei with Z in the range 50 to 82, and for the most part find the predicted trend well supported with $g_\pi \sim 1$ and $g_\nu \sim 0$. (In their analysis they allowed g_π and g_ν to be smoothly varying functions of N_π and N_ν respectively.) But there were a couple of notable exceptions. These concerned the $N = 86$ and $N = 88$ isotopes of Sm and Nd. The problem here can be traced to a subshell closure at $Z = 64$, which is particularly

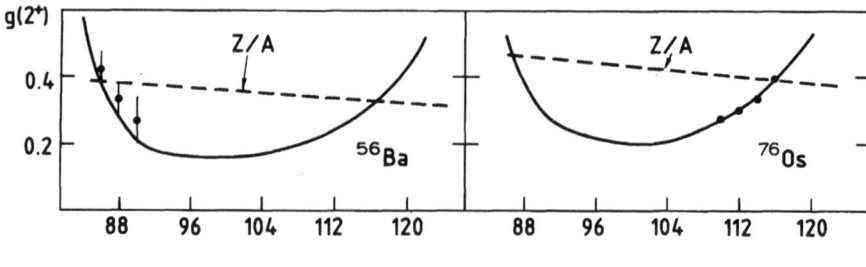

NEUTRON NUMBER

Fig. 5.5 The g-factors of the first 2^+ state in Ba and Os isotopes as a function of neutron number. The solid curves are the IBM-2 values with effective boson g-factors taken from Sambataro et al.[36] The dashed line represents the collective model of $g(2^+) = Z/A$.

pronounced for $N \leq 88$ but which rapidly disappears for larger neutron numbers. The trends in Sm isotopes can be improved by assuming a rigid $Z = 64$ shell closure for $N \leq 88$ and counting the number of valence protons from this value rather than from $Z = 50$. A slightly different analysis is given by Wolf et al.[37] They rewrite eqn (5.68) as

$$\frac{N}{N_\nu} g(2^+) = g_\nu + g_\pi \frac{N_\pi}{N_\nu} \tag{5.69}$$

then, on assuming that g_π and g_ν are constants, a linear relationship exists between $g(2^+)N/N_\nu$ and N_π/N_ν. To test this they use data not contaminated by the $Z = 64$ subshell closure. The result is shown in Fig. 5.6 where

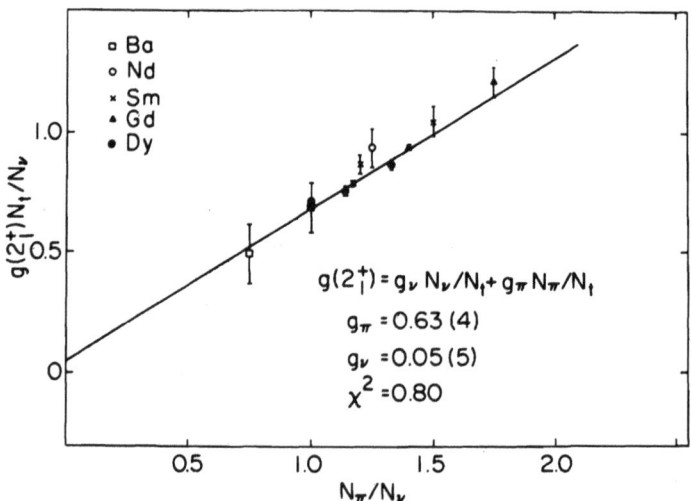

Fig. 5.6 Plot of the linear relationship between $g(2^+)N/N_\nu$ and N_π/N_ν. (From Wolf et al.[37])

it is seen that an excellent straight line fit is obtained with $g_\pi = 0.63 \pm 0.04$ and $g_\nu = 0.05 \pm 0.05$. Thus, while g_π is considerably reduced from its expected value of unity, g_ν remains consistent with zero. These values are then adopted for the analysis of the anomalous values of $g(2^+)$ that occur in the region of the $Z = 64$ subshell closure, which are then described in terms of changes in the effective number of proton bosons, N_π^{eff}, taking part in the collective motion.

5.11 Scissors mode

A major property of IBM-2 is the occurrence of 1^+ states. These states are missing in IBM-1 since they belong to representations which are not fully symmetric, $F < F_{\text{max}}$. Such states have been found experimentally in heavy deformed even–even nuclei at an excitation energy as low as 3 MeV. Classically, the introduction of separate collective degrees of freedom for neutrons and protons can be visualized in a two-rotor model.[38] The protons and neutrons are assumed to form separate rigid bodies of ellipsoidal shape. The excitation mode, then, is a small-amplitude oscillation in terms of the angle between the two symmetry axes of the proton and neutron distributions. This two-dimensional vibration breaks the overall axial symmetry but has a fundamental excitation mode with projection $K^\pi = 1^+$ on the symmetry axis. It leads to a rotation band of spins $1^+, 2^+, 3^+ \ldots$. This excitation mode has become known as the scissors mode, and should not be confused with the well known isovector giant dipole resonance in the Goldhaber–Teller picture in which the neutrons and protons carry out a collective translational oscillation.

In the classic picture of Lo Iudice and Palumbo[38] all the protons and all the neutrons participate in the collective vibration. In IBM-2 only the valence nucleons (i.e. those outside the major closed shells) are involved. The experimental strengths of M1 transitions appear to indicate that valence particles rather than the entire nucleus take part in the collective motion.

In IBM-2, we assume the ground state is a state of maximum F-spin, and the 1^+ excitation a state with $F = F_{\text{max}} - 1$. The $B(\text{M1}; 0^+ \to 1^+)$ transition rate then, as was the case with cross-band E2 transitions discussed in the last section, depends on the difference, $(g_\pi - g_\nu)^2$, of the boson g-factors. In the limit that $g_\pi = g_\nu$ the rate goes to zero as the M1 operator becomes of generator of the direct product group $U_\pi(6) \otimes U_\nu(6)$ and transitions between states of differing F-spin are forbidden. For magnetic transitions, however, $g_\pi \neq g_\nu$ and the expressions for the $B(\text{M1})$ rate for the three chains of subgroups, eqn (5.61), are as follows:[26]

$$B(\text{M1}; 0^+ \to 1^+) \begin{cases} = 0 & \text{U(5)} \\ = (3/4\pi)(g_\pi - g_\nu)^2 8 N_\pi N_\nu / (2N - 1), & \text{SU(3)} \\ = (3/4\pi)(g_\pi - g_\nu)^2 3 N_\pi N_\nu / (N + 1), & \text{O(6)} \end{cases} \quad (5.70)$$

with $N = N_\pi + N_\nu$. In the two-rotor model,[38] the $B(M1)$ rate is

$$B(M1; 0^+ \to 1^+) \cong 0.035 \delta A^{3/2},$$

where δ is a deformation parameter, defined in eqn (5.29), with a value of $\delta \simeq 0.25$ for rare-earth nuclei. This expression leads to a $B(M1)$ rate of order $17\mu_N^2$ for rare-earth nuclei and of order $32\mu_N^2$ for the actinides, between five and ten times larger than experiment. The experimental data together with IBM-2 predictions in the SU(3) limit are given in Table 5.4. With the canonical values of $g_\pi = 1$ and $g_\nu = 0$ for the boson g-factors (as expected microscopically from the g_L values of fermions and close to the values determined by Sambataro et al.[36]), the theory reproduces more or less the summed strength in three of the cases cited but underpredicts slightly on the strongest transition in ^{164}Dy. For the g-factors determined by Wolf et al.[37] in fits to magnetic moments of 2^+ states in the rare earths, $g_\pi = 0.63$ and $g_\nu = 0.05$, the summed strength is underpredicted by more than a factor of two.

The centroid energy of the 1^+ excitation scales roughly as $66\delta A^{-1/3}$ MeV, where δ is the deformation parameter, and is located at around 3 MeV in the rare earths and at 2.6 MeV in the actinides. The transition strength is concentrated into one or two collective states; there is little spreading. In the two actinide nuclei, the M1 strength is clustered into two groups. In ^{232}Th both groups are about equally excited. In ^{238}U, the lower group carries two-thirds of the strength.[39] In both nuclei the energy splitting between the two groups is roughly 0.2 MeV. In the rare earths, the nucleus ^{164}Dy also has two groups of 1^+ states; there the higher-lying group carries most of the strength.[40] The separation is larger, around 0.5 MeV. In ^{156}Gd six 1^+ states are found in a range of 0.3 MeV with most of the strength in one peak at 3.07 MeV.[41]

Table 5.4 Summed $B(M1; 0^+ \to 1^+)$ strength in nuclear magneton units seen in electron scattering and nuclear resonance fluorescence experiments compared to IBM-2 predictions in the SU(3) rotational limit. Experimental data from ^{156}Gd,[41] ^{164}Dy,[40] ^{232}Th,[39] ^{238}U.[39] In addition, weaker 1^+ excitations at around 3 MeV have been observed in ^{154}Sm, ^{158}Gd, ^{168}Er and ^{174}Yb[42]

	Experiment: $B(M1; 0^+ \to 1^+)$		Theory: IBM-2	
	(e, e')	(γ, γ')	$g_\pi = 1$, $g_\nu = 0$	$g_\pi = 0.63$, $g_\nu = 0.05$
^{156}Gd	2.3 ± 0.5	2.1 ± 0.5	2.9	1.0
^{164}Dy	5.17 ± 0.52	5.55 ± 0.53	3.9	1.3
^{233}Th	2.7 ± 1.1	2.58 ± 0.25	2.7	0.9
^{238}U	4.0 ± 1.7	3.19 ± 0.24	3.3	1.1

Theoretically the excitation energy is predicted in the two-rotor model[38] to be at $42\delta A^{-1/6}$ MeV, which is about 1.6 MeV too high. IBM-2, on the other hand, cannot really make a prediction since the excitation depends on the strength of the Majorana term in the Hamiltonian. If the experimental energy is used to adjust λ' of the Majorana term, eqn (5.60), which is responsible for the splitting of the symmetric and antisymmetric neutron–proton states, one obtains $\lambda' = 0.2$ MeV in reasonable agreement with values deduced from other phenomenological information.

5.12 Summary

In heavier nuclei and away from closed-shell configurations, it becomes impossible to mount a microscopic calculation based on individual nucleons. Thus it becomes more economical to introduce the idea of collective models and this chapter was concerned with elucidating the ways in which nuclear properties can be used as testing grounds for these models.

We began with the rigid-body model and obtained the standard rotational model, core–particle coupling and band-crossing results. We also discussed in detail how measurements of magnetic moments of the yrast states through the backbending region helped determine the physical nature of the band-crossing. Furthermore we saw that the newly discovered superdeformed bands (with major-to-minor axis ratio of two to one) have large intrinsic quadrupole moments consistent with rigid-body theory.

We said little in this chapter on vibrational nuclei, deferring detailed discussion on these to Chapter 7. We concentrated instead on the boson operator structure, where vibrational spectra (as indeed rotational spectra) were obtained as special limits to the interacting boson model (IBM). We quoted some general results for quadrupole moments and transitions in a model version known as IBM-1. A separate treatment of proton and neutron degrees of freedom is required to discuss magnetic properties. This was achieved by an extension of the model known as IBM-2. The systematics of nuclear g-factors of 2^+ states were found to be in good agreement with the IBM-2 predictions (and superior to the standard collective model predictions of Z/A). IBM-2 also provided a collective 1^+ excitation at around 3 MeV known as the scissor mode. This mode has been identified recently in photon and electron scattering experiments.

5.13 References

1. Bohr, A. and Mottelson, B. (1975). *Nuclear structure*, vol. 2. Benjamin, New York.

2. Brink, D. M. and Satchler, G. R. (1968). *Angular momentum.* Clarendon, Oxford.
3. Bunker, M. E. and Reich, C. W. (1971). *Rev. Mod. Phys.*, **43**, 348; Ogle, W., Wahlborn, S., Piepenbring, R., and Fredriksson, S. (1971). *Rev. Mod. Phys.*, **43**, 424.
4. Engeland, T. (1984). In *Collective phenomena in atomic nuclei* (ed. T. Engeland, J. Rekstad, and J. S. Vaagen). World Scientific, Singapore.
5. Nilsson, S. G. (1955). *K. Danske Vidensk. Selsk. Mat.-Fys. Medd.*, **29**, No. 16.
6. Prior, G., Boehm, F., Nilsson, S. G. (1968). *Nucl. Phys.*, **A110**, 257.
7. Bochnacki, Z. and Ogaza, S. (1966). *Nucl. Phys.*, **83**, 619.
8. Ring, P. and Schuck, P. (1980). *The nuclear many-body problem.* Springer, Heidelberg.
9. Johnson, A., Ryde, H., and Hjorth, S. A. (1972). *Nucl. Phys.*, **A179**, 753.
10. Garrett, J. D., Hagemann, G. B., and Herskind, B. (1986). *Ann. Rev. Nucl. Part. Sci.*, **36**, 419; Bengtsson, R. and Garrett, J. D. (1986). In *Collective phenomena in atomic nuclei* (ed. T. Engeland, J. Rekstad, and J. S. Vaagen). World Scientific, Singapore; de Voigt, M. J. A., Dudek, J., and Szymanski, Z. (1983). *Rev. Mod. Phys.*, **55**, 949; Diamond, R. M. and Stephens, F. S. (1980). *Ann. Rev. Nucl. Part. Sci.*, **30**, 383.
11. Stephens, F. S. (1975). *Rev. Mod. Phys.*, **47**, 43.
12. Frauendorf, S. (1981). *Phys. Lett.*, **100B**, 219.
13. Chen, Y. S. and Frauendorf, S. (1983). *Nucl. Phys.*, **A393**, 135.
14. Taras, P., et al. (1985). *Nucl. Phys.*, **A345**, 294.
15. Billowes, J., et al. (1986). *Phys. Lett.*, **B178**, 145.
16. Seiler-Clark, G., et al. (1983). *Nucl. Phys.*, **A399**, 211.
17. Dönau, F. and Frauendorf, S. (1983). In *High angular momentum properties of nuclei* (ed. N. R. Johnson), p. 143, Academic, New York.
18. Twin, P. J., et al. (1986). *Phys. Rev. Lett.*, **57**, 811.
19. Specht, H. J., Weber, J., Konecny, E., and Heunemann, D., (1972). *Phys. Lett.*, **B41**, 43.
20. Ragnarsson, I. and Åberg, S. (1986). *Phys. Lett.*, **B180**, 191.
21. Bentley, M. A., et al. (1987). *Phys. Rev. Lett.*, **59**, 2141.
22. Twin, P. J. (1988). In *Nuclei far from stability.* American Institute of Physics Conference Proceedings, No. 164 (ed. I. S. Towner), p. 499. American Institute of Physics, New York.
23. Haas, B., et al. (1988). *Phys. Rev. Lett.*, **60**, 503.
24. Bohr, A. (1954). *K. Danske Vidensk. Selsk. Mat.-Fys. Medd.*, **26**, No. 14.
25. Kumar, K. and Baranger, M. (1967). *Nucl. Phys.*, **A92**, 608.
26. Iachello, F. and Arima, A. (1987). *The interacting boson model.* Cambridge University Press.
27. Feshbach, H. and Iachello, F. (1973). *Phys. Lett.*, **45B**, 7; Feshbach, H. and Iachello, F. (1974). *Ann. Phys. (N.Y.)*, **84**, 211.
28. Janssen, D., Jolos, R. V., and Dönau, F. (1974). *Nucl. Phys.*, **A224**, 93.
29. Arima, A. and Iachello, F. (1975). *Phys. Rev. Lett.*, **35**, 1069.
30. Iachello, F. and Talmi, I. (1987). *Rev. Mod. Phys.*, **59**, 339; Casten, R. F. and Warner, D. D. (1988). *Rev. Mod. Phys.*, **60**, 389; Arima, A. and Iachello, F. (1984). *Adv. Nucl. Phys.*, **13**, 139.

31. Elliott, J. P. (1958). *Proc. R. Soc.*, **A245,** 128, 562.
32. Casten, R. F. and von Brentano, P. (1985). *Phys. Lett.*, **152B,** 22.
33. Otsuka, T., Arima, A., Iachello, F., and Talmi, I. (1978). *Phys. Lett.*, **76B,** 139.
34. Hamermesh, M. (1962). *Group theory and its applications to physical problems.* Addison-Wesley, Reading, Mass.
35. Bohle, D., *et al.* (1984). *Phys. Lett.*, **137B,** 27.
36. Sambataro, M., Scholten, O., Dieperink, A. E. L., and Piccitto, G. (1984). *Nucl. Phys.*, **A423,** 333.
37. Wolf, A., Warner, D. D. and Benczer-Koller, N. (1985). *Phys. Lett.*, **158B,** 7.
38. De Franceschi, G., Palumbo, F., and Lo Iudice, N. (1984). *Phys. Rev.*, **C29,** 1496; Lo Iudice, N. and Palumbo, F. (1978). *Phys. Rev. Lett.*, **74,** 1046; Lo Iudice, N. and Palumbo, F. (1979). *Nucl. Phys.*, **A326,** 193.
39. Heil, R. D., *et al.* (1988). *Nucl. Phys.*, **A476,** 39.
40. Bohle, D., *et al.* (1987). *Phys. Lett.*, **195B,** 326.
41. Bohle, D., *et al.* (1986). *Nucl. Phys.*, **A458,** 205.
42. Bohle, D., *et al.* (1984). *Phys. Lett.*, **148B,** 260.

6
SINGLE-PARTICLE AND SHELL-MODEL THEORIES OF QUADRUPOLE MOMENTS

We now turn our attention to electric quadrupole moments. These moments have traditionally been invoked as good indicators of collective effects in contrast to magnetic moments. The reason for that is simple and has to do with the difference in the nature of the two relevant operators. The magnetic dipole operator leads to opposite contributions from a state $|\psi\rangle$ and its time-reversed conjugate $|\bar\psi\rangle$ (since both currents and spin change sign in the $|\psi\rangle \to |\bar\psi\rangle$ conjugation). Thus one member of the paired-off nucleon cancels the contribution from the other. Such is not the case for the quadrupole operator since its contributions are invariant under time reversal. Thus if nuclear correlations can be induced by some deformation, a coherent build up of quadrupole moment can be expected. As we shall see, this coherence can lead to sizeable effects which will be examined in this chapter in the light of single-particle and various shell models. The next chapter will be devoted to a survey of the macroscopic models currently available for the study of quadrupole moments and nuclear deformations.

6.1 Single-particle quadrupole moments

As we have already seen in our study of magnetic moments, the value of any operator of rank not equal to zero vanishes for a system of completely filled levels. As a consequence the quadrupole moment of a system consisting of a proton outside a closed shell is given by the quadrupole moment of the valence proton. Calculating the expectation value of the quadrupole operator as given in eqn (1.6) in a shell-model orbit $|nljm\rangle$ yields

$$Q_{\text{s.p.}} = e\left(\frac{16\pi}{5}\right)^{1/2} \langle nl|\, r^2\, |nl\rangle \langle jj20\,|\,jj\rangle \langle l\tfrac{1}{2}j\|\, Y_2\, \|l\tfrac{1}{2}j\rangle \qquad (6.1)$$

$$= -e\frac{2j-1}{2(j+1)} \langle nl|\, r^2\, |nl\rangle. \qquad (6.2)$$

The negative sign of $Q_{\text{s.p.}}$ is of course consistent with the fact that the density of a state $|j, m=j\rangle$ is concentrated along the equatorial plane. For such a distribution $\langle 3z^2 \rangle < \langle r^2 \rangle$ which according to eqn (1.6) results

6.1 SINGLE-PARTICLE QUADRUPOLE MOMENTS

in a negative quadrupole moment. We also notice that in contrast to single-particle magnetic moments, $Q_{\text{s.p.}}$ depends on the exact nature of the radial integral $\langle nl| r^2 |nl \rangle$.[1]

In the case of n protons in configuration $|j^n; I = j\rangle$ with n odd, one obtains the resultant[2,3]

$$Q_n = Q_{\text{s.p.}}\left(1 - \frac{2n-2}{2j+1}\right). \tag{6.3}$$

Note that when n increases, Q_n should vary linearly with n. It is interesting to compare Fig. 6.1 showing the expected behaviour of Q_n between closed shells with the plot of Fig. 6.2 displaying the observed quadrupole moments for odd-A nuclei as a function of A. The agreement between theory and experiment observed near closed shells disappears rapidly as n increases and suggests the onset of collective deformation.

Equation (6.3) also clearly indicates that the quadrupole moment of a proton–hole state is equal to minus that of a proton state. If one considers ^{39}K as a $d_{3/2}$ proton–hole state, its quadrupole moment should be equal and opposite to that of ^{37}Cl (if we assume ^{37}Cl to be a $d_{3/2}$ proton outside an inert ^{36}S core.) Experimentally $Q(^{39}\text{K}) = 4.9 \text{ fm}^2$ whereas $Q(^{37}\text{Cl}) = -6.2 \text{ fm}^2$.[4] Note that according to eqn (6.3), $Q(^{39}\text{K})$ should be equal to 4.8 fm^2 if one used harmonic oscillator wavefunctions (with $\hbar\omega = 41A^{-1/3}$). Thus the quadrupole moment of ^{39}K is in excellent agreement with the single-particle estimate. In many cases, however, the quadrupole moment of odd-proton nuclei differs from the simple estimate

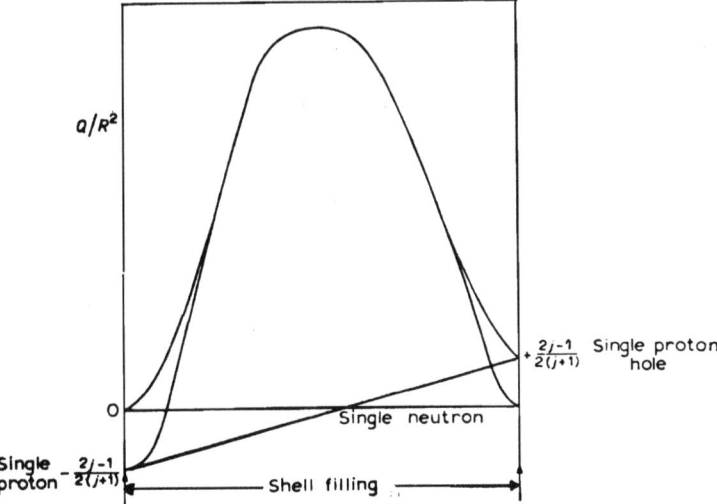

Fig. 6.1 The behaviour of the quadrupole moment whilst a shell is being filled. The straight line is essentially eqn (6.3) while the curve represents the more realistic situation.

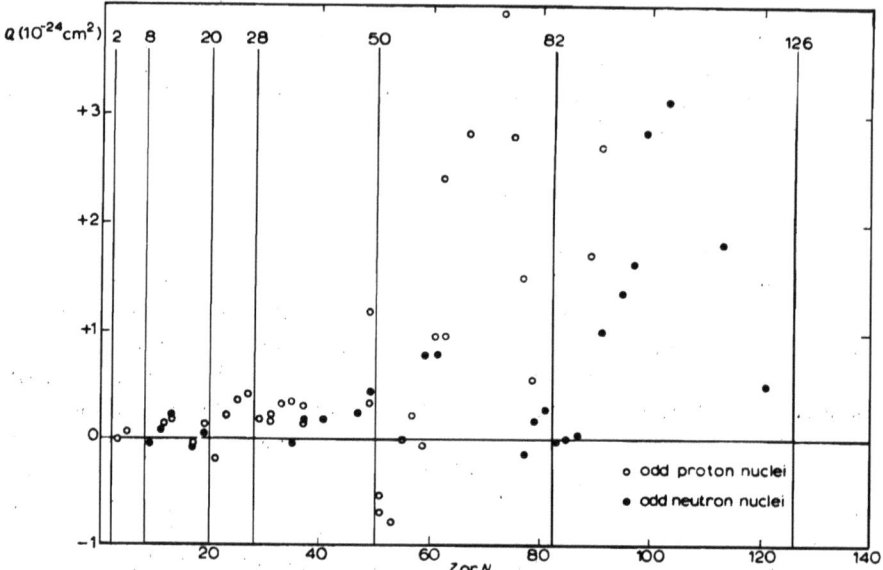

Fig. 6.2 Observed quadrupole moments for valence proton and neutron nuclei as a function of proton or neutron number.

and the discrepancy characterized by an effective charge (see Section 1.8). The charge on the odd-proton is not that of a bare proton but is modified presumably by the presence of other nucleons in the nucleus and is written as $e_{\text{eff}} = 1 + e_p$ in units of the bare proton charge e, where e_{eff} is known as the effective charge. Similarly odd-neutron nuclei would all have zero quadrupole moments in the extreme single-particle model. However, the effect of the other protons in the nucleus induces an effective charge, $e_{\text{eff}} = e_n$, on the valence nucleon. Thus, as we saw earlier in Section 1.8, the quadrupole moment of ^{17}O is nearly equal to the one estimated for ^{17}F.[5] Of course, a single neutron can also affect the proton distribution by shifting the nuclear centre of mass but the resulting neutron quadrupole moment $Q_{\text{s.p.}}(\text{neutron}) = -Z/A^2 \times Q_{\text{s.p.}}(\text{proton})$ is in general much smaller than experiment.[3] We are thus enticed to look at one-particle and one-hole systems as excellent candidates for the study of effective charges. Before concentrating on that project, we will briefly examine to what extent two-particle configurations can add to the scarce body of data provided so far by valence nucleon systems.

6.2 Quadrupole moments of two-particle systems

Let us examine the quadrupole moment of a two-proton or two-neutron state in the configuration, $|j^2; I = 2\rangle$. Using the two-particle

6.3 EFFECTIVE CHARGE AND THE SHELL MODEL

wavefunction,

$$|j_1 j_2; IM\rangle = \sum_{m_1 m_2} \langle j_1 m_1 j_2 m_2 | IM \rangle |j_1 m_1\rangle |j_2 m_2\rangle (1 + \delta_{j_1 j_2})^{-1/2}, \quad (6.4)$$

one obtains a result of opposite sign to that given by the single-particle moment (Yoshida and Zamick[6]). For instance

$$Q(f_{7/2}^2; I = 2) = -32/49 Q_{\text{s.p.}}(f_{7/2})$$

where $Q_{\text{s.p.}}(f_{7/2})$ is a single-particle moment for an $f_{7/2}$ nucleon (see, for example, de Shalit and Talmi[2]).

Thus a positive quadrupole moment is predicted for the first 2^+ state in ^{42}Ca, whereas Cline[6] has measured $Q(2^+) = -0.18$ b. This result was construed as good evidence that the ^{42}Ca $J = 2^+$ state is not a simple two-particle configuration but contains close to 50% admixture of 4p–2h correlations.[7] Other two-particle configurations can also play a role in increasing generally the $B(E2)$ transition rate between the 2^+ and the 0^+ ground state, but the effects on quadrupole moments are varied. Again in ^{42}Ca, Yoshida and Zamick[6] calculated that a 4% admixture of the $f_{7/2}p_{3/2}$ configuration in the basic $f_{7/2}^2$ state increased the $B(E2: 2^+ \to 0^+)$ by about 50%, but decreased the quadrupole moment from $+0.07$ to 0.02 b. This is a good example demonstrating the difficulty of controlling both rates and moments using a single parameter such as the effective charge.

6.3 Effective charge and the shell model

We have seen that in order to understand observed quadrupole moments and E2 transitions, one must introduce the concept of effective charge. Recall that we have introduced the definitions $e_{\text{eff}} = 1 + e_p$ for a proton and $e_{\text{eff}} = 1 + e_n$ for a neutron, where e_p and e_n are additional charges on a nucleon induced by the presence of other protons in the nucleus. Two mechanisms contribute to these polarization effects. The first one is linked to excitations in which a nucleon is excited from a closed-shell occupied orbital to an unoccupied orbital of energy $2\hbar\omega$ higher in the oscillator model. These are the particle–hole excitations that coherently make up the giant quadrupole resonance. Since shell-model calculations are frequently carried out in a model space in which the fully occupied closed-shell orbitals are ignored these calculations naturally omit the giant quadrupole resonances. The effective charge, in essence, is compensating for this omission. The second mechanism is a polarization *induced by valence particles*, that is nucleons in partially occupied orbitals. This is not a factor if one considers simply closed-shell-plus-one nuclei but is relevant for open-shell nuclei. We return to this second mechanism in the next section.

Consider the case of a closed-shell-plus-one nucleus. In the simplest approximation its ground-state wavefunction is given by the single-particle wavefunction of the valence nucleon; the core is inert. There will, however, be corrections. For instance there will be admixtures of configurations in which the core is excited, such as 2p–1h configurations. If we are interested in the E2 properties of the valence nucleon, then the most important admixture is the one in which the additional particle and hole are the coherent linear superposition of particle–hole excitations that make up the electric giant quadrupole resonance in the core. How important is this admixture in the ground-state wavefunction can be estimated in perturbation theory and in first order is given by $\langle 2\text{p}-1\text{h}| V |1\text{p}\rangle/\Delta E$, where V is the perturbing residual interaction and ΔE the unperturbed energies of the configurations, which in the oscillator model is $\Delta E \simeq 2\hbar\omega$. The larger this admixture the greater will be the additional charge induced on the valence nucleon. Both the particle and the hole in the admixed configuration must be protons as the electric quadrupole operator only acts on protons. Thus if the valence particle as well is a proton then the matrix element $\langle 2\text{p}-1\text{h}| V |1\text{p}\rangle$ will only depend on the isospin $T=1$ part of the interaction, V. Whereas if the valence particle is a neutron both the $T=0$ and $T=1$ parts of the interaction contribute to the matrix element. Since the $T=0$ part of the interaction is generally stronger than the $T=1$ part, we expect the core-polarization contribution to the effective charge for valence neutrons to be larger than for protons, i.e. e_n should be larger than e_p. As we will see later, this observation is indeed borne out by experiment.

First of all, since the effective charge, in principle, converts an operator into an effective operator to compensate for truncations in model space, i.e.

$$\langle \psi_\text{f}(\text{exact})| T |\psi_\text{i}(\text{exact})\rangle \equiv \langle \phi_\text{f}(\text{truncated})| T_\text{eff} |\phi_\text{i}(\text{truncated})\rangle, \quad (6.5)$$

it is clear that the effective charge required is a function of the model space selected. This is well illustrated by an example using ^{42}Ca data.[6] Within the $(\nu\text{f}_{7/2})$ configuration, the effective charge required to fit the $B(\text{E2}: 2^+ \rightarrow 0^+)$ is $e_\text{n} = (2.08 \pm 0.04)e$. If one now uses the complete (0f, 1p) shell, e_n drops by about 20% to $(1.73 \pm 0.03)e$. If one now adds the 4p–2h basis $(0\text{f}, 1\text{p})^4(0\text{d}, 1\text{s})^{-2}$ the required effective charge drops to $e_\text{n} = (0.95 \pm 0.02)e$.[3]

Of course, attempts to calculate effective charges within the shell model have had a long history. Several authors starting with Arima and Horie[7] have used a perturbation expansion for the effective operator T_eff and applied it through to second order in the residual interaction in the case of E2 transitions.[8–10] The set of diagrams relevant to that expansion are shown in Fig. 6.3 and have been discussed in detail by Towner.[11] Diagram (a) is the zeroth-order term and represents the bare operator T evaluated between single-particle oscillator states. The effective charge is

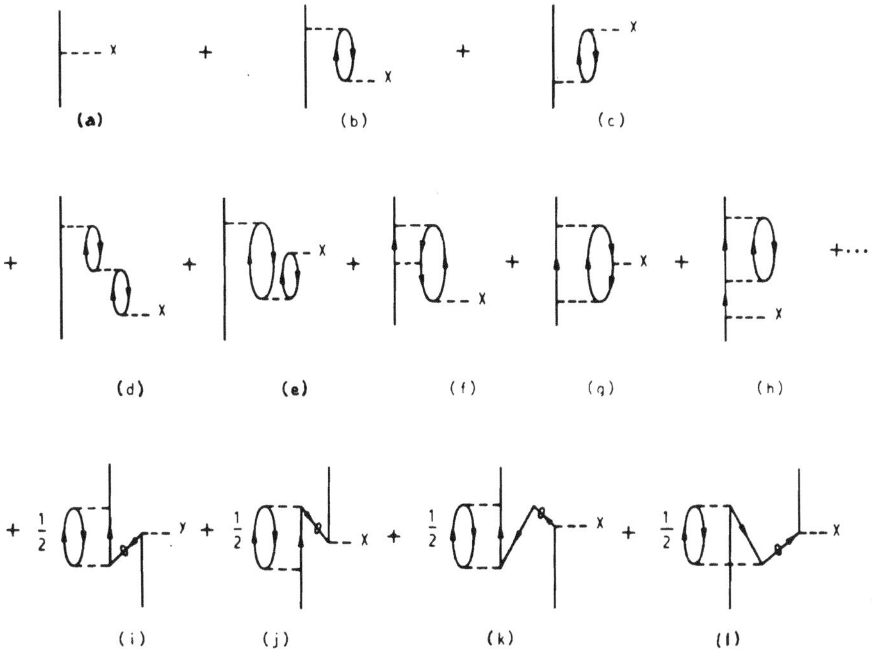

6.3 EFFECTIVE CHARGE AND THE SHELL MODEL

Fig. 6.3 Diagram expansion for the effective transition operator in closed-shell-plus-one nuclei. Diagram (a) represents the bare transition operator. Diagrams (b) and (c) are first-order corrections, and diagrams (d) to (h) are selected second-order corrections. Diagrams (i) to (l) are the folded diagrams that appear in second order. (From Towner.[11])

then given by the sum of all the other diagrams divided by this zeroth-order term. Diagrams (b) and (c) are the only first-order terms; they are known as the core-polarization diagrams and select out just the 2^+ particle–hole component in the residual interaction. In the ^{17}O calculation by Ellis and Siegel[10] whose results are shown in Table 6.1, e_n is found to be twice as large as e_p in the first order—both results are well

Table 6.1 First- and second-order core-polarization contributions to the effective charge in $A = 17$[10]

	$(1s_{1/2}\| e \|0d_{5/2})$		$(0d_{5/2}\| e \|0d_{5/2})$	
	Neutron	Proton	Neutron	Proton
First order	0.219	0.115	0.258	0.099
Total second order	0.035	−0.095	0.056	−0.057
Total first plus second	0.254	0.020	0.314	0.042
Experiment	0.54	0.81	0.44	—

short of experiment. The more complex second-order corrections were first calculated by Siegel and Zamick[9] and by Ellis and Siegel[10] whose results we also quote in Table 6.1. Adding first- and second-order contributions yields satisfactory results for e_n in that the second-order contribution is some five times smaller than the first, so that convergence order-by-order looks promising, but the summed effect is still too small by a factor of two compared with experiment. The proton results are more disappointing in that the second-order contribution is as large as the first and of opposite sign, so there seems to be no order-by-order convergence and the summed effect is smaller than experiment by an order of magnitude. Ellis and Mavromatis[12] later found that the contributions to the effective charge from Hartree–Fock diagrams which had been omitted earlier brought neutron and proton effective charges closer together but the total results were still too small by about a factor of three compared with experiment.

Historically, the inability to explain effective charges in closed-shell-plus-one nuclei in perturbation theory has been considered quite a set-back since, as we can see from eqn (6.5), this inability can be linked to an inability to construct an appropriate effective operator. This in turn reflects our lack of knowledge of the character of the effective interaction.

6.4 Mass dependence of the effective charge

The effective charge also is not strictly speaking a constant within a major shell but should also be mass dependent. This is because the effective charge originates as we have seen in the core-polarization diagrams (b) and (c) drawn in Fig. 6.3, which depict the correction to an electromagnetic transition matrix element in a closed-shell-plus-one nucleus, from first-order perturbation theory. We define the polarizability χ as the ratio of this correction to the single-particle matrix element; it is given by the expression

$$\chi = \frac{\langle a| \delta T_\lambda |a\rangle}{\langle a| T_\lambda |a\rangle} = 2 \sum_\alpha \frac{\langle 0| T_\lambda |\alpha\rangle \langle \alpha| V |a^{-1}a\rangle}{\langle 0| T_\lambda |a^{-1}a\rangle(-\varepsilon_\alpha)}, \qquad (6.6)$$

where a represents the single-particle valence orbital outside the closed shell and α are particle–hole states of spin and parity I^π, namely $|\alpha\rangle = |j_h^{-1}j_p: I^\pi\rangle$, with the I^π value given by the multipolarity of the one-body electromagnetic transition operator T_λ. In the case of the electric E2 operator, $I^\pi = 2^+$. Here ε_α is the energy of particle–hole state, which in a harmonic oscillator model is just $2\hbar\omega$ for E2 transitions. Finally V is the two-body residual interaction that excites the particle–hole state. The enhancement this correction brings to the single-particle matrix element $\langle a| T_\lambda |a\rangle$ relates to the effective charge, namely $e_{\text{eff}} = 1 + \chi$.

6.4 MASS DEPENDENCE OF THE EFFECTIVE CHARGE

Now the mass dependence of the effective charge follows from the following argument. Consider the case of ^{17}O. Contributions to eqn (6.6) arise from exciting a 0s particle to the 0d,1s shell and from a exciting a 0p particle to the 0f,1p shell. As one goes from shell closure at $A = 16$ towards shell closure at $A = 40$ by adding valence nucleons to the 0d,1s shell the $2\hbar\omega$ core excitation 0s$\rightarrow$0d,1s is gradually quenched by the Pauli exclusion principle from its full value at $A = 17$ to zero at $A = 39$. Likewise, a valence polarization from the $2\hbar\omega$ excitation 1s,0d$\rightarrow$2s,1d,0g is gradually introduced increasing from zero at $A = 17$ to its full value at $A = 39$. These two effects lead to a mass dependence in the effective charge. It is conceivable, of course, that these two effects being of opposite sign could mutually cancel leaving a constant polarization. However, as has been shown in schematic model calculation of Alexander et al.[13] the valence polarization is the larger effect and dominates the phenomenon.

In Fig. 6.4 we show the core-polarization graphs for two valence nucleons outside closed shells. Graph (a) is just the standard core-polarization mechanism acting on one of the valence nucleons with the other unaffected. Notice that for the particle–hole excitation being $p = 0d,1s$ and $h = 0s$, the particle p is not antisymmetrized with respect to the nucleon, a_1. This is corrected in graph (b), which restores the antisymmetrization of the valence nucleons. Graphs (b) and (c) lead to two-body electromagnetic transition operators in the valence space of sd-shell nucleons, whereas graph (a) leads to one-body operators.

For n particles in the sd shell, the off-diagonal matrix elements of a one-body operator is expected to depend on the particle number as $[n(N-n)]^{1/2}$, where n is the number of particles and $N-n$ the number

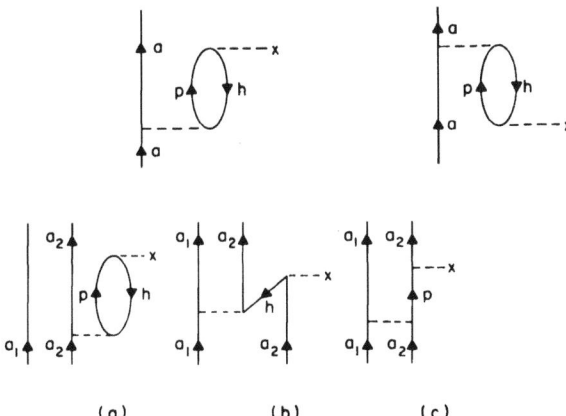

(a)   (b)   (c)

Fig. 6.4 First-order core-polarization graphs in closed-shell-plus-one and plus-two nuclei. Hermitian graphs and topologically equivalent graphs have been omitted. (From Alexander et al.[13])

of holes in the sd shell. Similarly for two-body operators, the matrix element is the sum of two pieces, one of which goes as $(N-n+2)[n(N-n)]^{1/2}$ and the other as $(n-2)[n(N-n)]^{1/2}$. For the effective charge, we are interested in the size of these core-polarization graphs relative to the size of the bare single-particle graph evaluated in the n-particle valence space. Thus graph (a), being an effective one-body graph, will, in taking the ratio, produce a polarization χ that is independent of particle number n, whereas graphs (b) and (c), being effective two-body graphs, will give contributions to χ that have pieces independent of n and pieces linear in n.

To get an estimate of how big this linear dependence on the mass number is, we appeal to the simple schematic model of Brown and Bolsterli.[14] The Brown–Bolsterli schematic factorization can in fact be realized with a zero-range interaction which we write as[13]

$$V(r) = t_0[1 + xP^\sigma]\delta(r_1 - r_2). \qquad (6.7)$$

It is an attractive interaction, $t_0 > 0$, central and contains a spin-exchange mixture for which x is positive and typically of order $x \sim \frac{1}{3}$. Here P^σ is the spin-exchange operator, $P^\sigma = \frac{1}{2}(1 + \boldsymbol{\sigma}_1 \cdot \boldsymbol{\sigma}_2)$. This interaction is just the central part of a Skyrme interaction,[15] omitting the repulsive velocity-dependent and density-dependent terms.

In Table 6.2 we give the calculated values for the isoscalar and isovector polarizabilities for a ^{16}O and ^{40}Ca core, and the effective

Table 6.2 Schematic model values for the proton and neutron effective charges and the corresponding isoscalar and isovector polarizabilities,[a] χ, in terms of the parameter[b] k that characterizes the strength of the $T=0$ part[c] of the zero-range interaction, eqn (6.7)

	^{16}O	^{40}Ca	$1/24(^{40}\text{Ca}-^{16}\text{O})$
e_p	$\frac{16}{9}k$	$4k$	$\frac{1}{24}(\frac{20}{9}k)$
e_n	$\frac{56}{9}k$	$14k$	$\frac{1}{24}(\frac{70}{9}k)$
e_n/e_p	$\frac{7}{2}$	$\frac{7}{2}$	
$\chi_{T=0}$	$8k$	$18k$	
$\chi_{T=1}$	$-\frac{40}{9}k$	$-10k$	

[a] $\chi_{T=0} = e_p + e_n$; $\chi_{T=1} = e_p - e_n$.
[b] $k = 3t_0 I/(2\hbar\omega)$, where I is a radial integral defined in Alexander et al.[13] and assumed to be a constant for all orbitals in the schematic model.
[c] With $x = 1/3$, the strength of the $T=1$ part of the zero-range interaction is $-5k/9$.

6.4 MASS DEPENDENCE OF THE EFFECTIVE CHARGE

charges they lead to. We also give the difference in these effective charges beween ^{40}Ca and ^{16}O cores that is responsible for the linear mass dependence in these quantities. It is quite clear that the *isoscalar effective charge increases in going from* ^{16}O *to* ^{40}Ca while the *isovector effective charge decreases*. The magnitude of these charges depends on the strength of the zero-range particle–hole force, but the ratio of isovector to isoscalar mass dependence of the effective charges is just the ratio of the strengths of the isovector to isoscalar particle–hole force, namely $-\frac{1}{3}(1+2x) \simeq -\frac{5}{9}$.

Finally, we would like to comment on the ratio of neutron and proton effective charges, e_n/e_p. This ratio has recently been analysed by Alexander et al.[13] using a parametrization involving all E2 isoscalar and isovector data in the sd shell and found to vary from 1.68 at ^{16}O to 2.24 at ^{40}Ca in a smooth way. These values are somewhat less than the value $e_n/e_p = 3$ deduced by Brown et al.[16] Theoretically the value of e_n/e_p is quite sensitive to the core-polarization model used. For example, in the schematic model evaluated in first order we have $e_n/e_p = 7/2$ (see Table 6.2). Actually the expression is

$$e_n/e_p = \frac{\chi_0 - \chi_1}{\chi_0 + \chi_1} = \frac{2+x}{1-x}, \tag{6.8}$$

where x is the spin-exchange mixture in the force. Here χ_0 and χ_1 are the isoscalar and isovector polarizabilities given, in first order, by the schematic model as

$$\chi_0 = k\sigma, \qquad \chi_1 = -\tfrac{1}{3}(1+2x)\chi_0, \tag{6.9}$$

where k and σ are defined in Alexander et al.,[13] but roughly k is proportional to the strength of the particle–hole interaction and σ to the number of protons in the core effective in the polarization calculation. The calculation is trivially extended beyond first order in the Tamm–Dancoff (TDA) and random phase (RPA) approximations in the schematic model by replacing the polarizabilities by

$$\chi_{\text{TDA}} = \frac{\chi}{1-\tfrac{1}{2}\chi}, \qquad \chi_{\text{RPA}} = \frac{\chi}{1-\chi}, \tag{6.10}$$

where χ is the first-order expression. The same replacements apply to both the isoscalar and isovector polarizabilities. Because these replacements have a larger influence on χ_0 than χ_1, the impact of higher orders is to reduce the value of e_n/e_p. Specifically for a typical value of $\chi_0 = \tfrac{1}{2}$, we obtain $e_n/e_p = 28/13 = 2.15$ in TDA and $e_n/e_p = 14/9 = 1.56$ in RPA. These estimates are well within the range of the experimentally deduced values.[13]

Two major conclusions suggest themselves from this review of E2 effective charges. The first is that the isoscalar effective charge, $e_0 = 1 + \chi_0$, shows a pronounced mass dependence as one goes through the sd shell; the second implies that valence neutrons are more effectively polarized than protons by core nucleons (with a ratio of neutron to proton polarization charge being of the order of two).

Several years ago Abbas and Zamick[17] studied the evolution of E2 effective charges in closed-shell nuclei ($A = 16, 40, 80, \ldots$) and concluded that in the large-A limit, one should expect an increase of isoscalar effective charge e_0 with A. Their conclusions, together with those of the schematic model discussed here in the context of the E2 experimental data,[18] present a coherent picture of a mass dependence for the effective charge. The isovector component of this picture could be tested further if more precise experimental data on isovector transitions become available.

Finally, one can also ask whether there is any state dependence to the effective charge. In principle there should be, although presumably it is weak or else the analysis of experimental data assuming state independence would be meaningless. In the schematic model the core-polarization calculations lead to state-independent effective charges because of the zero range nature of the interaction used and the assumption that all the radial integrals are state independent. More realistic core-polarization calculations, such as those shown in Table 6.1, produce state-dependent results. If one assumes that most of the E2 strength is concentrated in a $2\hbar\omega$ giant quadrupole state which can be characterized as a vibration of the nuclear surface (Bohr and Mottelson[1]) then states with high l should have larger values for e_p and e_n. This is because the wavefunction for these states will be more localized in the vicinity of the nuclear surface and consequently interact more strongly with a surface oscillation. Data by Astner et al.[19] in the Pb region seem to corroborate this contention. For instance $e_n(0i_{13/2}) \sim 0.96$ whereas $e_n(1g_{9/2}) \sim 0.84$ and $e_n(2p_{\frac{1}{2}}) \sim 0.75$.[19,20] One should note, however, that since the core-polarization mechanism is itself state dependent Astner et al.'s result might be reproducible by a core-polarization calculation using a finite-range force.

6.5 Electric quadrupole moments in the shell model

We now consider open-shell nuclei away from the closed shells and see to what extent the simple expedient of modifying the E2 quadrupole operator with an effective charge is successful in explaining the data. We first examine the various shell model calculations available for the sd shell which has been a region of predilection for theoretical testing for, among other reasons, its interesting transition from prolate to oblate deformation at around $A \sim 28$. An operator well suited to characterize nuclear deformation in even nuclei is the static quadrupole moment of the 2^+

6.5 ELECTRIC QUADRUPOLE MOMENTS

state defined as the expectation value of the quadrupole moment operator in the $I = M$ state.

The measurement of static quadrupole moments of 2_1^+ states in even–even nuclei not only provides a direct test of the nuclear wavefunctions but, when used in conjunction with a collective model, gives information on the *intrinsic shape* of these nuclei.

In general, most shell-model calculations we will review have been performed using phenomenological interactions.[12] In the conventional approaches experimental information on the energy levels is used for the determination of the effective interaction. For this purpose the matrix elements of the Hamiltonian are written as a linear combination of parameters, e.g. single-particle and two-body matrix elements, relative matrix elements, or Talmi integrals. Optimal values for these parameters are then determined iteratively with a least-squares fitting routine to a selected set of energy levels. The number of parameters that enters in such a calculation thus depends not only on the model space considered, but also on the general structure assumed for the effective Hamiltonian.[22]

In general, shell-model calculations of the sd shell vary substantially with respect to degree and method of model-space truncation and assumptions concerning the two-body matrix elements of the residual interaction. The conventional shell-model calculations of Wildenthal et al.,[23] Krewald et al.,[24] and Van Hienen et al.[25] all assume an inert ^{16}O core, and permit no excitations beyond the 1s0d shell. Various degrees of truncation involving the $0d_{5/2}$, $1s_{1/2}$, and $0d_{3/2}$ orbits are then used, and various 'empirical' effective interactions adopted. Recently, Singhal et al.[26] and Brown and Wildenthal[27] have used the Lanczos algorithm for matrix diagonalization and allowed access to the full 1s0d shell space for the active nucleons outside an inert ^{16}O core.

In spite of this diversity of detail, the results of the calculations are impressively consistent and on the whole they represent the variation of $Q(2^+)$ in the 1s0d shell very well. In particular the sharp transition from prolate to oblate deformation occurring at $A = 28$ is generally well reproduced (see Fig. 6.5).

Recently Carchidi et al.[28] have attempted to surmount the difficulties and inconsistencies inherent in truncated shell model calculations by calculating quadrupole moments and E2 transitions using the full 1s0d shell space. These calculations are internally consistent in that every state is calculated from a model Hamiltonian generated from the same set of two-body matrix elements and single-particle energies, in that the effective charges used in the model E2 operator are independent of state and mass, and in that the single-particle radial wavefunctions are parametrized as functions of the mass in a systematic manner.

The salient results of calculated and experimental static quadrupole

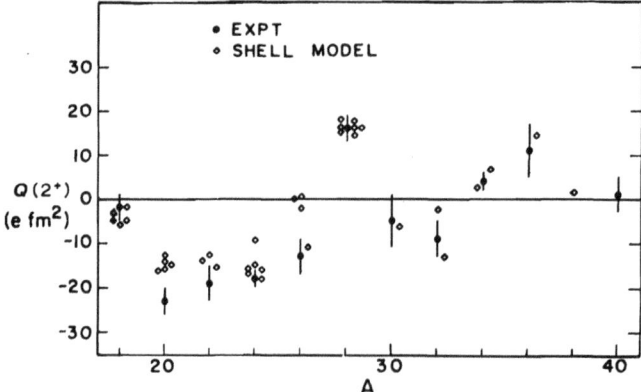

Fig. 6.5. Adopted experimental values for $Q(2^+)$ compared with the results of shell-model calculations.

moments of $I = 2^+$ states in even nuclei were a *small* negative value for ^{18}O and *large* negative values for ^{20}Ne, ^{22}Ne, and ^{24}Mg reflecting, presumably, strong stable prolate intrinsic states. A smaller negative value for ^{26}Mg and a large positive one for ^{28}Si indicated a rapid shape transition from prolate to oblate between $A = 24$ and 28.

The predicted negative value of $Q(2_1^+, {}^{18}\text{O})$, (and hence the prolate shape of its intrinsic state in the context of the rotational model) was construed as being the result of strong mixing between the $(d_{5/2}^2$ (61%) and $d_{5/2}^1 s_{1/2}^1$ (32%) configurations.

For ^{20}Ne, the model wavefunction exhibited extensive configuration mixing. The largest probability (21%) component has a $d_{5/2}^3 s_{1/2}^1$ configuration and the second largest (14%) is $d_{5/2}^4$. The value of Q is enhanced by a factor of 2.4 over the single component, $d_{5/2}^3 s_{1/2}^1$, estimate. As can be seen from Fig. 6.5 the experimental value of Q is significantly larger than the predicted value. It is also significantly larger than the 'dynamic moment' estimated from the measured $B(E2)$ of the $0_1^+ \rightarrow 2_1^+$ transition ($Q = -15.4 \pm 1.1 e$ fm^2). This alternative experimental value only slightly exceeds the predicted value. In sum, Carchidi *et al.*'s shell-model wavefunction (and all previous ones for that matter) underestimate the quoted experimental value of the quadrupole moment of ^{20}Ne (but the significance of the underestimation is uncertain).

For ^{24}Mg, the model wavefunction has for its largest (14%) component the $d_{5/2}^8$ configuration. This component yields neutron and proton E2 matrix elements whose values are smaller than those of the $d_{5/2}^3 s_{1/2}^1$ component which is the leading term in the ^{20}Ne(2_1^+) model wavefunction. The coherent amplification from configuration mixing is 20% greater than was obtained in ^{20}Ne and yields a factor of four enhance-

6.5 ELECTRIC QUADRUPOLE MOMENTS

ment of Q over the jj limit, $d_{5/2}^8$, value. The theoretical values of $Q(2_1^+, {}^{24}\text{Mg})$ are 10% smaller than the experimental value, but lie within its estimated uncertainty.

The leading component of the model wavefunction for ${}^{26}\text{Mg}(2_1^+)$ is $d_{5/2}^{10}$. This term is the proton–hole analogue of the $d_{5/2}^2$ neutron component of the ${}^{18}\text{O}$ wavefunction; ${}^{26}\text{Mg}$ would have a quadrupole moment of the same magnitude but opposite in sign, to that of ${}^{18}\text{Ne}$, the mirror state of ${}^{18}\text{O}$. The configuration mixing predicted for ${}^{26}\text{Mg}$ is, however, completely different from that calculated for ${}^{18}\text{O}$–${}^{18}\text{Ne}$. The predicted value of Q is enhanced by a factor of 2.4 over the $d_{5/2}^{-2}$ estimate by the effects of configuration mixing and yields a result consistent with experiment. Note that the observed decrease in the experimental moment of ${}^{26}\text{Mg}$ relative to that of ${}^{24}\text{Mg}$ is accurately reproduced by the shell-model results.

We now come to ${}^{28}\text{Si}$ where the shell-model wavefunction is dominated by the one-hole–one-particle $d_{5/2}^{11}s_{1/2}^1$ configuration (followed by a variety of three-hole–three-particle components). All of these components yield signs for the value of the one-body densities and the E2 matrix elements which are opposite to those obtained for ${}^{20}\text{Ne}$, ${}^{22}\text{Ne}$, ${}^{24}\text{Mg}$, and ${}^{26}\text{Mg}$. Thus in the language of the collective model, the wavefunction for ${}^{28}\text{Si}$ has emerged with an *oblate* deformation rather than the prolate shapes which characterized the lighter even–even nuclei.

In ${}^{30}\text{Si}$, the configuration of the leading (25%) component, $d_{5/2}^{12}s_{1/2}^1d_{3/2}^1$, reflects the 'shell effects' of both the approximate 'filling' of the $d_{5/2}$ subshell at $A=28$ and the impossibility of forming an $I=2$ state within the confines of the next, $j=1/2$, subshell. The next most important components have $d_{5/2}^{11}s_{1/2}^3$ (9%), $d_{5/2}^{11}s_{1/2}^1d_{3/2}^2$ (7%), etc., configurations. Configuration mixing results in a competition between the 'particle' (neutron) nature of the leading component and the 'hole' nature of the admixed components.

The model wavefunction for ${}^{32}\text{S}(2_1^+)$ has for its leading components the $d_{5/2}^{12}s_{1/2}^3d_{3/2}^1$ (21%) and $d_{5/2}^{12}s_{1/2}^1d_{3/2}^3$ (15%) configurations. This dominant 'particle-like' structure is reflected in the positive values of the one-body transition densities. The predicted value of $Q(2_1^+, {}^{32}\text{S})$ is definitely prolate and is enhanced by a factor of 2.6 over the value of the jj limit configuration $s_{1/2}^3d_{3/2}^1$. The search for an experimental value for $Q(2_1^+, {}^{32}\text{S})$ has been difficult, but recent work suggests a large negative value which is somewhat larger than the shell-model prediction.

Finally in ${}^{36}\text{Ar}$, the shell-model wavefunction has the $d_{3/2}^4$ configuration as its main component (34%), with the second largest component being $s_{1/2}^3d_{3/2}^5$ (20%). The $d_{3/2}^4$ configuration yields a zero E2 matrix element. The off-diagonal contributions between these two leading terms have the same sign as the diagonal contribution from the $s_{1/2}^3d_{3/2}^5$ component. The resultant value predicted for $Q(2^+, {}^{36}\text{Ar})$ is small and positive, agreeing in sign and magnitude with the only existing experimental value.

In conclusion, features of the observed region of strong prolate deformation extending from ^{20}Ne through to ^{26}Mg are well reproduced by the shell model, with the exception of the magnitudes of the ^{20}Ne and, to a lesser degree, ^{22}Ne moments. (The large experimental values for these two states may also reflect problems in the reorientation-effect experiments.) The sharp transition to oblate deformation at ^{28}Si is unambiguously predicted by the shell model as is the rapid transition back to prolate at ^{32}S. Finally, the predicted change from a large negative moment to a small positive moment in going from ^{32}S to ^{34}S is in agreement with the available experimental evidence.

In the case of odd-A nuclei, the overall trends of quadrupole data, as seen in Fig. 6.6 is also correctly reproduced by the shell-model calculations. One should note in particular the excellent agreement between theory and the three values determined in muonic atom measurements (^{23}Na, ^{25}Mg, ^{27}Al). Also the predictions for the 'true' single $d_{5/2}$ proton hole in ^{17}F and the 'quasi'-single $d_{5/2}$ proton hole state in ^{27}Al differ by a factor of 2, in rough agreement with experiment. The predictions for the quasi-single $d_{3/2}$ neutron state in ^{33}S agree with experiment and are likewise about twice as large in magnitude as predicted for the 'true' single $d_{3/2}$ neutron hole state in ^{39}Ca.

It would also be of great interest to test experimentally the trend predicted by the shell model for the ground-state quadrupole moments of

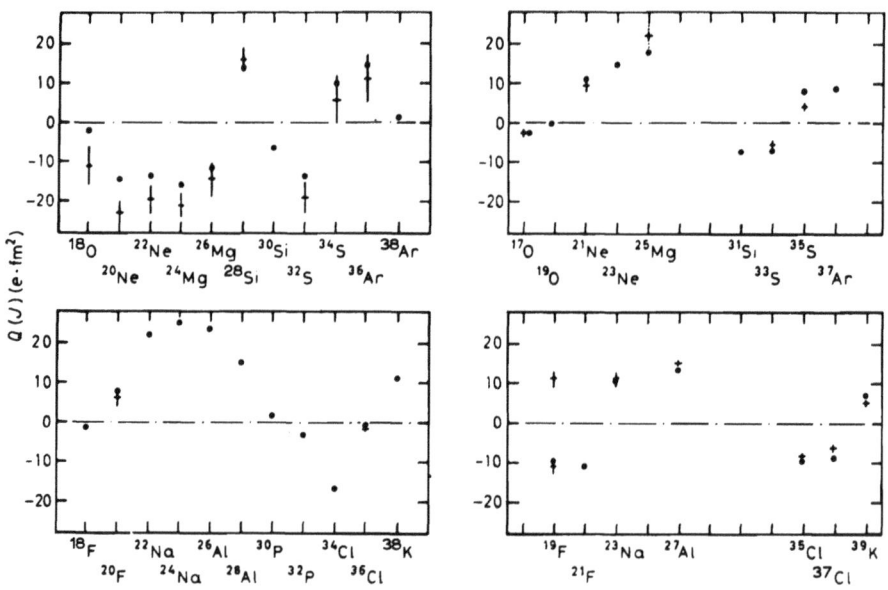

Fig. 6.6 Comparison of calculated and observed electric-quadrupole moments for the sd-shell nuclei.

6.5 ELECTRIC QUADRUPOLE MOMENTS

doubly odd nuclei. The only two measurements available so far[30] are in good agreement with theory as can be seen from Fig. 6.6.

Insight into the fashion by which the many components of realistic shell-model wavefunctions combine to produce coherent amplification of E2 matrix elements between certain states may be obtained by a decomposition due to Wildenthal of the internal constitution of E2 matrix elements. In Fig. 6.7, the summed value of the various *single-component* contributions $\alpha_i \alpha_j \langle \phi_i | E2 | \phi_j \rangle$ to the quadrupole matrix element of ^{20}Ne 2_1^+ are plotted as the number of basis vectors ϕ_i increases from one to the dimension limit. Note that any single component $\alpha_i \phi_i$ of the wavefunction makes a very small contribution to the total value. Also obvious is the fact that nearly all contributions to the sum have the same

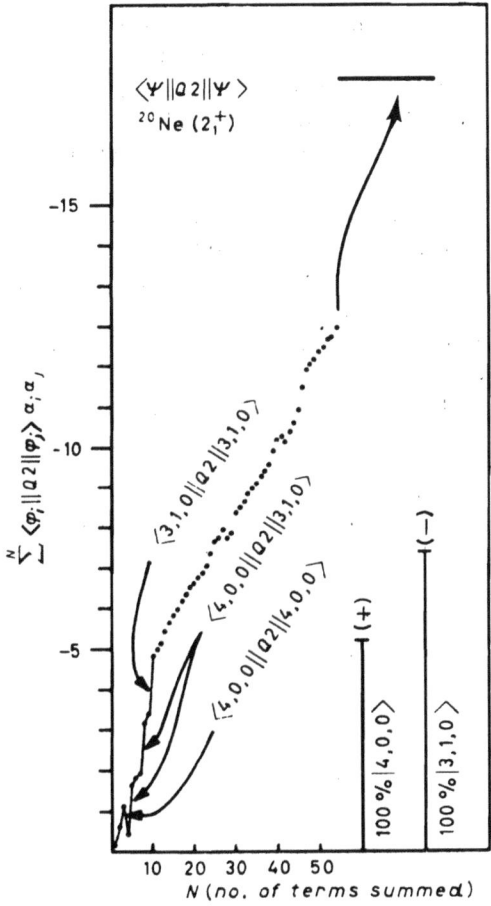

Fig. 6.7 Depiction of internal constitution of electric-quadrupole matrix element in jj coupling shell-model representation. (From Wildenthal.[29])

sign. The vertical bars to the right indicate the magnitudes of the quadrupole moments which would be obtained if various plausible configurations contained 100% of the wavefunction strength. It is seen that the value obtained with the realistic, highly mixed wavefunction is about 3 times larger than these typical one-component estimates. This amplification comes about via the many small in-phase contributions from off-diagonal, $\phi_i \neq \phi_j$, matrix elements. The non-zero values of these terms for the E2 operator produce, when the phases of the wavefunction are appropriate, coherent amplification as the many terms in the wavefunction expansion come to have significant amplitudes.

6.6 Shell-model calculations for the p-shell

With regard to the lighter nuclei of the p shell, van Hees et al.[32] have recently presented a new method which goes beyond the traditional shell model introduced by Cohen and Kurath[33] in that it uses nuclear moments in the fitting procedure.

The experimental data used in van Hees et al.'s fitting procedure consisted of 146 energy levels and all 27 experimentally known magnetic dipole moments and 12 electric quadrupole moments of the $A = 4$–16 nuclei. These 185 pieces of experimental data were used to determine a total of 28 parameters, i.e. four nucleon g-factors, two effective nucleon charges, 21 Talmi integrals, and $\hbar\omega$. The number of 21 Talmi integrals which enter into the calculations is a consequence of the restriction to include only the $0\hbar\omega$ and $1\hbar\omega$ configurations into the harmonic-oscillator shell-model space.

As is seen in Table 6.3, the agreement between calculated and observed quadrupole moments is much improved compared with the

Table 6.3 Electric quadrupole moments (e fm^2) of $A = 6$–16 nuclei

Nucleus	J^π	Expt.[32]	Cohen and Kurath[33]	Van Hees et al.[32]
^{8}Li	1^+	-0.06 ± 0.01	-0.73	-0.27
^{7}Li	$\frac{3}{2}^-$	-3.70 ± 0.08	-3.94	-3.92
^{8}Li	2^+	2.4 ± 0.2	2.42	2.59
^{9}Li	$\frac{3}{2}^-$	-3.3 ± 0.7	-4.00	-3.97
^{9}Be	$\frac{3}{2}^-$	5.3 ± 0.3	4.32	4.66
^{10}B	3^+	8.47 ± 0.06	8.49	8.48
^{11}B	$\frac{3}{2}^-$	4.07 ± 0.03	4.50	4.48
^{11}C	$\frac{3}{2}^-$	4.43	3.24	3.15
^{12}B	1^+	134 ± 0.14	1.89	1.62
^{12}C	2^+	6 ± 3	7.26	7.12
^{13}B	$\frac{3}{2}^-$	470 ± 0.46	3.97	4.00
^{14}N	1^+	1.56	1.02	1.34

earlier calculations by Cohen and Kurath.[33] The main differences between the present values and the ones obtained with the Cohen and Kurath wavefunctions occur for ^{6}Li and ^{14}N. The quadrupole moments of these $T=0$ nuclei are the ones which depend most sensitively on the interaction used. It is in particular the $T=0$ tensor part of the interaction which affects these two quadrupole moments rather strongly.

The proton and neutron effective charges ($e_p = 1.25$ and $e_n = 0.48$) are nearly the same in both the Cohen–Kurath and van Hees versions of the shell model and result from the fact that in both shell-model calculations the E2 operator and the Hamiltonian have large matrix elements between the $0\hbar\omega$ states *inside* and the $2\hbar\omega$ states *outside* the model space.

6.7 Expansion of the shell model and nuclear moments

As we have seen in the two previous sections, the shell model often provides great insight into the structure of nuclear moments. Yet despite the spectacular success of several calculations, a number of discrepancies between predicted and observed moments remain. And at present the question of whether these discrepancies should be attributed to restrictions of model space, to non-nucleonic effects, or to a combination of both remains unanswered.

Recently the question of the influence of the model space has been investigated in an ambitious manner by Wolters et al.[32] These authors have for the first time presented a systematic treatment of normal parity states in the $A = 4$–16 region by including the full $(0+2)\hbar\omega$ space. (The $2\hbar\omega$ part of this space contains all configurations where two particles can each make a jump of $1\hbar\omega$ between two neighbouring major shells or one particle can make a $2\hbar\omega$ jump.)

Apart from handling large matrices, Wolters et al. faced the problem of finding an appropriate effective nucleon–nucleon interaction. It turned out that 671 two-body matrix elements were involved in the evaluation of the effective Hamiltonian for the $(0+2)\hbar\omega$ model space but the assumption of translational invariance reduced this amount to 51. As far as the choice of an interaction was concerned, Wolters et al. attempted to reduce the number of parameters by using a completely phenomenological effective interaction which was assumed to have the following structure:

$$V(r_1 - r_2, p_1 - p_2, \sigma_1, \sigma_2, \tau_1, \tau_2) = \sum_{iT} U_i^{(T)}(r_1 - r_2) W_i P_T,$$

where the operator W_i denotes one of the four types of interaction, i.e. central spin–singlet, central spin–triplet, tensor, and two-body spin–orbit, and P_T is the isospin projection operator. The matrix elements of $U_i^{(T)}(r)$ between harmonic-oscillator radial wavefunctions were then written as a linear combination of a finite number of Talmi integrals. These were then considered as parameters to be determined empirically from a fit of calculated level energies to the corresponding experimental values. (The only other parameter entering in the calculations was $\hbar\omega$, which determined the kinetic energy part of the Hamiltonian as well as the scaling of the harmonic oscillator.)

In the calculation of nuclear spectra, Wolters et al. obtained a small but significant overall improvement of the calculated energies in the $(0+2)\hbar\omega$ model space with respect to those in the $0\hbar\omega$ model space. The most spectacular results, however, were obtained for quadrupole moments which are depicted in Fig. 6.8. There one observes a large difference between the corresponding theoretical values obtained in each space. In general the values in the $(0+2)\hbar\omega$ model space agree much better with the experimental data than those obtained in the $0\hbar\omega$ space. (The $0\hbar\omega$ calculations even produced a vanishing moment for ^{14}N.) Another important point concerned the effective charge needed in the least-squares fit to the experimental quadrupole moments. This charge was reduced by a factor of two due to the inclusion of the $2\hbar\omega$ space (from $e_n = 0.29$ to 0.14).

Thus one may conclude that the extension of the model space to include the $2\hbar\omega$ configurations as well as the $0\hbar\omega$ configurations without expanding the number of parameters leads to a significantly better description of quadrupole moments of the p-shell nuclei. Some future steps could require the determination of the Talmi integrals in the

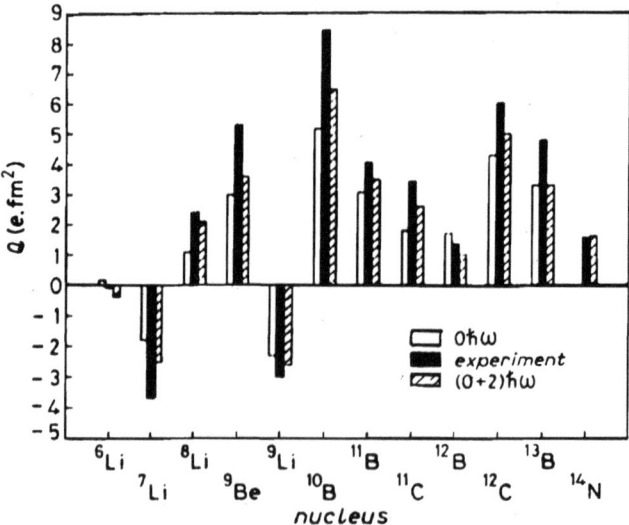

Fig. 6.8 Quadrupole moments of p-shell nuclei, calculated in the $0\hbar\omega$, $(0+2)\hbar\omega$ model spaces and compared with experiment. (From van Hees.[32])

$(0+2)\hbar\omega$ model space as well as applications of the same techniques to the beginning of the sd shell where we have encountered some difficulties in comparing shell-model results with experiment for the Ne and Mg isotopes.

6.8 The multiconfiguration Hartree–Fock model

Finally we come to discuss the band-mixing Hartree–Fock model considered for a long time a powerful competitor to the shell model in the investigation of sd shell nuclei.[34,35]

In this model the low-lying nuclear eigenstates are assumed spanned by states of good angular momentum I, projected from several low-lying Hartree–Fock intrinsic states. If the Hartree–Fock gap is small, single and double nucleon excitations above the Fermi level are included in the non-orthogonal basis in which the nuclear Hamiltonian is diagonalized to produce the eigenstates of the system. This technique not only offers a massive truncation of the conventional shell-model approach but offers more insight into the role of intrinsic shapes and their interaction which is of great importance in interpreting static quadrupole moments. Morrison[36] has recently used this technique to calculate the static quadrupole moments of even nuclei in the sd shell (see Fig. 6.9). His results, which generally compare well with those of the untruncated shell model,[28] offer some insight in the competition between the different

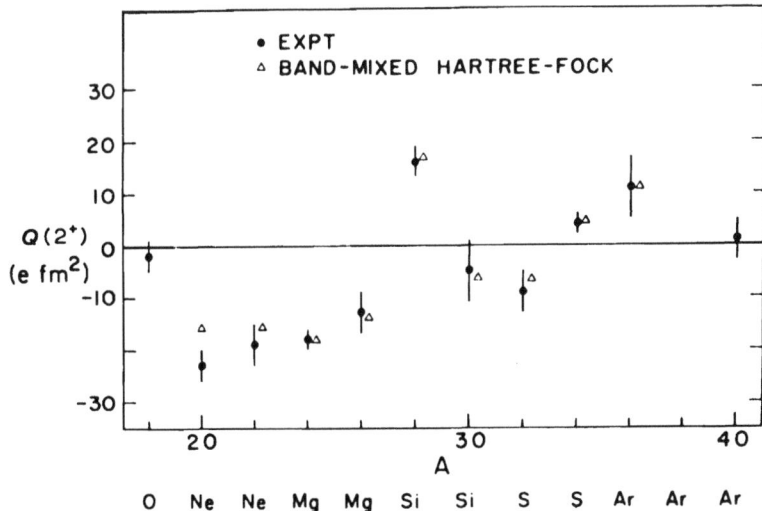

Fig. 6.9 Static quadrupole moments of the 2_1^+ states as a function of A, calculated in the band-mixing Hartree–Fock model and compared with experiment. (From Morrison.[36])

nuclear shapes, as seen through static quadrupole moments. A few cases deserve special attention.

1. ^{24}Mg. A single prolate intrinsic shape give a $Q(2^+)$ value of $Q = -18.3e$ fm^2 [37].

2. ^{26}Mg. Three low-lying Hartree–Fock solutions were found by Morrison (Fig. 6.10); one oblate (Ob) and two prolate (Pr$_{1,2}$) of which the oblate and one of the prolate solutions exhibited a small Hartree–Fock gap and were almost degenerate. The individual and mixed spectra (3D) are compared with experiment in Fig. 6.10. Overall agreement is poor and the calculated $Q(2^+)$ of $-9.9e$ fm^2 shows cancellation from the prolate-only value of $-19.9e$ fm^2, due to oblate mixing. Inclusion of single and double (to include pairing effects) nucleon excitation above the Fermi level increased the basis dimension to 9 and the resultant spectrum (9D) compares favourably both with experiment and a full (sd) basis calculation of Cole et al.[37] The calculated $Q(2_1^+)$ of -13.9 fm^2 is in excellent agreement with experiment.[38] Thus we see here that the inclusion of configuration interaction effects (through nuclear excitation) has reduced the degree of shape mixing and restored the prolate shape as dominant.

3. ^{28}Si. A single oblate Hartree–Fock state gives $Q(2_1^+) = +16.7e$ fm^2

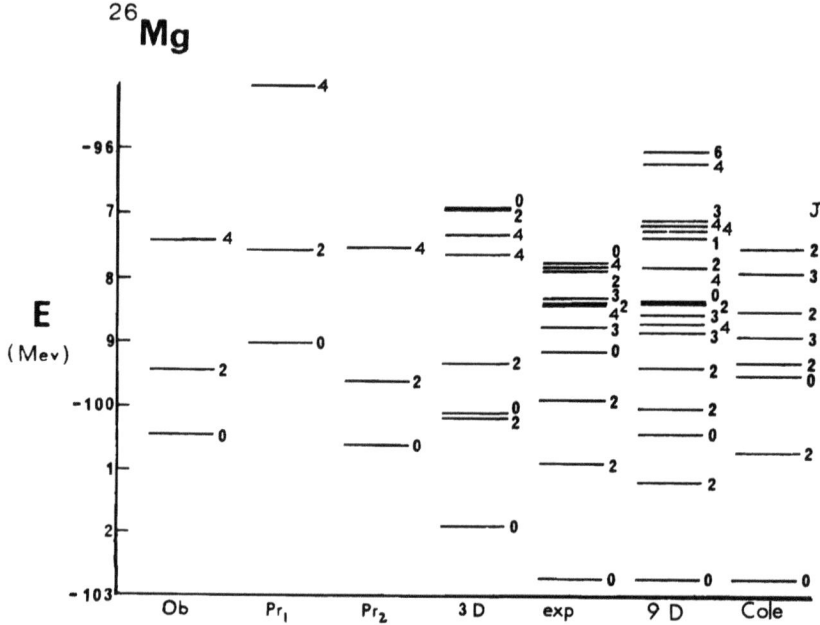

Fig. 6.10 Experimental and calculated spectrum for ^{26}Mg.[36]

in good agreement with the full (sd) basis shell-model result $(+17.6e\ \text{fm}^2)$.[37]

4. ^{32}S. Three Hartree–Fock solutions were found, one spherical, one oblate, and a prolate solution which gave the lowest 2^+ state after projection. After mixing, the calculated $Q(2_1^+)$ was $-6.9e\ \text{fm}^2$, reflecting the small deformation of the intrinsic states. Shell-model calculations were performed using the same interaction allowing 1, 2, and 3 holes in the $d_{5/2}$ orbit giving $Q(2_1^+)$ of -8.1, -9.4, and $-10.2e\ \text{fm}^2$, respectively. All Hartree–Fock solutions had large gaps, therefore the discrepancy is likely to be due to large configuration interaction effects as shown by the sensitivity to the shell-model space. Both methods unambiguously support a prolate shape and are in reasonable agreement with experiment.[38]

5. ^{36}Ar. One oblate and one prolate solution were found with the oblate one dominating the 0_1^+–2_1^+ transition after projection and mixing. The calculated values of $Q(2_1^+)$ were $+12.2e\ \text{fm}^2$ (oblate only) and $+10.8e\ \text{fm}^2$ (after mixing) which are to be compared with the full shell-model value of $Q = +7.8e\ \text{fm}^2$ and with experiment.[38]

In general, the band-mixing Hartree–Fock model suggests that variation of $Q(2_1^+)$ as A increases is due to transition from prolate to oblate intrinsic shapes as the $d_{5/2}$ shell fills preferentially due to the two-nucleon interaction. This effect can be viewed as the start of a competition between shape and shell effects which is exacerbated in the second half of the shell because of the 'stability' of spherical shapes in ^{28,30}Si. The transition nucleus ^{26}Mg exhibits pairing effects forestalling the prolate–oblate transition which becomes complete in ^{28}Si.

6.9 Summary

In this first chapter devoted specifically to quadrupole moments, we have focused on E2 transitions and moments as tests of microscopic models. We have discussed how the effective charge concept arose in the shell model to compensate for truncation in model space and that the concept itself had a long history. Indeed, the inability to explain effective charges in closed-shell-plus-one nuclei in perturbation theory still poses a vexing problem to theorists today. On a more practical level we have seen that the shell model is, however, capable of impressive results in reproducing correctly the magnitudes and signs of $Q(2^+)$ values of even nuclei in the sd shell. Some interesting avenues are also now being explored beyond the conventional limits of the shell model and the question of the influence of the model space on quadrupole moments is one of them. In that context, the recent extension of the shell-model space by Wolters et al. to include $2\hbar\omega$ excitations proved most rewarding in improving the results of $Q(2^+)$ values in p-shell nuclei.

Finally, we examined in detail the results and predictive power of the band-mixing Hartree–Fock model, for long considered a simpler competitor to the shell model. The results were particularly transparent in the study of the variation of $Q(2^+)$ in the sd shell where the prolate to oblate transition was interpreted as due to the preferential filling of the $d_{5/2}$ shell under the influence of the two-nucleon interaction. This effect was then viewed as the start of a competition between *shape* and *shell* effects which will later dominate the search for stability among the different Hartree–Fock solutions in the upper sd shell. Finally, this chapter prepared the ground to launch a more extensive study of quadrupole moments in the collective model, a context which will become necessary for the study of heavier nuclei.

6.10 References

1. Bohr, A. and Mottelson, B. R. (1975). *Nuclear structure,* Vol. I. Benjamin, New York.
2. de Shalit, A. and Talmi, I. (1963). *Nuclear shell theory.* Academic Press, New York.
3. Lawson, R. D. (1980). *Theory of the nuclear shell model.* Oxford University Press.
4. Endt, P. M. and Van der Leun, C. (1973). *Nucl. Phys.,* **A214,** 1.
5. Pal, M. K. (1983). *Theory of nuclear structure p.* 46. Scientific and Academic Editions, New York.
6. Yoshida, S. and Zamick, L. (1972). *Ann. Rev. Nucl. Sci.,* **22,** 121; Towsley, C. W., Cline, D., and Horoshko, R. N., (1972). *Phys. Rev. Lett.,* **28,** 368.
7. Arima, A. and Horie, H. (1954). *Prog. Theor. Phys.,* **11,** 509.
8. Federman, P. and Zamick, L. (1969). *Phys. Rev.,* **177,** 1534.
9. Siegel, S. and Zamick, L. (1969). *Phys. Lett.,* **28B,** 450; (1970). *Nucl. Phys.,* **A145,** 89.
10. Ellis, P. J. and Siegel, S. (1971). *Phys. Lett.,* **34B,** 177.
11. Towner, I. S. (1977). *A shell model description of light nuclei.* Clarendon, Oxford.
12. Ellis, P. J. and Mavromatis, H. A. (1971). *Nucl. Phys.,* **A175,** 309.
13. Alexander, T. K., Castel, B., and Towner, I. S. (1985). *Nucl. Phys.,* **A445,** 189.
14. Brown, G. E. and Bolsterli, M. (1959). *Phys. Rev. Lett.,* **3,** 472.
15. Vautherin, D. and Brink, D. M. (1972). *Phys. Rev.,* **C5,** 626.
16. Brown, B. A., Wildenthal, B. H., Chung, W., Massen, S. E., Bernas, M., Bernstein, A. M., Miskimen, R., Brown, V. R., and Madsen, V. A. (1982). *Phys. Rev.,* **C26,** 2247.
17. Abbas, A. and Zamick, L. (1980). *Phys. Rev.,* **C21,** 738; (1980). *Phys. Rev.,* **C22,** 1755.
18. Alexander, T. K., Ball, G. C., Forster, J. S., Davies, W. G., Mitchell, I. V., and Mak, H.-B. (1982). *Phys. Rev. Lett.,* **49,** 438; Endt, P. M. (1979). *At. Data Nucl. Data Tables,* **23,** 3.
19. Astner, G., *et al.* (1972). *Nucl. Phys.,* **A182,** 219.
20. Blomquist, J. (1973). *J. Phys. Soc. Japan Suppl.,* **34,** 223.

21. Wildenthal, B. H. (1983). *Prog. Part. Nucl. Phys.*, **11,** 5.
22. de Shalit, A. and Feshbach, H. (1974). *Theoretical nuclear physics*, Vol. 1. John Wiley, New York.
23. Wildenthal, B. H., Halbert, E. C., McGrory, J. B., and Kuo, T. T. S. (1971). *Phys. Rev.*, **C4,** 1966.
24. Krewald, S., *et al.* (1974). *Nucl. Phys.*, **A228,** 524.
25. Van Hienen, J. F. A., Glaudemans, P. W. M., and Van Lidth de Jeude, J. (1974). *Nucl. Phys.*, **A225,** 119.
26. Singhal, R. P., Kelvin, D., Knight, E. A., Watt, A., and Whitehead, R. R. (1979). *Nucl. Phys.*, **A323,** 91.
27. Brown, B. A. and Wildenthal, B. H. (1983). *Phys. Rev.*, **C28,** 2397.
28. Carchidi, M., Wildenthal, B. H., and Brown, B. A. (1986). *Phys. Rev.*, **C34,** 2280.
29. Wildenthal, B. H. (1977). In *elementary modes of excitations in nuclei*, Enrico Fermi School of Physics, Varenna (ed. A. Bohr and R. A. Broglia), p. 453. Amsterdam, North Holland.
30. Horie, H. and Sugimoto, K. (ed.) (1973). *Nuclear moments and nuclear structure*, Physical Society of Japan.
31. Wildenthal, B. H. (1978). *Nucleonika*, **23,** 459.
32. van Hees, A. G. M., Wolters, A. A., and Glaudemans, P. W. M. (1988). *Nucl. Phys.*, **A476,** 61; Wolters, A. A., van Hees, A. G. M., and Glaudemans, P. W. M. (1988). *Europhys. Lett.*, **5,** 7.
33. Cohen, S. and Kurath, D. (1965). *Nucl. Phys.*, **73,** 1.
34. Goeke, K., Müther, H., and Faessler, A. (1973). *Nucl. Phys.*, **A201,** 49.
35. Lee, H. C. and Cusson, R. Y. (1972). *Ann. Phys. (N.Y.)*, **72,** 353.
36. Morrison, I. (1980). *Phys. Lett.*, **91B,** 4.
37. Cole, B. J., Watt, A., and Whitehead, R. R. (1975). *J. Phys. G: Nucl. Phys.*, **1,** 213.
38. Christy, A. and Haüsser, O. (1972). *At. Data Nucl. Data Tables*, **11,** 281.
39. Schwalm, D., Warburton, E. K., and Olness, J. W. (1977). *Nucl. Phys.*, **A293,** 425.

7
QUADRUPOLE MOMENTS AND COLLECTIVE MODEL THEORIES

Large quadrupole moments and E2 transition rates, characteristic of deformed systems, are often well described using simple collective models. These models which were actually first devised for this purpose have also shown some success in describing vibrational excitations and in particular anharmonic excitations. We will examine them in this chapter in an evolution having a dual character, from the phenomenological to the microscopic and from the vibrational to the rotational. We thus first concentrate on collective excitations in spherical nuclei.

7.1 The harmonic vibration model

In Section 5.5, we briefly introduced vibrational models principally to introduce β and γ vibration bands as common features in many nuclear spectra. Here we discuss the vibrational model in more detail. Recall that the Hamiltonian is written in terms of deformation amplitudes $\alpha_{\lambda\mu}$ as

$$H = \tfrac{1}{2}\sum_\mu [D_\lambda |\dot\alpha_{\lambda\mu}|^2 + C_\lambda |\alpha_{\lambda\mu}|^2], \tag{7.1}$$

where D_λ and C_λ are the mass parameter and the restoring force parameter respectively. If the canonical variable to $\alpha_{\lambda\mu}$ is denoted by

$$\pi_{\lambda\mu} = \frac{\partial H}{\partial \dot\alpha_{\lambda\mu}} = D_\lambda \dot\alpha^*_{\lambda\mu} = D_\lambda(-)^\mu \dot\alpha_{\lambda-\mu} \tag{7.2}$$

then the Hamiltonian becomes

$$H = \tfrac{1}{2}\sum_\mu [1/D_\lambda |\pi_{\lambda\mu}|^2 + C_\lambda |\alpha_{\lambda\mu}|^2]. \tag{7.3}$$

The Hamiltonian can then be quantized by imposing the commutation relation

$$[\alpha_{\lambda\mu}, \pi_{\lambda\mu'}] = i\hbar\delta_{\mu\mu'}.$$

The Hamiltonian is expressed in a diagonal form by introducing boson

7.1 THE HARMONIC VIBRATION MODEL

operators

$$b^\dagger_{\lambda\mu} = \left(\frac{\omega_\lambda D_\lambda}{2\hbar}\right)^{1/2}\left[\alpha_{\lambda\mu} - \frac{i}{\omega_\lambda D_\lambda}(-)^\mu \pi_{\lambda-\mu}\right]$$
$$b_{\lambda\mu} = \left(\frac{\omega_\lambda D_\lambda}{2\hbar}\right)^{1/2}\left[(-)^\mu\alpha_{\lambda-\mu} + \frac{i}{\omega_\lambda D_\lambda}\pi_{\lambda\mu}\right] \qquad (7.4)$$

where
$$\hbar\omega_\lambda = \hbar C_\lambda/D_\lambda$$

and the commutation relation is given by

$$[b_{\lambda\mu}, b^\dagger_{\lambda\mu'}] = \delta_{\mu\mu'}.$$

As we saw earlier (cf. eqn (5.47)) the vibrational Hamiltonian can then be simply written as

$$H = \hbar\omega_\lambda \sum_\mu (n_{\lambda\mu} + \tfrac{1}{2}). \qquad (7.5)$$

In the harmonic vibration model the electric transition operator is written in terms of the charge density $\rho(r)$ as

$$T(E\lambda, \mu) = e(4\pi/2\lambda + 1)^{1/2} \int \rho(r) r^\lambda Y_{\lambda\mu}(\theta, \phi)\, d^3r, \qquad (7.6)$$

with the charge density normalized as

$$Z = 4\pi \int \rho(r) r^2\, dr.$$

Integrating now by parts and exploiting the fact that in the harmonic approximation the Hamiltonian and the electric transition operator are constructed out of the same quantities, one obtains[1]

$$T(E\lambda, \mu) = \frac{2\lambda + 1}{4\pi} Ze\langle r^\lambda \rangle \alpha^*_{\lambda\mu} + O(\alpha^2)$$
$$= \frac{2\lambda + 1}{4\pi} Ze\langle r^\lambda \rangle \left(\frac{\hbar}{2\omega_\lambda D_\lambda}\right)^{1/2} [b_{\lambda\mu} + (-)^\mu b^\dagger_{\lambda-\mu}]. \qquad (7.7)$$

This operator must change the number of quanta by one, so the quadrupole moment of the one-phonon quadrupole vibrational state ($\lambda = 2$, $n_\mu = 1$) vanishes

$$Q(2^+) = 0. \qquad (7.8)$$

The E2 transition rate from a one- to a zero-phonon state is simply written as

$$B(E2: 2^+ \to 0^+) = \left[\frac{5}{4\pi} Ze\langle r^2 \rangle\right]^2 \frac{\hbar}{2\omega_2 D_2}. \qquad (7.9)$$

This model is of course far too simple to describe real nuclei. It was soon realized that anharmonic effects were necessary to account for experimental data, like the energy splitting of the two-phonon states, cross-over transitions, and large $Q(2^+)$ moments. It was in fact the discovery by de Boer et al.[2] of large quadrupole moments in the Cd isotopes which prompted numerous theorists to develop anharmonic models.

7.2 The anharmonic vibration model

The first exploratory models of anharmonic vibrations were based on adding to the vibrational Hamiltonian several low-rank terms consistent with spherical symmetry and time reversal.[3-5] In the Hamiltonian written in terms of phonon operator $b^\dagger_\mu$ and b_μ, the following terms were empirically added

$$[b^\dagger b^\dagger]_0, [bb]_0, [bbb]_0, [b^\dagger b^\dagger b]_0, \ldots$$

This indicated that the phonon number is no longer a good quantum number. The wavefunctions for the ground state and 2^+ states were written as

$$|0_0\rangle = a_0|0\rangle + a_1[b^\dagger b^\dagger]_0|0\rangle + \ldots,$$
$$|2_1\rangle = a_1^1 b^\dagger_\mu|0\rangle + a_2^1[b^\dagger b^\dagger]_{2\mu}|0\rangle + \ldots, \qquad (7.10)$$
$$|2_2\rangle = a_1^2 b^\dagger_\mu|0\rangle + a_2^2[b^\dagger b^\dagger]_{2\mu}|0\rangle + \ldots.$$

Extensive investigations were conducted by Sørensen and others[3-5] in order to fit the ^{114}Cd data using the following Hamiltonian

$$H = w^{21}[b^\dagger b]_0 + w^{30}[b^\dagger b^\dagger b^\dagger]_0 + w^{31}[b^\dagger b^\dagger b]_0$$
$$+ w^{40}[b^\dagger b^\dagger]_0[b^\dagger b^\dagger]_0 + \sum_I w^{41}_I[[b^\dagger b^\dagger]_I[b^\dagger b]_I]_0$$
$$+ \sum_I w^{42}_I[[b^\dagger b^\dagger]_I[bb]_I]_0 + \text{h.c.}$$

Although Sørensen included up to seven phonons in the construction of the vibrational states wavefunction, it soon became apparent that this method could not easily provide simultaneous fit to the quadrupole moment, the energy levels of the two phonon states, and their decay. Later calculations[6] showed that adding fourth-order terms modified the energy spectrum completely and also had strong effects on quadrupole moments of 2^+ states (see Fig. 7.1). This suggested that fairly extensive calculations were in general necessary, well beyond the apparent simplicity implied by this phenomenological model.

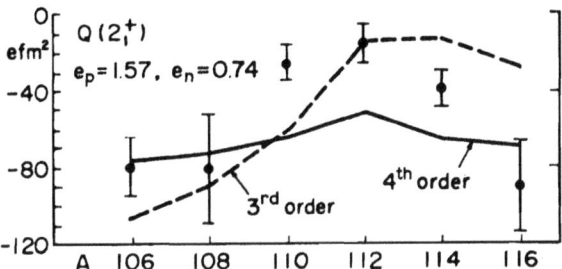

Fig. 7.1 Third- and fourth-order contributions to the static quadrupole moment of $^{106-116}$Cd nuclei calculated in the anharmonic vibration model of Sørensen.[3]

7.3 Anharmonicity and the particle-vibration coupling model

A good example of a slightly more complex model which still retains a degree of intuitive appeal is afforded by the particle-vibration model. We describe here a technique used to describe anharmonic features in vibrational nuclei by incorporating two-particle excitations coupled to surface vibrations. The following Hamiltonian has been adopted by most authors[7-15]

$$H = H_{SM} + \tfrac{1}{2}C_2 \sum_{\mu=-2}^{2} |\alpha_{2\mu}|^2 + \tfrac{1}{2}D_2 \sum_{\mu=-2}^{2} |\dot{\alpha}_{2\mu}|^2 + H_{pair}$$
$$\mp k \sum_{i,\mu} \alpha_{2\mu} Y_{2\mu}^{\mu *}(\theta_i, \phi_i). \quad (7.12)$$

The first term represents the motion of the particles or holes in a shell-model potential. The next two terms represent the free quadrupole vibrations, and the last two terms are the residual interaction between the shell-model particles and the particle–surface interaction. The form of this interaction has its origin in the interaction between an average quadrupole field and the motion of a particle in this average field.[8] In the above k denotes the strength of the interaction, θ_i and ϕ_i are the two particle coordinates and the + and − signs refer to the particle–surface or hole–surface interaction respectively.[11]

The operators for the electric moments and transitions consist of two parts, one describing the component associated with particles (or holes) and the other the one associated with the vibrator[11]

$$T(E2) = \sum_{i=1}^{k} (\pm) e_{eff} r_i^2 Y_{2\mu}(\theta_i, \phi_i) + \frac{3}{4\pi} Ze^2 R_0^2 \alpha_{2\mu}. \quad (7.13)$$

In one of the first applications of this model to studies of anharmonicity, Alaga[11] applied the model to ^{114}Cd as two proton holes coupled to the surface vibration of a ^{116}Sn core. The diagonalization of the Hamiltonian

(7.12) showed that the low-lying states of ^{114}Cd contain a very large contribution of $g_{9/2}^{-2}$ configuration. It was 85% for the ground state, 90% for the 2_1 state, and 61% for the 2_2 state. The ground state itself had 72%, 20%, 7%, and 1% of the zero-, one-, two-, and three-phonon states, respectively. Thus the vibration-like wavefunctions resulting from the coupling of holes to the vibrations contained more structure and more flexibility than Sørensen's anharmonic model discussed in the previous section. If we now consider the electromagnetic properties of the coupled system, it is clear that they will appear as a result of the competition between the holes and the vibrator.

Assuming a proton effective charge of $e_p = 2e$, Alaga calculated the $g_{9/2}^{-2}$ contribution to the quadrupole moment to be $Q(2^+) = -0.10e$ b while the contribution of the vibrator part was $Q(2^+) = -0.22e$ b with a total which thus compared well with Kleinfeld's data of $Q(2^+) = -0.36 \pm 0.5e$ b.[16]

From this we obtain an idea of how the quadrupole moment is created in this model. Recalling that $Q(2^+) = 0$ in the harmonic vibrator limit, we see that the quadrupole moment of the two proton holes seems to act as a generator of the additional quadrupole moment of the vibrator. In all cases considered by Alaga,[11] the sign of the quadrupole moment of the vibrator was the same as that of the two-particle contributions. Of course, one is reminded by looking at the form of the Hamiltonian that $Q(2^+)$ still depends on the coupling strength k and the pairing interaction. Systematic calculations by Lopac[12] showed that for ^{114}Cd and the $g_{9/2}^{-2}$ proton configuration alone, the variation of the pairing constant (within reasonable limits) does not affect the essential properties of the quadrupole moment: it is negative and growing in absolute value with increasing coupling strength. Figure 7.2 provides a good illustration of these results for five different values of the pairing strength G.

In conclusion, the simplicity of these calculations and their rather intuitive description of anharmonic effects seem to justify the continuing popularity of the particle-vibration model.[13-15] In the next section we turn to the study of a higher vibrational mode and to a more stringent test of this model.

7.4 Quadrupole moments of octupole vibrational states: the 3⁻ state in ^{208}Pb

The measurement of the quadrupole moment of the lowest 3⁻ state in ^{208}Pb has presented an interesting challenge over the years. At first the large quadrupole moment ($Q(3^-) = -1.0 \pm 0.5\,b$) obtained by Barnett and Phillips[17] could not be explained by any available model. We briefly discuss here the theoretical implications of a large quadrupole moment for the octupole state.

7.4 MOMENTS OF OCTUPOLE VIBRATIONAL STATES   217

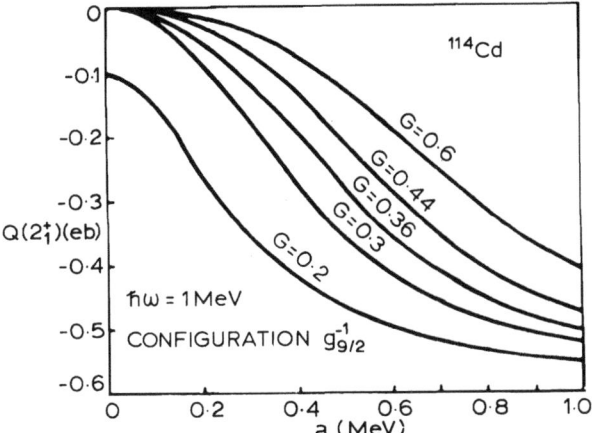

Fig. 7.2 Dependence of the quadrupole moment of even Cd isotopes, calculated by coupling two holes in the $g_{9/2}^{-1}$ configuration to the harmonic vibrator, on the coupling constant for five different values of the pairing strength G. (From Lopac.[12])

In a microscopic analysis, the 3^- state in ^{208}Pb is known to be expressed mainly by a superposition of many 1p–1h configurations, none of which has a predominant component, corresponding to the fact that the state is highly collective. The ^{208}Pb(p, p')^{208}Pb reaction, for instance, exciting the 3^- state shows resonances at the analogue states in ^{209}Bi of all the single-particle neutron states in ^{209}Pb (namely, $1g_{9/2}$, $0i_{11/2}$, $0j_{15/2}$, $2d_{5/2}$, $3s_{1/2}$, $1g_{7/2}$, and $2d_{3/2}$). Thus, one can see that the 3^- state in ^{208}Pb has in fact all these neutron single-particle components in appreciable amounts.

The static quadrupole moment of the 3^- one-phonon state was studied by Hamamoto[18,19] using the particle-vibration model described earlier. In her calculation, Hamamoto considered the second-order contributions to $Q(3^-)$ arising from decomposing the $|3^-\rangle$ phonon state as p–h states coupled to 2^+ vibrations, i.e.

$$|3^-\rangle = |p \otimes h\rangle + |(p \otimes 2^+) \otimes h\rangle + |p \otimes (h \otimes 2^+)\rangle + |(ph) \otimes 2^+\rangle + \ldots \quad (7.14)$$

We see that in lowest order, the 3^- state is just a linear combination of particle–hole states. Second-order graphs will correspond to the p–h states coupled to the 2^+ vibration (last three terms in Eqn 7.14) and third order will include components like $|(ph) \otimes (2^+ \otimes 2^+)\rangle$, etc. The various polarization graphs, which are associated with the second-order expansion of the 3^- state are illustrated in Fig. 7.3, where the dash-dotted lines express possible quadrupole phonons of the ^{208}Pb core. We see from these graphs that all the vertex functions are the same. Thus, by

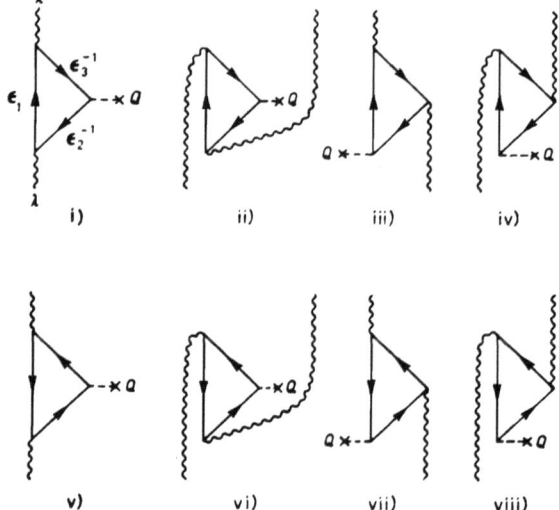

Fig. 7.3 Graphs which contribute to the quadrupole moment of the one-phonon state $\lambda \equiv |3^-\rangle$. These graphs result from all possible recouplings of the (ph) states to 2^+ vibrations to form the octupole state λ.

evaluating the energy denominators

$$\frac{1}{[\hbar\omega_\lambda - (\varepsilon_1 - \varepsilon_2)][\hbar\omega_\lambda - \hbar\omega_2 - (\varepsilon_1 - \varepsilon_3)](-\hbar\omega_2)}$$
$$+ \frac{1}{[\hbar\omega_\lambda - (\varepsilon_1 - \varepsilon_2)][\hbar\omega_\lambda - \hbar\omega_2 - (\varepsilon_1 - \varepsilon_3)][\hbar\omega_\lambda - (\varepsilon_1 - \varepsilon_3)]}$$
$$+ \frac{1}{[\hbar\omega_\lambda(\varepsilon_1 - \varepsilon_2)][\hbar\omega_\lambda - \hbar\omega_2 - (\varepsilon_1 - \varepsilon_3)][\hbar\omega_\lambda - (\varepsilon_1 - \varepsilon_3)]}$$
$$+ \frac{1}{(-\hbar\omega_2)[\hbar\omega_\lambda - \hbar\omega_2 - (\varepsilon_1 - \varepsilon_2)]\hbar\omega_\lambda - (\varepsilon_1 - \varepsilon_3)]}$$
$$= \frac{-2}{[\hbar\omega_\lambda - (\varepsilon_1 - \varepsilon_2)][\hbar\omega_\lambda - (\varepsilon_1 - \varepsilon_3)](\hbar\omega_2)}, \quad (7.15)$$

it is explicitly seen that these graphs yield a contribution equal to the static polarizability multiplied by the contribution of the graph (i) in Fig. 7.3. Therefore, it is sufficient to consider only graph (i) in Fig. 7.3 if the vertex Q contains the static polarization charge of the relevant particle in addition to the bare charge. For this information Hamamoto looked at the effective charges derived from the observed quadrupole moments of odd-A nuclei in the lead region.

Hamamoto also showed that the main contribution to $Q(3^-)$ comes from the graphs (i) and (v) of Fig. 7.3 (as can be immediately seen by considering the magnitudes of the energy denominators). The absolute magnitude of the contribution of the former, which has a plus sign in most cases, is smaller than that of the latter, which has a minus sign in most cases, since the expectation values of r^2 in hole states are, roughly speaking, smaller than those in particle states. Consequently, the total calculated quadrupole moment of the octupole phonon has a minus sign. By using $e_p = 1.5$ and $e_n = 1$ derived from neighbouring nuclei, one gets the calculated quadrupole moment of -0.13 b. i.e. nearly one order of magnitude smaller than the experimental value known at the time.

Guidetti et al.[20] went then a step further and examined a variety of particle–hole models. Having concluded that particle–hole space either in TDA or RPA is incapable of explaining the large observed quadrupole moment, they considered the inclusion of particle–hole states *coupled* to a 2^+ phonon, i.e. up to and including second-order graphs, thus adopting the same space as considered previously by Hamamoto. The problem was thus to find the state

$$|\lambda\rangle = \sum_a x_a |\alpha\rangle, \qquad (7.16)$$

where λ represents the octupole state, and x_a are amplitudes to be determined in the given basis $\{\alpha\}$ of p–h states coupled to 2^+ phonons, such that the quadrupole matrix element,

$$q = \langle\lambda\| Q \|\lambda\rangle, \qquad (7.17)$$

was an extremum for a given octupole transition matrix element,

$$\langle\lambda\| Q3 \|0\rangle = C, \qquad (7.18)$$

subject to normalization

$$\langle\lambda | \lambda\rangle = \sum_a x_a^2 = 1.$$

Thus Guidetti et al. had to solve the variational equation

$$\delta[\langle\lambda\| Q \|\lambda\rangle - \nu\langle\lambda | \lambda\rangle - \mu\langle\lambda\| Q3 \|0\rangle] = 0, \qquad (7.19)$$

with ν and μ acting as Lagrange multipliers.

Now that second-order graphs were included, various phonon–phonon interaction matrix elements became necessary to complete the variational analysis. These of course depended on matrix elements such as $\langle 2^+\| Q |0^+\rangle$ and $\langle 2^+\| Q \|2^+\rangle$. The former could be obtained from the $B(E2)$ value,[21] which yielded

$$\langle 2^+\| Q \|0^+\rangle = \pm 1.738 e \text{ b}.$$

Unfortunately the static quadrupole moment of the 2^+ state had not been measured. Two extreme estimates of $\langle 2^+\| Q \|2^+\rangle$ were thus considered. The simple vibrational model for the 2^+ state which would predict

$$\langle 2^+\| Q \|2^+\rangle = 0,$$

and the rotational model value, which relates the static moment to the $B(E2)$ transition, i.e.

$$\langle 2^+ \| Q \| 2^+ \rangle = \mp 7.264 e \text{ b}.$$

Figure 7.4 shows the extremum curves obtained assuming the vibrational and rotational values for $\langle 2^+ \| Q \| 2^+ \rangle$. As can be seen from these, the only phonon–phonon interaction scenario consistent with Barnett and Phillips' $Q(3^-)$ result requires a value of $\langle 2^+ \| Q \| 2^+ \rangle$ close to the rotational limit and quite inconsistent with the vibrational hypothesis required among other things to explain the splitting of the $(h_{9/2} \times 3^-)$ septuplet in ^{209}Bi.[19] In conclusion, this calculation showed that using the *same space* as in Hamamoto's core–particle model but performing a *variational calculation* instead did not change drastically the conclusion, i.e. the inability of a nuclear structure calculation to yield a large $Q(3^-)$ value while remaining consistent with the vibrational nature of the other observables in the Pb region.

In 1977, Joye *et al.*,[22] at the Australian National University remeasured the 3^- state quadrupole moment by Coulomb excitation and found $Q(3^-) = -0.42 \pm 0.32 e$ b in final agreement with most calculations (see Fig. 7.5).

Few problems in nuclear structure physics had aroused as much attention as this dramatic discrepancy between experiment and theory.

7.5 Anharmonic vibrations and the time-dependent Hartree–Fock theory

We will now conclude our study of static quadrupole moments as tests of vibrational models by presenting a microscopic description of $Q(2^+)$ moments using time-dependent Hartree–Fock equations. This method has been used extensively in a variety of many-body problems.[25–28] Its application to vibrational excitations has been shown to be particularly promising in extensive works by Meyer[25] and Speth.[26]

Meyer and Speth have shown that a direct treatment of anharmonicity is possible by taking into account all second-order contributions to the single-particle density matrix following from time-dependent Hartree–Fock (TDHF) equations. Much attention had been paid to the fact that the random-phase approximation fails to give large $Q(2^+)$ moments because interaction processes describing the scattering of quasi-particles by quasi-particle pairs were neglected.[1] The complete second-order treatment includes these processes and yields a strong enhancement of the quadrupole moment by a mechanism which may be described as a second-order response of the system to the first-order harmonic vibration.

7.5 ANHARMONIC VIBRATIONS

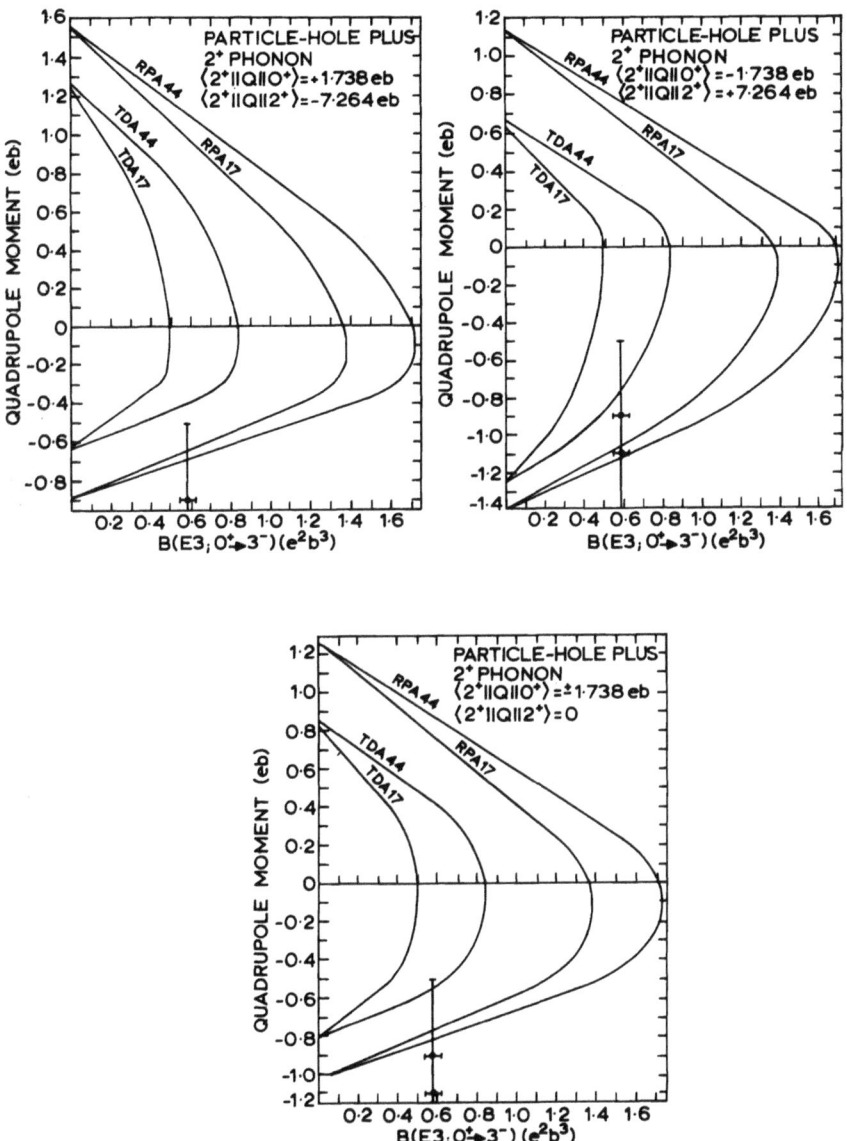

Fig. 7.4 Extremum curves for the particle–hole plus 2^+ phonon model. The three sets of curves correspond to the different assumed quadrupole matrix elements of the 2^+ states shown in the figures. The curves are also labelled by the dimensions of the particle–hole space and according to whether it is treated in TDA or RPA. Only the curves on the right-hand sides indicate a possible consistency between theory and experiment, but these correspond to a rotational model value for $\langle 2^+ \| Q \| 2^+ \rangle$ which contradicts other observables in neighbouring odd-A nuclei. (From Guidetti et al.[20])

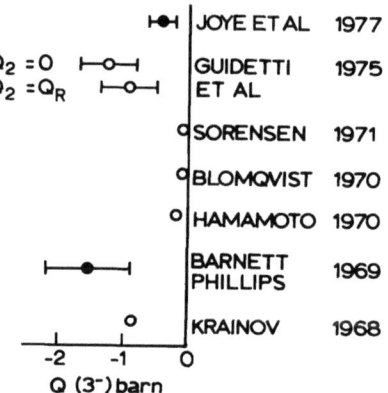

Fig. 7.5 Comparison of experimental and theoretical values for $Q(3^-)$ displayed in chronological order. Krainov,[24] Hamamoto[23] and Blomquist are all particle–core coupling calculations, while Sørensen and Guidetti et al. have used anharmonic and RPA type calculations respectively.

According to the semiclassical features of the TDHF method[29] the quadrupole moment of the system is obtained in a time-dependent form as

$$\langle Q \rangle = \mathrm{Tr}\, Q\rho = Q_0 + 2Q_t \cos \omega t + Q_2 + 2Q_{2t} \cos 2\omega t + \ldots . \quad (7.20)$$

Here, Q is the Y_{20} component of the quadrupole tensor, ρ the single-particle density and the amplitudes can be identified with matrix elements of Q as follows

$$Q_0 = \langle \psi_0 | Q | \psi_0 \rangle,$$
$$Q_1 = \langle \psi_0 | Q | \psi_1 \rangle,$$
$$Q_2 = \langle \psi_1 | Q | \psi_1 \rangle,$$

and

$$Q_{2t} = \sqrt{2} \langle \psi_0 | Q | \psi_2 \rangle,$$

where ψ_0, ψ_1, and ψ_2 denote the normalized wavefunctions of the ground state, the first, and second 2^+ state, respectively. For spherical nuclei, $Q_0 = 0$ and the term $Q_t \cos \omega t$ describes the harmonic quadrupole oscillation with frequency ω following from RPA, whereas the first anharmonic terms $(Q_2 + 2Q_{2t} \cos 2\omega t)$ arise in second order.

The single particle density matrix ρ is calculated from the TDHF equations[29,30]

$$i\dot{\rho} = [T + V\rho, \rho], \quad (7.21)$$
$$\rho = \rho^2, \quad (7.22)$$

7.5 ANHARMONIC VIBRATIONS

where

$$(T + V\rho)_{\alpha\beta} = T_{\alpha\beta} + \sum_{\gamma,\delta} V_{\alpha\gamma,\beta\delta}\rho_{\delta\gamma} \quad (7.23)$$

is the self-consistent Hamiltonian operator with the kinetic energy T, V is the two-particle interaction, and $\alpha, \beta, \gamma, \ldots$ denote single-particle states obtained from diagonalization of the time-independent HF problem, i.e.

$$(T + V\rho^{(0)})_{\alpha\beta} = \varepsilon_\alpha \delta_{\alpha\beta},$$

$$\rho^{(0)}_{\alpha\beta} = n_\alpha \delta_{\alpha\beta}, \quad n_\alpha = \begin{cases} 1, & \varepsilon_\alpha \leq \varepsilon_F \\ 0, & \varepsilon_\alpha > \varepsilon_F \end{cases}.$$

In the ground state, which is assumed to be non-degenerate, the particles fill the lowest levels up to the Fermi level with energy ε_F. In the following, occupied levels (hole states) are denoted by i, j, k, and unoccupied levels (particle states) by m, n, p. One is now looking for time-dependent solutions which describe slow, but anharmonic, vibrations. The density matrix ρ and eqns (7.21) and (7.22) are expanded with respect to the parameter of adiabaticity

$$\rho = \rho^{(0)} + \rho^{(1)} + \rho^{(2)} + \ldots.$$

In first order, trying a solution in the form

$$\rho^{(1)}_{mi} = X^{(1)}_{mi} e^{-i\omega t} + Y^{(1)}_{mi} e^{i\omega t} = (\rho^{(1)}_{im})^*, \quad (7.24)$$

where X and Y denote the forward- and backward-going amplitudes, the well known RPA equations are then obtained

$$(\varepsilon_m - \varepsilon_i - \omega)X^{(1)}_{mi} + \sum_{n,j} (V_{mj,in}X^{(1)}_{nj} + V_{mn,ij}Y^{(1)*}_{nj}) = 0,$$

$$(\varepsilon_m - \varepsilon_i + \omega)Y^{(1)}_{mi} + \sum_{n,j} (V_{mj,in}Y^{(1)}_{nj} + V_{mn,ij}X^{(1)*}_{nj}) = 0. \quad (7.25)$$

Since solution (7.24) corresponds to a pure harmonic vibration, there is no first-order contribution to the static quadrupole moment; the time average $\text{Tr}\, Q\rho^{(1)} = 0$. Thus the first non-zero contributions to the quadrupole moments of one-phonon states arise in second order.

One can now expand ρ in eqn (7.23) to second order

$$\rho^{(2)} = \rho^{(0)}\rho^{(2)} + \rho^{(2)}\rho^{(0)} + \rho^{(1)}\rho^{(1)}$$

and obtain

$$\rho^{(2)}_{mn} = \sum_j \rho^{(1)}_{mj}\rho^{(1)}_{jn} = W^{(2)}_{mn} + X^{(2)}_{mn}e^{-2i\omega t} + Y^{(2)}_{mn}e^{2i\omega t},$$

$$\rho^{(2)}_{ij} = -\sum_n \rho^{(1)}_{in}\rho^{(1)}_{nj} = W^{(2)}_{ij} + X^{(2)}_{ij}e^{-2i\omega t} + Y^{(2)}_{ij}e^{2i\omega t}. \quad (7.26)$$

Due to the non-linearity of the basic eqns (7.21) and (7.22), $\rho^{(2)}$ consists now of two parts: one being independent of time (the Ws) and another oscillating with the two-phonon frequency ω (the Xs and Ys). Using eqn (7.23), one verifies easily that

$$\text{Tr}' QW^{(2)} = Q_2^{RPA}, \quad (7.27)$$

$$\text{Tr}' QX^{(2)} = \text{Tr}' QY^{(2)} = Q_{2t}^{RPA}, \quad (7.28)$$

where the prime at Tr' indicates that only particle–particle (pp) and hole–hole (hh) states are included under the trace. Thus the RPA results of eqns (7.25) have been regained. But it is seen that all particle–hole contributions to Q_2 and Q_{2t} are cut off in the approximation. This is the reason why RPA fails to reproduce the experimental values for the quadrupole moments of the 2_1^+ states.

The complete second-order treatment shows that the p–h contributions $\rho_{mi}^{(2)}$ are by no means negligible. They are calculated from the equation of motion (7.21) that reads, expanded to second order,

$$i\dot{\rho}_{mi}^{(2)} = (\varepsilon_m - \varepsilon_i)\rho_{mi}^{(2)} - [V\rho^{(2)}, \rho^{(0)}]_{mi} + [V\rho^{(1)}, \rho^{(1)}]_{mi}. \quad (7.29)$$

This equation describes the response of the system to the driving term

$$[V\rho^{(1)}, \rho^{(1)}]_{mi} = \sum_{n,p,j} (V_{mj,pn}\rho_{nj}^{(1)}\rho_{pi}^{(1)} + V_{mn,pj}\rho_{jn}^{(1)}\rho_{pi}^{(1)})$$

$$- \sum_{n,k,j} (V_{kj,in}\rho_{mk}^{(1)}\rho_{nj}^{(1)} + V_{kn,ij}\rho_{mk}^{(1)}\rho_{jn}^{(1)}).$$

This term only involves interaction matrix elements that correspond to the scattering of a particle (hole) under emission (annihilation) of a particle–hole pair. All these processes are neglected in the RPA, but must be taken into account when calculating phonon expectation values. Meyer[25] calculated the second-order contribution to $\langle Q \rangle = \text{Tr} Q\rho$ given as

$$\text{Tr} Q\rho^{(2)} = Q_2 + 2Q_{2t} \cos 2\omega t,$$

with the static moment determined by

$$Q_2 = \sum_{m,n} Q_{mn} W_{nm}^{(2)} + \sum_{i,j} Q_{ij} W_{ji}^{(2)} + \sum_{m,i} Q_{mi}(W_{mi} + W_{mi}^*). \quad (7.30)$$

He then used this equation for the static quadrupole moment in a simple two-level model using a quadrupole–quadrupole interaction. As is shown in Fig. 7.6, a strong enhancement of the static quadrupole moment of one-phonon states was obtained by the second-order response mechanism. Increasing the strength of the quadrupole interaction, x, the deformation of the vibrational state sets in much earlier than the RPA phase transition. At the same time, the ratio Q_{2t}/Q_2 is reduced by 50% compared with the RPA in agreement with experiment.

Birbrair[31] used graphical methods and Speth[26] the Green function

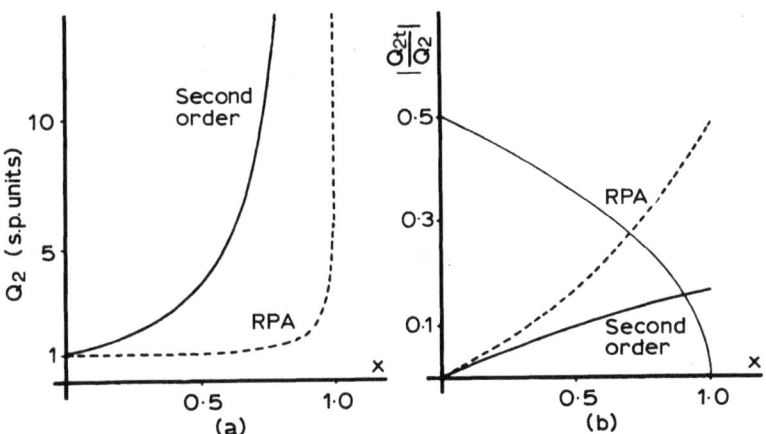

Fig. 7.6 (a) The static quadrupole moment of the one-phonon state plotted as a function of the interaction strength x. Broken curve, the RPA result; solid curve, the complete second-order results.

(b) Ratio $Q_{2'}/Q_2$ plotted as a function of x. Broken curve, RPA results; solid curve, second-order results.

Whereas the RPA results for Q_2 nearly keep to the single particle (s.p.) value up to $x = 0.9$, the complete second-order result grows rapidly with x reaching a value of five times the single-particle value at $x = 0.6$ and eight times at $x = 0.7$. These enhancement factors are sufficient to explain the large static moments observed in experiment. At the same time the ratio $Q_{2'}/Q_2$ is reduced; compared with experiment, this is again the right trend.

formalism to calculate second-order responses. In spite of the difference in methods, these two authors reached essentially the same conclusion as Meyer, namely that enhancement to $Q(2^+)$ values in the vibrational method are due to second-order responses of the system to first-order harmonic vibrations.

With this model we conclude here our study of purely vibrational modes and consider now three techniques specifically designed to analyse rotational motion and its possible coupling to vibrational excitations.

7.6 The sum-rule method

Recently Cline[32,33] has proposed an appealing method to correlate E2 data with the aim of isolating quadrupole collectivity in the nucleus. The method is based on a model-independent sum rule which exploits the rotational invariance of zero-coupled products of spherical tensor operators. We recall that electromagnetic multipole operators are spherical tensors and thus zero-coupled products of such operators can be formed that are rotationally invariant, i.e. these products are identical in the intrinsic and laboratory frames. Consider for example the electric quadrupole tensor which when rotated into the principal axis frame can be defined in terms of two intrinsic quadrupole moments, Q_{20} and Q_{22}. These have been expressed by Cline in terms of two parameters Q and δ. That is

$$\left(\frac{5}{16\pi}\right)^{1/2} Q_{20} = \langle r^2 Y_{20}\rangle \equiv Q\cos\delta \tag{7.31}$$

$$\left(\frac{5}{16\pi}\right)^{1/2} Q_{22} = \langle r^2 Y_{22}\rangle \equiv (Q/\sqrt{2})\sin\delta \tag{7.32}$$

where Q is positive and $0° \leq \delta \leq 60°$.

The expectation value in a nuclear state $|s\rangle$ of some of the invariant products of E2 operators can easily be expressed in terms of the intrinsic parameters Q and δ using simple recoupling algebra,[34]

$$\langle s|\,\{E2 \times E2\}^0\,|s\rangle = (1/\sqrt{5})\langle s|\,Q^2\,|s\rangle,$$

$$\langle s|\,\{[E2 \times E2]^2 \times E2\}^0\,|s\rangle = -(2/35)^{1/2}\langle s|\,Q^3\cos(3\delta)\,|s\rangle, \tag{7.33}$$

and

$$\langle s|\,\{[E2 \times E2]^0[E2 \times E2]^0\}\,|s\rangle = \tfrac{1}{5}\langle s|\,Q^4\,|s\rangle.$$

Cline also showed that these invariants can be expressed in terms of summations of products of single-particle E2 reduced matrix elements using an intermediate state expansion. For example,

$$\langle s|\,\{E2 \times E2\}^0\,|s\rangle = \sum_r U(ss22;0r)\langle s\|\,E2\,\|r\rangle\langle r\|\,E\,\|s\rangle, \tag{7.34}$$

$$\langle s|\,\{[E2 \times E2]^2 \times E2\}^0\,|s\rangle = \sum_{rt} U(ss22;0t)U(st22;2r)\langle s\|\,E2\,\|r\rangle$$

$$\langle r\|\,E2\,\|t\rangle\langle t\|\,E2\,\|s\rangle \tag{7.35}$$

where U-coefficients are defined in eqn (1.17) and the Wigner–Eckart theorem in eqn (1.4).

This sum rule (7.34) corresponds to a sum of all $B(E2, s \to r)$ values to all final states r. Of course the method should be particularly useful for the interpretation of numerical data in terms of a few collective variables. Rotational model assumptions can be used to relate the $Q-\delta$ E2 distributions to the familiar $\beta-\gamma$ shape distribution if so desired. A good example of this is given in Figs 7.7 and 7.8 which analyse shape transitions occuring in the Os and Pt isotopes. For instance Fig. 7.7 shows that the centroids $\langle Q^2\rangle$ and $\langle \cos 3\delta\rangle$ are almost constant for the excited states of ^{192}Os, which implies a strong correlation of the E2 properties consistent with rotation-like quadrupole collective motion. Similar results were obtained for ^{186,188,190}Os and ^{194}Pt. The expectation values of both

7.7 CRANKING MODEL AND QUADRUPOLE COLLECTIVITY

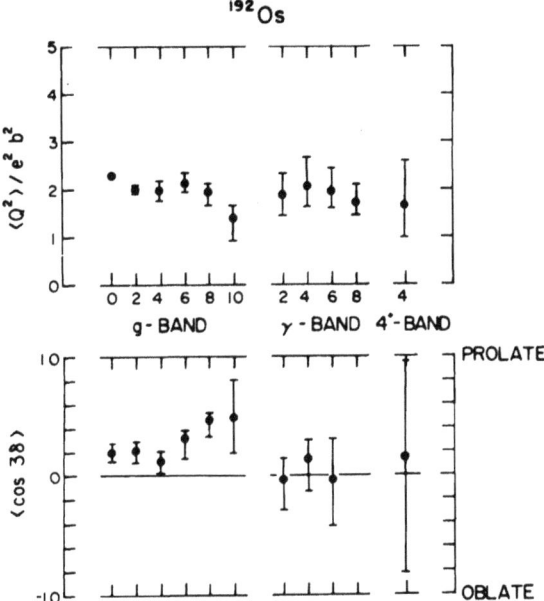

Fig. 7.7 Plot of centroids Q^2 and $\cos 3\delta$ versus spin for the ground band, γ band, and 4^- band head in ^{192}Os. (From Cline.[32])

the centroids and widths for the ground states of these nuclei are shown in Fig. 7.8. The centroids $\langle Q^2 \rangle$ illustrate the gradual reduction in deformation with increasing mass while the centroids $\langle \cos 3\delta \rangle$ correspond to triaxial deformation throughout a prolate-to-oblate shape transition.

The mass dependence of the softness in asymmetry (δ) and the moderate stiffness in magnitude (Q^2) of the quadrupole deformation are apparent. These results demonstrate that the E2 data for the low-lying levels are correlated well using only quadrupole collective degrees of freedom throughout a prolate-to-oblate shape transition and illustrate very adeptly the power of the sum-rule technique.

7.7 The cranking model and quadrupole collectivity in strongly deformed nuclei

Departing further from vibrations one should remember that quadrupole collectivity is associated, however, in its most spectacular effects with rotational motion. One of the most successful models, the cranking model, has its origin in the pioneering work of Inglis.[35] Instead of considering rotational motion macroscopically, as discussed earlier, Inglis evaluated the effect of rotational motion on the intrinsic structure. In

228 QUADRUPOLE MOMENTS AND COLLECTIVE MODEL THEORIES

GROUND STATE E2 INVARIANTS

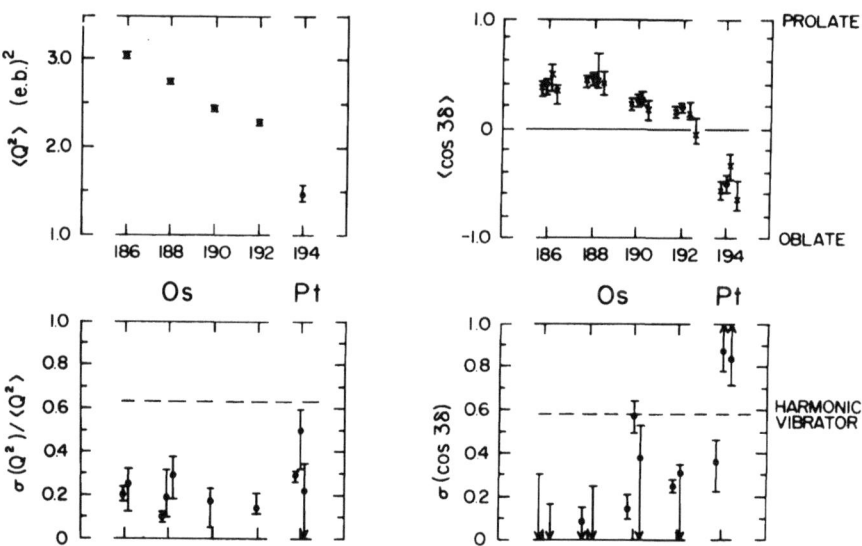

Fig. 7.8 Centroids (upper) and fluctuation widths (lower) for the magnitude and asymmetry of the intrinsic frame E2 properties of the ground states of ^{186}Os, ^{188}Os, ^{190}Os, ^{192}Os, and ^{194}Pt. Four values of cos 3δ are shown for each state; one derived using the third-order invariant (solid point) and three from the fifth-order invariants (crosses).[32]

Inglis' model the deformed potential is cranked with constant angular velocity ω along a fixed axis, say the x-axis. In the rotating coordinate system, a Coriolis term $-\omega \mathscr{I}_x$ is then added to the Hamiltonian

$$H_\omega = H_0 - \omega \mathscr{I}_x \qquad (7.36)$$

where H_0 is a sum of deformed single-particle Hamiltonians. In the early version of the model, Inglis calculated the first-order effect on the wavefunction by perturbation and evaluated the expectation value of the cranked Hamiltonian. The results were written as

$$E = E_0 + \tfrac{1}{2}\mathscr{I}\omega^2 \qquad (7.37)$$

where the moment of inertia was expressed in terms of the matrix elements of $\mathscr{I}_x$,

$$\mathscr{I}(\text{cranking}) = 2 \sum_{mi} |\langle m| \mathscr{I}_x |i\rangle|^2 (\varepsilon_m - \varepsilon_i)^{-1}. \qquad (7.38)$$

Here m and i denote the eigenstates of the unperturbed Hamiltonian H_0 with eigenenergies ε_m and ε_i. In evaluating the moment of inertia the effect of pairing interaction is important and BCS wavefunctions are

7.7 CRANKING MODEL AND QUADRUPOLE COLLECTIVITY

generally used.[35,36] This is done using a more general starting Hamiltonian where H_0 stands for a sum of deformed single-particle Hamiltonian plus a pairing interaction, like eqn (5.22) for example. In recent versions of the model, the Coriolis term $-\omega \mathcal{I}_x$ has been interpreted as a constraint to keep the expectation value of $\mathcal{I}_x$ at a fixed value.

The effect of pairing correlations becomes generally less important for higher rotational states because the Coriolis term tends to cancel the pairing effect.

A remarkable result discovered about 12 years ago and confirmed in recent experiments has been that for strongly deformed nuclei, the measured ground state $B(E2)$ value and intrinsic quadrupole moments obey the simple rigid spheroidal rotor relation (see eqn (5.14)) with remarkable accuracy to high-spin states in the band. The relationship naturally breaks down in the region of band crossings. In ^{232}Th for instance, a combination of Doppler shift attenuation and Coulomb excitation experiments[37,38] show that the intrinsic *transition* quadrupole moment

$$Q_0(I+1 \to I-1) = (\tfrac{4}{5})Ze^2R_0^2\beta(1 + 1/6\beta), \qquad (7.39)$$

where β is the deformation parameter (eqn (5.30)), remains remarkably constant up to $I = 28^+$ (see Fig. 7.9). Similar effects have been observed in ^{168}Er and ^{248}Cm.[39]

Also in these nuclei, the moment of inertia usually exhibits a smooth ω dependence of the form $\mathcal{I}(\omega) = \mathcal{I}_1 \omega^2$.[40] This dependence is obtained from data by noting that

$$\hbar\omega = dE/dI = \tfrac{1}{2}E_\gamma(I+1 \to I-1)$$

and also

$$\hbar\omega = dE/dI = \frac{1}{2}\frac{\hbar^2}{2\mathcal{I}}[(I+1)(I+2) - I(I-1)]. \qquad (7.40)$$

This angular momentum dependence of the moment of inertia can be ascribed either to a weakening of the pairing correlation at high spin, i.e. Coriolis anti-pairing, or to shape variation due to centrifugal stretching. This latter explanation seems inconsistent with the remarkable constancy of the transition quadrupole moments for strongly deformed nuclei illustrated in Fig. 7.9. Thus Coriolis anti-pairing seems to be the reason for the observed spin dependence of the moment of inertia.[32]

The interacting boson model[41] is also reasonably successful in correlating data on collective behaviour for low-spin states in deformed nuclei. However, the limited number of $L = 0$ and 2 bosons employed leads to band termination and concomitant retardation of the $B(E2)$ values, which is in conflict with the $B(E2)$ data for ^{232}Th and for many other strongly deformed rare-earth and actinide nuclei. These results indicate that higher-angular-momentum bosons as well as excitation into

230 QUADRUPOLE MOMENTS AND COLLECTIVE MODEL THEORIES

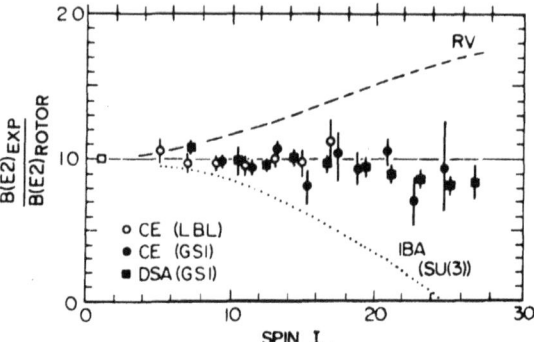

Fig. 7.9 The plot shows the measured values of $B(E2: I+1 \to I-1)$, proportional to $Q_0(I+1 \to I-1)$, normalized to the axially symmetric rigid rotor value for the ground band of ^{232}Th.[37,38] The rotation–vibration model values are those that result from attributing change in the ground band moment of inertia to centrifugal stretching. The IBA calculation assumes 12 bosons outside a ^{208}Pb core and the SU(3) limit.

the next valence shell need to be included in this model to account for the properties of higher-spin states.

7.8 The Kumar–Baranger theory

We will now conclude this chapter by describing a theory which, despite some shortcomings, has been very successful in describing nuclei in transitional regions. In the late 1960s, Kumar and Baranger[42,43] developed a model consisting of a pairing-plus-quadrupole interaction acting in a relatively small space of a few neutron and proton shells. The method first aimed at generating a self-consistent Hartree–Fock field. This was easily achieved because of the form of the quadrupole interaction chosen. Next, the time dependence of deformation parameters α_μ was taken into account using the time-dependent Hartree–Fock equation[40]

$$[H, \rho] = i\hbar \dot\rho(\alpha_\mu, \dot\alpha_\mu). \tag{7.41}$$

The density matrix ρ was then expanded in terms of the time-independent Hartree–Fock solution taking into account up to the first time derivative of α_μ. The density, which is a function of α_μ and $\dot\alpha_\mu$, was easily solved for by the perturbation theory. Next the collective energy was evaluated using the density matrix and the result expressed as

$$H(\alpha_\mu, \dot\alpha_\mu) = V(\beta, \gamma) + \tfrac{1}{2}\sum_{\mu\nu} B_{\mu\omega}(\alpha_\mu)\dot\alpha_\mu\dot\alpha_\nu \tag{7.42}$$

where the higher-order terms in $\dot\alpha_\mu$ were neglected. The first term

7.8 THE KUMAR-BARANGER THEORY

represents the potential energy of the collective motion. From the second term the rotational and vibrational kinetic energy term were obtained by expressing α_μ in terms of Euler angles θ_i and the intrinsic deformation β and γ. Once the potential and mass parameters were obtained, the Hamiltonian was diagonalized by using eigenfunctions coupling harmonic vibrations and rotations.[41]

Baranger and Kumar calculated energy spectra and electromagnetic moments up to 4^+ states in transitional nuclei W, Os, and Pt. They succeeded in reproducing the change of energy spectra, $B(E2)$, and $Q(2^+)$ values as the deformation changes sign at around $A = 192$ (see Fig. 7.10). Later Kumar[44] improved the model and included up to $J = 6^+$ states in applying the theory to ^{152}Sm which he fitted very well for both energy levels and $Q(2^+)$ moments.

Kumar then suggested that, although a complete treatment of collective quadrupole motion must include the specific dependence of the potential and mass parameters on β and γ, one could possibly use, as an intermediate step, the prolate-oblate energy difference as a test of calculated equilibrium deformations. He thus proposed a simple empirical relation between the value of the quadrupole moment and the

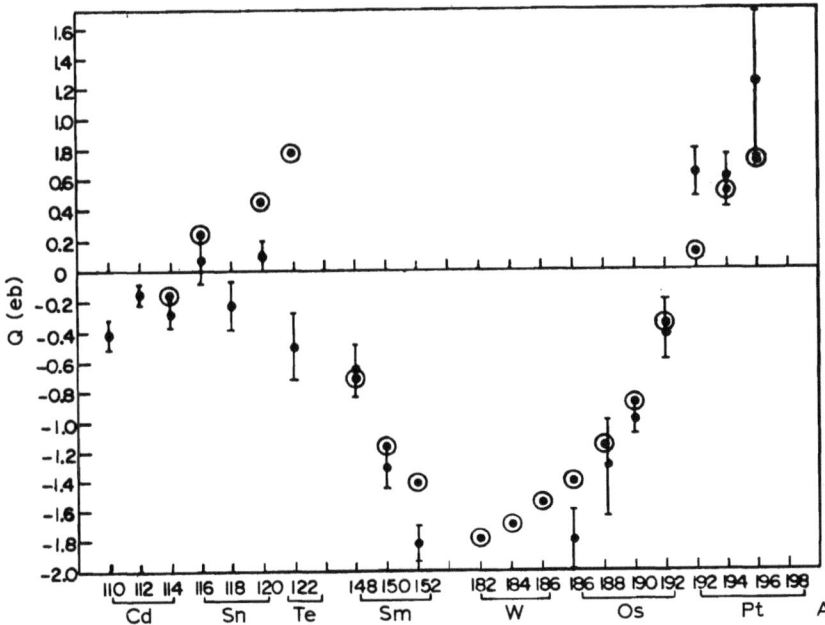

Fig. 7.10 Comparison of theoretical $Q(2^+)$ values with experiment for nuclei in the transitional region $112 \leq A \leq 198$. Theoretical values are shown by open circles and are taken from the work of Kumar and Baranger.[42,43]

232 QUADRUPOLE MOMENTS AND COLLECTIVE MODEL THEORIES

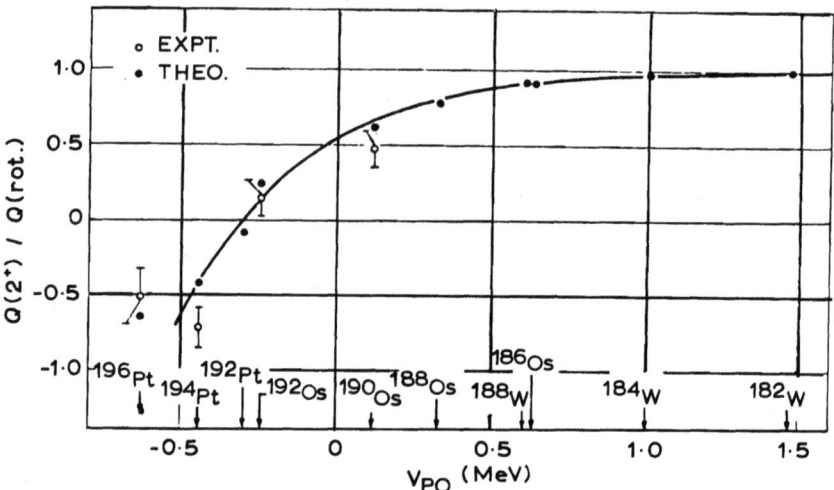

Fig. 7.11 The quadrupole moment ratio $Q(2^+)/Q(\text{rot.})$ plotted agains the prolate–oblate energy difference. The smooth curve has been drawn to guide the eye for positive values of V_{PO}. (From Kumar.[47])

prolate–oblate energy difference V_{PO}, of the calculated Hartree–Fock solutions.[45] In Fig. 7.11, the observed values of $Q(2^+)/Q(\text{rot.})$ and those calculated by Kumar are plotted against V_{PO}. The smooth curve corresponds to the relation

$$Q_{2^+}/Q_R = 1 - 0.45 \exp(-2.6 V_{PO}).$$

This curve deviates from the expected behaviour at low values of V_{PO}; the net quadrupole moment is non-zero even when V_{PO} equals zero! The reason invoked is that the mass parameter favours prolate shapes and hence increases the domain of 'prolate'-type quadrupole moments. Although Kumar considered only a limited number of cases, the theory seemed to work well. Later Cline[46] applied the relation to more extended cases and found reasonable correlations between $Q(2^+)$ and V_{PO}. Kumar[47] also proposed a method for determining the shape of nuclei independent of the nuclear model by using sum rules over the quadrupole moments and $B(E2)$. This method is very reminiscent of Cline's sum-rule analysis which we have described in detail earlier.

7.9 Summary

In this chapter a broad review of collective model theories has been presented. All of them had in common that they were either inspired or were otherwise intimately connected with nuclear quadrupole moments. The logical order of presentation of the chapter followed a dual purpose, going from phenomenological to microscopic and from vibrational to rotational characters.

The first two models discussed were inspired by the discovery of strong anharmonic effects in vibrational nuclei. We saw that macroscopic models based on boson expansion methods and microscopic calculations based on core–particle coupling required very extensive labour to reproduce the large quadrupole moments observed in the cadmium isotopes and other transitional nuclei. We then focused on the rich history of the $Q(3^-)$ moment in ^{208}Pb as an example demonstrating how important quadrupole moments appear in the eyes of theoretical practitioners.

Anharmonic vibrations also represent a challenge for time-dependent theories. In that context we examined how anharmonicity inspired Meyer and others to go beyond ordinary RPA and investigate how second-order responses to first-order harmonic vibrations are instrumental in producing the desired enhancement of static quadrupole moments of one-phonon states.

Finally, in our evolution toward deformed nuclei, rotational spectra were examined in the context of the cranking model and the more microscopic theories of Kumar and Baranger. We also saw that after years of extensive microscopic calculations, Kumar finally proposed a method for determining quadrupole moments (and thereby nuclear shapes) independently of nuclear models by introducing prolate–oblate potential energy differences or using sum rules over E2 transitions and moments. The last method had strong analogies with Cline's empirical analysis of collective spectra based on E2 invariants. This demonstrated well the ability of nuclear moments to connect phenomenological and microscopic attempts at describing the collective features of transitional and deformed nuclei.

7.10 References

1. Bohr, A. and Mottelson, B. R. (1975). *Nuclear structure*, Vol. II. Benjamin, New York.
2. de Boer, J. and Eichler, J. (1968). *Adv. Nucl. Phys.*, **1**, 1.
3. Sørensen, B. (1970). *Nucl. Phys.*, **A142**, 392; (1967). *Nucl. Phys.*, **A97**, 1.
4. Sørensen, B. (1971). *Phys. Lett.*, **35B**, 101.
5. Marshalek, E. (1971). *Nucl. Phys.*, **A161**, 401; Janssen, D., et al. (1971). *Nucl. Phys.*, **A171**, 726; Hage-Hassan, M. and Lambert, M. (1972). *Nucl. Phys.*, **A188**, 545.
6. Sørensen, B. (1972). In *Nuclear moments and nuclear structure*, p. 399. Physical Society of Japan.
7. Bohr, A. and Mottelson, B. R. (1953). *K. Danske Vidensk Selsk. Mat.-Fys. Medd.*, **27**, 16.
8. Mottelson, B. R. (1977). In *Elementary modes of excitation in nuclei*, Course LXIX Enrico Fermi School of Physics, p. 31. North Holland, Amsterdam.

9. Alaga, G. and Ialongo, G. (1967). *Nucl. Phys.*, **A97**, 600.
10. Alaga, G. (1975). In *Problems in vibrational nuclei* (ed. G. Alaga, V. Paar, and I. Sips), p. 344. North Holland, Amsterdam.
11. Alaga, G. (1969). In *Cargese lectures in physics*, Vol. 5, (ed. M. Jean), p. 579. Gordon and Breach, New York.
12. Lopac, V. (1975). In *Problems in vibrational nuclei* (ed. G. Alaga, V. Paar, and I. Sips), p. 396. North Holland, Amsterdam.
13. Sips, I. (1972). In *The structure of nuclei*, p. 339. IAEA, Vienna.
14. Paar, V. (1975). In *Problems in vibrational nuclei* (ed. G. Alaga, V. Paar, and I. Sips), p. 25. North Holland, Amsterdam; Rybarska, W. (1972). In *The structure of nuclei*, p. 357. IAEA, Vienna.
15. Castel, B. (1979). In *Neutron capture gamma ray spectroscopy* (ed. R. E. Chrien and W. R. Kane). Plenum Press, New York.
16. Kleinfeld, A. (1975). In *Problems in vibrational nuclei* (ed. G. Alaga, V. Paar, and I. Sips), p. 363. North Holland, Amsterdam.
17. Barnett, A. R. and Phillips, W. R. (1969). *Phys. Rev.*, **186**, 1205.
18. Hamamoto, I. (1970). *Nucl. Phys.*, **A155**, 362.
19. Hamamoto, I. (1975). In *Problems of vibrational nuclei* (ed. G. Alaga, V. Paar, and I. Sips), p. 54. North Holland, Amsterdam.
20. Guidetti, M., Rowe, D. J., and Chow, H. (1975). *Nucl. Phys.*, **A238**, 225.
21. Ziegler, J. F. and Peterson, G. A. (1968). *Phys. Rev.*, **165**, 1337.
22. Joye, A. M. R., et al. (1977). *Phys. Rev. Lett.*, **38**, 807.
23. Hamamoto, I. (1977). In *Elementary modes of excitation in nuclei*, Course LXIX Enrico Fermi School of Physics, p. 268. North Holland, Amsterdam.
24. Krainov, V. P. (1968). *Phys. Lett.*, **27B**, 341.
25. Meyer, J. (1971). *Z. Phys.*, **242**, 332.
26. Speth, J. (1970). *Z. Phys.*, **239**, 249.
27. Baranger, M. and Veneroni, M. (1978). *Ann. Phys.*, (New York), **114**, 123.
28. Villars, F. (1977). *Nucl. Phys.*, **A285**, 269.
29. Ring, P. and Schuck, P. (1980). *The nuclear many-body problems*, p. 485. Springer, Berlin.
30. Kisslinger, L. S. and Sorensen, R. A. (1963). *Rev. Mod. Phys.*, **35**, 853.
31. Birbrair, B. L. (1970). *Phys. Lett.*, **32B**, 165.
32. Cline, D. (1986). *Ann. Rev. Nucl. Part. Sci.*, **36**, 683.
33. Cline, D. (1987). In *Interacting Bose-Fermi systems in nuclei*, (ed. F. Iachello) pp. 241-55. Plenum Press, New York.
34. Cline, D. (1972). In *Nuclear moments and nuclear structure*, p. 377. Physical Society of Japan.
35. Inglis, D. R. (1954). *Phys. Rev.*, **96**, 1959; (1956). *Phys. Rev.*, **103**, 1786.
36. Belyaev, S. T. (1961). *Nucl. Phys.*, **24**, 322.
37. Guidry, M., et al. (1976). *Nucl. Phys.*, **A266**, 228.
38. Ower, H., et al. (1982). *Nucl. Phys.*, **A388**, 421.
39. Czosnyka, T., et al. (1986). *Nucl. Phys.*, **A458**, 123.
40. Ring, P. and Schuck, P. (1980). *The nuclear many-body problems*, p. 129. Springer, Berlin.
41. Arima, A. and Iachello, F. (1981). *Ann. Rev. Nucl. Sci.*, **31**, 75.
42. Kumar, K. and Baranger, M. (1968). *Nucl. Phys.*, **110**, 490, 529; Baranger, M. and Kumar, K. (1965). *Nucl. Phys.*, **62**, 113.

43. Kumar, K. and Baranger, M. (1968). *Nucl. Phys.*, **A112,** 241, 273.
44. Kumar, K. (1971). *Phys. Rev. Lett.*, **26,** 269.
45. Kumar, K. (1970). *Phys. Rev.*, **C1,** 369.
46. Cline, D. (1972). In *Proceedings of the Orsay Colloquium on Intermediate Nuclei.* North Holland, Amsterdam.
47. Kumar, K. (1972). *Phys. Rev. Lett.*, **28,** 249.

8

NUCLEAR MOMENTS: FUTURE DIRECTIONS

In recent years, progress in knowledge of nuclear moments has developed along several branches beyond the magnetic dipole and electric quadrupole moments which have been the focus of our attention so far. We devote this chapter to a discussion of what we consider are promising new directions in investigations involving nuclear moments. We first discuss E4 moments and transitions before turning to monopole transitions and concluding with an appraisal of a current relativistic model.

8.1 Hexadecapole transitions and moments

The experimental and theoretical study of hexadecapole (E4) transitions and moments is still in an early state of growth, in marked contrast to quadrupole moments which, as we have seen in the last two chapters, have been studied extensively. There is no reason, however, why these E4 moments should not turn out to be as fertile and interesting as their E2 counterparts. Indeed, recent results reported by several groups indicate that this might actually be the case.

Several years ago, Bertsch[1] gave a simple but elegant explanation of the systematics of E4 moments in light nuclei. When nucleons are added to a deformed core they will fill orbits as close to the z-axis (symmetry axis) as possible; hence at the beginning of a shell there would be a polar 'bulge' whereas near the end of the shell there would be an equatorial 'pinch' which would have positive and negative E4 moments respectively. Hartree–Fock calculations by Castel and Parikh[2] corroborated these intuitive predictions.

So far all experimental measurements of E4 moments have been measurements of the *transition* matrix elements *not* of static moments. On the strength of existing data, mostly from proton and alpha inelastic scattering, deducing the existence of static E4 moments is analogous to deducing a static quadrupole moment because of a non-zero $B(E2; 0^+ \to 2^+)$. The procedure has therefore been to invoke a nuclear model, like the rotational model.

Recently Brown *et al.* have carried out extensive shell-model calculations of hexadecapole moments in the sd shell.[3,4] These authors find that the concept of an effective charge can be used for E4 transitions in the sd

8.1 HEXADECAPOLE TRANSITIONS AND MOMENTS

shell nuclei with an isoscalar effective charge $e_0 = 2$. They, however, find an exception in the E4 transition from the ground state to the first 4^+ state in ^{24}Mg. This state is weakly excited, as compared with the second 4^+ state, in inelastic electron scattering, and they are unable to obtain a reasonable fit to the inelastic form factor for this transition. Jaqaman and Zamick[5] have investigated carefully the role that open-shell effects would have in modifying the shell-model results. Their investigations involved using a Hartree–Fock programme with good axial symmetry. In general, good agreement is obtained between rotational Hartree–Fock, shell-model values, and experiment in the lower half of the sd shell but complete disagreement in the upper half. The Hartree–Fock results give values of $B(E4)$ of $28\,e\,\text{fm}^4$ for ^{32}S and $31\,e\,\text{fm}^4$ for ^{36}Ar. The corresponding numbers in the shell model are 230 and $190\,e\,\text{fm}^4$. The data support a large value for ^{32}S (either $465\,e\,\text{fm}^4$ by one determination with 20 MeV protons or $145\,e\,\text{fm}^4$ with 60 MeV protons). Of course, a basic difference between the two approaches is that, in the shell model, a fixed and rather large isoscalar effective charge is used but that in the collective model, since E4 transitions depend on E2 deformations *in a quadratic manner*, one would expect the isoscalar effective charge to be mass dependent. We should also keep in mind that ^{32}S has been a difficult system to analyse in terms of its quadrupole deformation and its equilibrium transition between oblate and prolate Hartree–Fock solutions. The fact remains that Hartree–Fock and shell-model results agree as far as E2 data of ^{32}S are concerned but not as far as E4 properties are concerned. Accurate E4 inelastic scattering data in this mass region would be most welcome.

There is also some evidence for E4 deformation beside inelastic scattering. There are several examples where spectroscopy of odd-*A* nuclei can provide some evidence for hexadecapole deformations. Casten and Cizewski[6] have shown that fragmentation of strengths in single-nucleon transfer reactions to levels in odd-*A* tungsten nuclei are consistent with Nilsson model calculations if those included hexadecapole deformations. Also, recently, Baker[7] reviewed earlier axially *asymmetric* rotor calculations for odd-*A* heavy transitional nuclei. Baker was able to show that including an hexadecapole deformation to an axially *symmetric* core provided as good or better description of the spectra. The E4 deformations needed were, furthermore, in excellent agreement with those deduced from inelastic scattering experiments.

Also in heavier nuclei, some of the evidence for static E4 moments originates not from experiment directly but again as a result of calculation. This is the case for Strutinsky-type calculations of potential energy surfaces which by including an E4 degree of freedom reproduce both the trend and magnitude of experimentally observed E4 moments in the $A = 184$ region and in the actinides.[8,9] The hexadecapole degree of freedom can also have an important effect in *lowering* the potential energy minimum more on the prolate side than on the oblate side. Thus in the transition from osmium to platinum the inclusion of the E4 deformation correctly predicts the prolate–oblate crossing, while, without hexadecapole deformation, the prolate–oblate transition is predicted to occur near $A = 182$, in contradiction to experiment.

Finally, we should mention that hexadecapole properties have also been discussed as constraints on the interacting boson model.[10] Baker[11] recently showed that E4 properties in the Gd and Pt regions could not be understood without consideration of g bosons. Morrison[12] subsequently demonstrated that by considering an sdg IBM, the systematics of E4 matrix elements could be qualitatively reproduced. Viewed microscopically, the 4^+ states strongly excited *inelastically* must be mainly composed of two quasiparticle configurations. In a boson model, 4^+ states of such structure can only be obtained by the inclusion of g bosons. Morrison also showed, however, that in the $160 < A < 200$ region the mass dependence of the E4 operator mirrors exactly that of the quadrupole operator and shows a turning point at $A = 186$ in agreement with experiment. According to Morrison, the qualitative shape of the E4 mass dependence came as a result of taking into account the Pauli corrections to the E4 operators in the shell model of $A \sim 186$ nuclei and not, *per se*, by the inclusion of g bosons.

8.2 Monopole transitions and moments

We now turn to monopole transitions and moments which have recently seen a resurgence of interest as tests of our knowledge of nuclear bulk properties. We realize that, simply speaking, the vanishing of the radiation field for a monopole ($\lambda = 0$) decay simply expresses the physical statement that each photon carries off at least one unit of angular momentum. Within the nucleus, however, E0 transitions can occur via the Coulomb coupling of the atomic electrons and the nuclear protons. But whether the transition from a 0_2^+ to 0_1^+ state proceeds by pair emission (electromagnetic creation of a positron–electron pair) or by conversion electron (where the excess nuclear energy is given to one of the electrons thereby expelled from the atom), the relevant nuclear structure information is contained within the monopole strength parameter ρ, defined as[13]

$$\rho = \langle 0_2^+ | T(E0) | 0_1^+ \rangle / eR^2, \qquad (8.1)$$

where R is the nuclear radius.[13,14]

Of course, while protons are usually considered, neutrons can also be incorporated through the use of an effective charge, which results in $T(E0) = e \sum_j r_j^2$ as the form of the monopole operator. Further, because

8.2 MONOPOLE TRANSITIONS AND MOMENTS

the monopole (E0) transition probability, $B(E0)$, is related to ρ by $B(E0) = e^2 R^4 \rho^2$, the quantity ρ^2 is commonly cited. A good example of this is given by Kantele[15] who showed that a typical 1 MeV E0 transition in ^{164}Gd having a ρ^2 value of 0.01 has a partial half-life for the $0_2^+ \to 0_1^+$ E0 K-conversion transition of 1 ns (which is much slower than the competing 1 W.u. $0_2^+(2) \to 2_1^+(1)$ E2 transition by about two orders of magnitude).

Recently the emergence of experimental techniques such as superconducting solenoidal magnets[16] and mini-orange spectrometers[17] and the ability to measure lifetimes of a fraction of a nanosecond have allowed the determination of the absolute rate of E0 transitions, i.e. the determination of the monopole strength. This information in turn has been exploited extensively as test of nuclear models. We present here two examples: the first one deals with a collective model interpretation of shape coexistence while the second illustrates a more microscopic test of nuclear bulk properties.

8.2.1 Monopole strengths as a measure of nuclear shape mixing

In the vibrational model, the E0 operator for harmonic quadrupole vibrational motion in nuclei can be written to lowest order as

$$T(E0) = \frac{3}{4\pi} ZeR^2 \sum_n |\alpha_n|^2, \tag{8.2}$$

where the α_n denote the spherical harmonics expansion coefficients for the nuclear surface in the collective model.[13] Thus, the expectation value of the E0 operator *within* a deformed state gives a measure of the nuclear radius related to this deformed state. Since the wavefunctions corresponding to 0^+ states with largely different deformation are only poorly overlapping, only that overlap region will contribute when calculating non-diagonal matrix elements of the E0 operator. This will normally result in a small value for ρ^2.

Recently Heyde and Meyer[18] have suggested that in nuclei with a decreasing potential energy barrier, mixing will increase and modify the E0 transition matrix element in a major way. In their argument Heyde and Meyer have adopted a schematic model where maximal mixing between spherical and deformed basis states is allowed. They write for $|0_f^+\rangle$ and $|0_i^+\rangle$

$$|0_f^+\rangle = 1/\sqrt{2}\,\{|\text{sph}\rangle - |\text{def}\rangle\},$$
$$|0_i^+\rangle = 1/\sqrt{2}\,\{|\text{sph}\rangle + |\text{def}\rangle\}, \tag{8.3}$$

which results in a value for ρ^2 equal to

$$|\rho_{i \to f}|^2 = \tfrac{1}{4}(\langle \beta^2 \rangle)^2 (3/4\pi)^2 Z^2. \tag{8.4}$$

Here $\langle \beta^2 \rangle = (\langle \text{def}| \sum er^2 |\text{def}\rangle$ and all matrix elements involving $|\text{sph}\rangle$ have been set to zero.

These authors then use a schematic model to show that whenever in a series of isotopes (or isotones) one considers a spherical ground state and an excited intruder configuration (with large β), if large mixing occurs in the crossing region point, large E0 rates will be seen there. (Here crossing means the proximity of spherical and deformed states plotted systematically as a function of mass number.) Heyde and Meyer then examined specific properties of E0 strengths in the $A \simeq 100$ region from the data which are now available through a combination of in-beam spectroscopy and (t, p) reactions.[19] The new E0 data plotted as a function of neutron number N indicated a striking increase in E0 strength when approaching $N = 60$. Since for the even–even Sr, Zr, and Mo nuclei, it is always in the $N = 60$ nucleus that the first excited 0^+ level occurs at its lowest excitation energy, the largest mixing between the ground state and this low-lying 0^+ state is expected there. This was then construed by Heyde and Meyer as an illustration that large E0 transitions can arise as a result of strong mixing between states with different shapes and that the observation of large E0 strength in itself is not evidence for shape coexistence between spherical and deformed states within a particular nucleus. Heyde and Meyer also showed that the difference of a factor of 10 in the monopole strength observed between the $N = 58$ and $N = 60$ isotones can be understood within the context of dynamical deformation. That is, the $N = 60$ nuclei have the deformed state as their ground state and the spherical state as a low-lying excited state. This is in contrast to the $N = 58$ nuclei where the deformed state not only occurs at a higher energy for the same isotope but exists because of promoted pairs that produce a dynamical deformation.

8.2.2 Nuclear incompressibility and its relation to monopole matrix elements

There are only two properties of infinite nuclear matter that can be obtained rather directly from nuclei. These are the binding energy per nucleon ($E/A = -16$ MeV) obtained by extrapolation from known nuclear masses, and the saturation density ($\rho_0 = 0.166$ fm^{-3}) obtained from the density of nucleons inside medium and heavy nuclei. But besides these two parameters there is interest in the complete equation describing the dependence of the energy on the density, the equation of state. This equation should naturally exhibit a minimum at the saturation density. How sharp this minimum is depends on the second derivative of the energy; a quantity known as the nuclear incompressibility. If the nuclear incompressibility, K, is small, say $K < 200$ MeV, the equation of state is said to be *soft*, while if $K > 200$ MeV, the equation is *stiff*. It is of great current interest, as we will see, to determine a value for the nuclear

8.2 MONOPOLE TRANSITIONS AND MOMENTS

incompressibility in nuclear matter. The best hope rests on an extrapolation from data in finite nuclei on the excitation energies of the isoscalar monopole giant resonances (so-called breathing modes) and from data on isotope shifts in neighbouring nuclei. Both these quantities depend on the matrix elements of the monopole operator, $T(\text{E0}) = \sum_{i=1}^{A} r_i^2$.

In infinite nuclear matter, the nuclear incompressibility is defined[20] as

$$K_\infty = k_F^2 \frac{d^2(E/A)}{d^2 k_F},$$

where k_F is the Fermi momentum and E/A is the binding energy per nucleon. In finite nuclei, the definition is

$$K_A = b^2 \frac{d^2(E/A)}{d^2 b},$$

where b is a parameter that has the dimension of length and characterizes the size of the system. The quantity E/A is the energy of the nucleus in its ground state when a small monopole field is applied such that its size would vary and, in this way, E/A would become a function of b.

Theoretical estimates in nuclear matter using realistic 'hard-core' nucleon–nucleon potentials lead to values for K_∞ ranging from 100 to 200 MeV,[21] while the phenomenological density-dependent forces lead to values in the range 200 to 400 MeV.[22] From the liquid drop model of Myers and Swiatecki[23] a value of $K_\infty = 240$ MeV can be inferred. The most commonly accepted value for K_∞ extrapolated from experimental data in finite nuclei on the breathing modes[20,24] and from isotope shifts[25] is $K_\infty = 210 \pm 30$ MeV. But it should be remembered that both the breathing modes and the isotope shifts are related to the surface of the nucleus. For this reason the extrapolation from finite nuclei to nuclear matter is not straightforward as surface, Coulomb, and asymmetry ($N \neq Z$) effects significantly modify K from its nuclear matter value, K_∞, to its value K_A in a finite nucleus. The problem is similar to the one of extracting the nuclear matter parameter, E/A, from the knowledge of nuclear binding energies. One major difference is, however, that in the latter case more than a thousand experimental data are known, whereas in the former only a couple of dozen of resonance energies are available.[24] (A recent analysis[26] of breathing-mode energies in a series of Sn and Sm isotopes suggest a larger value for the incompressibility of nuclear matter of $K_\infty = 300 \pm 25$ MeV.)

In recent years new and conflicting constraints have been placed on the nuclear incompressibility. For example, it plays a role in the analysis of pion multiplicity and multifragmentation following relativistic heavy-ion collisions.[27a] In this work a very stiff equation of state with $K_\infty \geq 200$ MeV is required in the analysis to get good agreement with data. On the other hand, a very soft equation of state is required by Baron et al.[27b] with $K_\infty \leq 180$ MeV in order that matter can be dispersed in supernova explosions. Clearly the incompressibility is an important quantity in many-body physics of fermion systems.

To get some understanding of the complexities involved, we will go through some of the simple model estimates that have been discussed by Zamick.[28] The method is to use the Hartree–Fock procedure with phenomenological zero-range effective interactions. Saturation in nuclear matter comes about either by the density dependence or the velocity dependence in the effective interaction. With *density* dependence the interaction becomes less and less attractive and eventually repulsive as the density increases. With *velocity* dependence, the attraction decreases when the relative momentum of the interacting nucleons increases.

The nuclear incompressibility depends very much upon the saturation mechanism. Consider the following simple force that saturates because of its density dependence:

$$V = (-t_0 + \tfrac{1}{6}t_3 \rho^\sigma(R))\delta(r) \tag{8.5}$$

with $R = \tfrac{1}{2}(r_1 + r_2)$ and $r = r_1 - r_2$. The density dependence is characterized by an exponent, σ, that will be left as a variable so that we can study its influence on the incompressibility.

Following Zamick,[28] we restrict our discussion to harmonic-oscillator wavefunctions characterized by only one parameter, b, the oscillator length parameter. The kinetic energy then scales as b^{-2} while the first term of the potential energy in eqn (8.5) scales as b^{-3} and the second term as $b^{-3(1+\sigma)}$. Since b is the only variational parameter in the Hartree–Fock theory, we can from dimensional analysis alone write down the form of the expression for the total energy

$$E = -E_B = a/b^2 + B/b^3 + C/b^{(3+3\sigma)}, \tag{8.6}$$

where a, B, and C are independent of b, and E_B is the (positive) binding energy. In the oscillator model, the value of a can be identified, since the kinetic energy is

$$a/b^2 = \frac{\hbar^2}{2Mb^2} \sum_h (2n_h + l_h + \tfrac{3}{2}) \equiv A\bar{T}, \tag{8.7}$$

where the sum h is a sum over all the occupied orbitals for the nucleus with mass number, A. Here n_h and l_h are the oscillator quantum numbers of the orbitals and $\bar{T}$ the average kinetic energy of a nucleon. Recall that the mean-square radius of the nucleus in the oscillator model is

$$\langle r^2 \rangle = \frac{1}{A} \sum_h (2n_h + l_h + \tfrac{3}{2})b^2$$

so that $\bar{T} = (\hbar^2/Mb^4)\langle r^2 \rangle$. Using typical values, namely $b \simeq A^{1/6}$ fm and $\langle r^2 \rangle = 3R^2/5$ with $R = r_0 A^{1/3}$ and $r_0 = 1.2$ fm we obtain $\bar{T} \simeq 18$ MeV.

Now, at equilibrium in the Hartree–Fock calculation, the energy is a minimum, $\partial E/\partial b = 0$, giving a second equation

$$0 = -2a/b^3 - 3B/b^4 - (3 + 3\sigma)C/b^{(4+3\sigma)}. \tag{8.8}$$

Equations (8.6) and (8.8) can be regarded as two equations for two unknowns, B and C. Solving for them, we obtain

$$B = -[(1 + \sigma)E_B + A\bar{T}(1/3 + \sigma)]b^3/\sigma,$$
$$C = [E_B + A\bar{T}/3]b^{(3+3\sigma)}/\sigma.$$

Finally the expression for the nuclear incompressibility is

$$K \equiv \frac{1}{A}b^2\frac{\partial^2 E}{\partial^2 b} = \bar{T}(1 + 3\sigma) + 9(E_B/A)(1 + \sigma). \tag{8.9}$$

In nuclear matter taking $\bar{T} = 18$ MeV (as above) and the accepted value for the binding energy per nucleon of $E_B/A = 16$ MeV leads to the expression $K_\infty = 162 + 198\sigma$ MeV. Thus we see that the incompressibility depends linearly on the exponent in the density-dependent term of the effective interaction. This parameter is not well known. A reasonable range would be 0 to 1, so that K_∞ would range from 160 to 360 MeV. In finite nuclei the kinetic energy is about the same, but the binding energy per nucleon E_B/A is closer to 8 MeV. So we see why the nuclear incompressibility is larger in nuclear matter than in conventional finite nuclei.

Now let us repeat the analysis but add a velocity-dependent term to the effective interaction

$$V = -t_0\delta(r) + \tfrac{1}{6}t_3\rho^\sigma(R)\delta(r)$$
$$+ \tfrac{1}{2}t_1[k^2\delta(r) + \delta(r)k^2] + t_2 k \cdot \delta(r)k, \tag{8.10}$$

where $k = (\nabla_1 - \nabla_2)/2$. This interaction is typical of Skyrme interactions[22] widely used in Hartree–Fock calculations of nuclei. The velocity-dependent terms scale as b^{-5}, so eqn (8.6) becomes modified to

$$E = -E_B = a/b^2 + B/b^3 + C/b^{(3+3\sigma)} + D/b^5, \tag{8.11}$$

and following the same analysis the incompressibility becomes

$$K = \bar{T}(1 + 3\sigma) + 9(E_B/A)(1 + \sigma) + 2D(2 - 3\sigma)/(Ab^5). \tag{8.12}$$

Note that for $\sigma = 2/3$ this reduces to eqn (8.9) showing that the velocity-dependent term is equivalent to a density-dependent term of the form $\rho^{2/3}(R)\delta(r)$. Now the velocity dependence of the effective interaction can be thought of as a modification of the kinetic energy. That is for a nucleon in nuclear matter the kinetic energy operator is replaced by

$$\hbar^2\nabla^2/2M \to \hbar^2\nabla^2/2M^*$$

in which the nucleon mass, M, is replaced by an effective mass, M^*. Thus in eqn (8.11), the parameter D is determined by

$$a/b^2 + D/b^5 = (M/M^*)a/b^2,$$

and the incompressibility becomes

$$K = -3\bar{T}(1-3\sigma) + 9(E_B/A)(1+\sigma) + 2(2-3\sigma)\bar{T}M/M^*. \quad (8.13)$$

Typical of a wide range of Skyrme interactions[22] is the one known as SIII for which $\sigma = 1$ and $M^* = 0.76M$. Then from eqn (8.13) with $\bar{T} = 18$ MeV and $E_B/A = 16$ MeV, the incompressibility is $K_\infty = 350$ MeV. Another parameter set, known as SGIII, has recently[29] been put together which in Hartree–Fock theory gives an equally good description of conventional properties of finite nuclei (binding energies, mean square radii, single-particle energies, and density profiles) but which nonetheless leads to a low incompressibility preferred by supernova theories.[27b] The parameters are $\sigma = 1/15$ and $M^*/M = 1$ yielding $K_\infty = 180$ MeV. Clearly theories fitted to these bulk properties alone are not capable of giving a definitive prediction for the incompressibility.

We turn now to ways to get information on the incompressibility in finite nuclei from data on the breathing mode and isotope shifts. First the excitation energy of the isoscalar monopole giant resonance can be obtained in an RPA calculation with the monopole operator $T(E0) = \sum_i r_i^2$. The resonance energy extracted from experiment is then identified with some definite average quantity that can be obtained from an RPA calculation. Treiner et al.[24] suggest that these average quantities come from the RPA sum rules and define moments as follows

$$m_k = \sum_n E_n^k |\langle n| T |0\rangle|^2$$

where $|0\rangle$ is the ground state of the nucleus, E_n the excitation energy of state $|n\rangle$, and T the monopole operator. Among the various mean energies $\bar{E}_k$ that can be defined

$$\bar{E}_k = (m_k/m_{k-2})^{1/2}$$

$\bar{E}_3$ and $\bar{E}_1$ have particularly been used. If the strength function is distributed in a very narrow energy region, all the values of $\bar{E}_k$ are very close to each other and there is no ambiguity in the comparison with experimental values. We will identify the experimental resonance energy, E_M, with the mean value $\bar{E}_3$, since the moment, m_1, can be related to the mean square radius and m_3 to the incompressibility:[20,24]

$$m_1 = \frac{2\hbar^2}{M} \langle r^2 \rangle \simeq \frac{6\hbar^2}{5M} r_0^2 A^{2/3},$$

$$m_3 = \frac{1}{2}\left(\frac{2\hbar^2}{M}\right)^2 K_A.$$

8.2 MONOPOLE TRANSITIONS AND MOMENTS

Hence
$$E_M = \bar{E}_3 = [(\hbar^2/M)K_A/\langle r^2\rangle]^{1/2},$$
$$= [48 \text{ MeV}]^{1/2} K_A^{1/2} A^{-1/3}. \quad (8.14)$$

Most of the information on the giant monopole resonance comes from inelastic hadron scattering (p, d, ^{3}He, α) with projectile energies of the order of 100 MeV.[30] The peak energies scale roughly as $E_M \simeq 80 A^{-1/3}$ leading to an incompressibility $K_A \simeq 130$ MeV. Note this is a finite nucleus value; it differs from K_∞ in nuclear matter by approximately 100 MeV, a huge number. In K_∞ only volume effects contribute, whereas in K_A surface effects come into play and are by no means negligible.

It is a similar story for the isotope shift, which is a measure of the difference in radii in adjacent isotopes. Consider Hartree–Fock theory with oscillator basis functions and suppose that for the A-particle system the equilibrium energy occurs with the oscillator parameter $b = b_A$ while in the $A - 1$ system it occurs at $b = b_{A-1}$. The mean square radii of the two nuclei are

$$A\langle r^2\rangle_A = \sum_i (2n_i + l_i + \tfrac{3}{2})b_A^2 \equiv N_A b_A^2,$$

$$(A-1)\langle r^2\rangle_{A-1} = \sum_i{}' (2n_i + l_i + \tfrac{3}{2})b_{A-1}^2 \equiv N_{A-1} b_{A-1}^2,$$

where the prime indicates a sum over the $A - 1$ nucleons. Let n_h and l_h be the quantum numbers of the Ath nucleon, then

$$N_A = N_{A-1} + (2n_h + l_h + 3/2)$$

and
$$[A\langle r^2\rangle_A - (A-1)\langle r^2\rangle_{A-1}] = A\langle r^2\rangle_A \frac{(b_A^2 - b_{A-1}^2)}{b_A^2} + r_h^2, \quad (8.15)$$

where $r_h^2 = (2n_h + l_h + \tfrac{3}{2})b_{A-1}^2$ is the square radius of the orbital, h. We will assume that the difference between b_A and b_{A-1} is small and write $b_A = b + \tfrac{1}{2}\Delta b$ and $b_{A-1} = b - \tfrac{1}{2}\Delta b$ and keep terms only to first order in Δb. It is convenient to characterize the isotope shift by a monopole polarization charge, e_h, defined as the difference in the square radius of the A and $A - 1$ systems divided by the square radius of the last nucleon

$$e_h = [A\langle r^2\rangle_A - (A-1)\langle r^2\rangle_{A-1}]/r_h^2.$$

With this definition, eqn (8.15) can be written in more compact form

$$\frac{1}{A}(e_h - 1) = \frac{\langle r^2\rangle_A}{r_h^2}\frac{2b\Delta b}{b^2}. \quad (8.16)$$

That is, the monopole polarization charge depends linearly on the shift in the size parameter, Δb, which in turn can be expressed in the terms of the incompressibility.

In Hartree–Fock theory, the total energy of A and $A-1$ particle systems is related by

$$E_A = E_{A-1} = \varepsilon_h + \Delta_h, \tag{8.17}$$

where ε_h is the Hartree–Fock single-particle energy and Δ_h is a positive rearrangement energy. This equation is a reminder of the fact that the total energy in a Hartree–Fock calculation is *not* equal to the sum of the single-particle energies, $E_A \neq \sum_h \varepsilon_h$, but rather is given by eqn (8.19) below. Only if it so happened that the Hartree–Fock orbitals obtained in a Hartree–Fock calculation of the $A-1$ system coincided with those of the A-particle system would the rearrangement energy be zero. Now at equilibrium

$$\left.\frac{\partial E_A}{\partial b}\right|_{b=b_A} = 0, \quad \left.\frac{\partial E_{A-1}}{\partial b}\right|_{b=b_{A-1}} = 0,$$

while the quantity we want is the variation of E_{A-1} at the value $b = b_A$. For this, a Taylor series expansion can be used

$$\left.\frac{\partial E_{A-1}}{\partial b}\right|_{b=b_A} = \left.\frac{\partial E_{A-1}}{\partial b}\right|_{b=b_{A-1}} + (b_A - b_{A-1})\left.\frac{\partial^2 E}{\partial^2 b}\right|_{b=b_{A-1}}$$

$$= \frac{(b_A - b_{A-1})}{b_{A-1}^2}(A-1)K_{A-1}$$

$$\simeq \frac{\Delta b}{b^2} A K_A$$

where we have assumed the incompressibility K does not vary significantly between $A-1$ and A-particle systems. Thus on differentiating eqn (8.17) with respect to b at the equilibrium value $b = b_A$,

$$-\frac{\Delta b}{b^2} A K_A = \frac{\partial \varepsilon_h}{\partial b} + \frac{\partial \Delta_h}{\partial b},$$

and eliminating Δb in favour of the monopole polarization charge, eqn (8.16), we obtain

$$\frac{1}{A}(e_h - 1)r_h^2 = -\frac{2b\langle r^2 \rangle_A}{A K_A}\left(\frac{\partial \varepsilon_h}{\partial b} + \frac{\partial \Delta_h}{\partial b}\right). \tag{8.18}$$

It remains to find expressions for $\partial \varepsilon_h / \partial b$ and $\partial \Delta_h / \partial b$ in Hartree–Fock theory. First $\partial \varepsilon_h / \partial b$ can be found from the Hartree–Fock expression for the total energy[31]

$$E_A = \tfrac{1}{2}\sum_h T_h + \tfrac{1}{2}\sum_h \varepsilon_h. \tag{8.19}$$

8.2 MONOPOLE TRANSITIONS AND MOMENTS

Differentiating with respect to b at the equilibrium value, b_A, gives

$$\sum_h \frac{\partial \varepsilon_h}{\partial b} = -\sum_h \frac{\partial T_h}{\partial b} = \frac{2}{b} A \bar{T}$$

where we again use the fact that the kinetic energy scales as b^{-2} so that $\partial T_h/\partial b = -2T_h/b$. We also introduce the average kinetic energy of the nucleon, $A\bar{T} = \sum_h T_h$. Substituting in eqn (8.18), we obtain

$$\frac{1}{A}\sum_h (e_h - 1)r_h^2 = -\frac{4\langle r^2 \rangle_A}{K_A}\left(\bar{T} + \frac{b}{2A}\sum_h \frac{\partial \Delta_h}{\partial b}\right). \tag{8.20}$$

If the rearrangement energy is neglected for a moment, then the mean of the sum of the polarization charges is negative definite, indicating that the average isotope shift is negative. Measurements usually gave positive values, but these are only made on orbitals near the Fermi surface so that it is not necessarily a contradiction.

Now, Hartree–Fock calculations with density-dependent interactions lead to positive values of the rearrangement energy. Indeed if the density-dependent interaction is of the form $\rho(R)\delta(r)$ then Δ_h scales as b^{-6} and $b\partial\Delta_h/\partial b = -6\Delta_h$. This term, being negative, has the tendency to change the sign of the right-hand side of eqn (8.20) from negative to positive. Being more specific, Zamick[28] shows that for the density-dependent force, eqn (8.10), the rearrangement energy is

$$\sum_h \Delta_h = \sigma C/b^{(3+3\sigma)}, \tag{8.21}$$

where C was introduced in eqn (8.11). Solving for C, as before, in terms of the binding energy E_B and effective mass M^* for a density- and velocity-dependent zero-range force gives

$$C = (E_B + A\bar{T} - \tfrac{2}{3}A\bar{T}M/M^*)b^{(3+3\sigma)}/\sigma.$$

Finally, differentiating the expression for the rearrangement energy, eqn (8.21) and inserting in eqn (8.20) leads to the final equation relating the average isotope shift to the incompressibility

$$\frac{A^{-1}\sum_h (e_h - 1)r_h^2}{A^{-1}\sum_h r_h^2} = \frac{4}{K_A}[\tfrac{3}{2}(1+\sigma)E_B/A + \tfrac{1}{2}(1+3\sigma)\bar{T} - (1+\sigma)\bar{T}M/M^*]. \tag{8.22}$$

The left-hand side of the equation has been written in this particular form to stress that it is the average value of $e_h - 1$ that is involved with r_h^2 being the weights. We write this as $\overline{(e_h - 1)}$. Using the Skyrme force, SIII, with $\sigma = 1$ and $M^*/M = 0.76$ for a finite nucleus with $E_B/A = 8$ MeV and $\bar{T} = 18$ MeV, we obtain from eqn (8.13) $K_A = 205$ MeV and hence $\overline{(e_h - 1)} = 0.25$. However, the expression, eqn (8.22), is very sensitive to M^*/M and can even change sign for $M^*/M < 0.6$. From an experimental point of view, mean isotope shifts are generally not available. The usual measurement involves the isotope shift for a particular orbital near the Fermi surface, such as the measurement of the charge density differences between ^{208}Pb and ^{207}Pb obtained in elastic electron scattering.[25] These density distributions are well reproduced in Hartree–Fock-style calculations with Skyrme interactions with a nuclear incompressibility of $K \simeq 230$ MeV. A change in K by 30 MeV, it is stated,[25] would lead to serious disagreement with the measured isotopic difference. The importance of eqn (8.22), however, is as a reminder that the isotope shift depends not only on the nuclear incompressibility but also on the effective mass and the form of the density-dependent interaction.

8.3 Relativistic theories of nuclear moments

Throughout this book our approach to nuclear structure has followed the traditional non-relativistic one based on the Schrödinger equation with nucleons interacting with static two-body potentials. Exact solutions for finite nuclei are complicated; nevertheless, the shell model has historically been a remarkably good starting point for understanding nuclear structure. Here the nucleons move in a self-consistent single-particle potential arising from their interactions with all the other nucleons. Hartree–Fock theory forms the basis for describing this situation in the non-relativistic many-body problem.

In spite of the success of the conventional non-relativistic approach there have been some exciting developments using a relativistic theory for nuclear systems. The simplest non-trivial approximate solution to the relativistic many-fermion problem is the Hartree approximation. (Recently the Fock, or exchange, terms have been computed, but they are not essential for this discussion.) The Hartree solutions have been extensively studied for infinite nuclear matter and for closed-shell nuclei. In both cases, the Hartree approximation leads to a high degree of symmetry in the system under consideration and simplifies the solutions. Many interesting results have emerged. For example, the calculations contain a uniquely relativistic contribution to the saturation mechanism of infinite nuclear matter resulting in saturation occurring at a lower density for fixed binding energy per nucleon. (Non-relativistic theories in the simplest Breuckner–Hartree–Fock calculations notoriously fail to get saturation in nuclear matter at the right density for the right binding energy per

8.3 RELATIVISTIC THEORIES OF NUCLEAR MOMENTS

nucleon.) There is also a direct and quantitative link between the saturation properties of nuclear matter and the strong nucleon–nucleus spin–orbit interactions. Indeed this has led to an explanation of the large nuclear effective single-particle spin–orbit potential and to an understanding in detail of the spin observables measured in intermediate energy proton–nucleus elastic scattering experiments. Furthermore these Hartree calculations accurately reproduce many of the properties of the ground states of closed-shell nuclei with relatively few free parameters. These and other findings are discussed in detail in Serot and Walecka[32] and references contained therein.

Here we will confine our discussion to nuclear moments in closed-shell-plus-one nuclei as obtained in the relativistic Hartree approximation. For many years these calculations posed a problem. They failed completely to reproduce the magnetic moments, especially the isoscalar ones for which large enhancements were predicted. For closed-shell-plus-one nuclei where the traditional isoscalar Schmidt values are in agreement with the observed moments this failure is particularly glaring, given the other successes of the relativistic approach. It was subsequently realized that the failure represented a self-consistency problem in the Hartree approximation. The first calculations were performed for the A nucleons comprising a closed-shell nucleus. Wavefunctions for both occupied and unoccupied orbitals are obtained. The magnetic moment of an extra-core nucleon was then evaluated using these wavefunctions and a relativistic operator. What is omitted is the impact that the extra-core nucleon has on the core itself. The correct treatment requires the Hartree calculation to be done for the $A+1$ occupied orbitals. Likewise the electromagnetic current must be calculated consistently for $A+1$ nucleons. The current on an extra-core nucleon is not just that of a single nucleon, but in relativistic theories contains a contribution from the core as well. When these considerations are correctly treated the difference between relativistic and non-relativistic theories becomes rather small, indeed the corrections are of the order of the nucleon binding energy divided by the nucleon mass. The 'enhanced' isoscalar moments were merely an artefact of an inconsistent treatment of the relativistic Hartree problem for an $A+1$ system.

For the rest of this section, we will amplify on these remarks, indicate where the large enhancements for isoscalar moments came from, and discuss the resolution of the problem. We will examine these issues within the framework of the simple $\sigma + \omega$ model of Walecka.[33] The model is characterized by a strong attractive scalar mean field and an almost equally strong repulsive vector mean field. These mean fields are generated from potentials representing the exchange of scalar, σ, and vector, ω, mesons. The calculation depends on two dimensionless parameters C_σ and C_ω where $C_i = (g_i/m_i)M$ and g_i is the meson–nucleon coupling constant, m_i the meson mass, and M the nucleon mass. These two quantities are adjusted to reproduce the binding energy and saturation density in infinite nuclear matter. The success of the Walecka model is its ability to reproduce single-particle separation energies and, in particular, the large spin–orbit splitting evident in nuclear spectra. This comes about because the spin–orbit potential depends on the difference of the scalar and vector potential strengths while the central (binding) potential depends on the sum. In Walecka's model a typical strength for the scalar potential is about -400 MeV while the vector strength is about 350 MeV. The sum is the relatively small (-50 MeV) nuclear potential commonly adopted in non-relativistic nuclear models.

Such strong scalar and vector potentials demand a relativistic treatment. In particular such potentials have a significant impact on the lower component of the Dirac four-spinor describing a nucleon in a nuclear medium. The single-particle spinors are solutions to the Dirac equation:

$$[i\gamma \cdot p - \gamma_4(E - V_\omega) + M + V_\sigma]\psi = 0, \tag{8.23}$$

which in nuclear matter reduces to the free Dirac equation with modified energy and mass: $E^* = E - V_\omega$ and $M^* = M + V_\sigma$. With the scalar potential, $V_\sigma \simeq -400$ MeV and the vector potential, $V_\omega \simeq 350$ MeV, it is seen that E^* and M^* attain only about 60% of their value in free space. The lower component in the Dirac spinor $\psi = (G, F)$ in a nuclear medium is now

$$F_{medium} = \frac{E + M}{E^* + M^*} F_{free} \simeq 1.7 F_{free} \tag{8.24}$$

compared with the free-space value. Hence the matrix element of an 'odd' Dirac operator that couples an upper and lower component will typically be enhanced by a factor of the order of 1.7. An example of this is the contribution to the magnetic moment operator arising from the Dirac (non-anomalous) coupling γ. Recall in Chapter 4, eqn (4.14), the single-nucleon electromagnetic current is written

$$J_\mu = F_1 \gamma_\mu - \frac{F_2}{2M} \sigma_{\mu\nu} k_\nu \tag{8.25}$$

where the first term is known as the Dirac current and the second term the anomalous current. Note that $\sigma_{\mu\nu} k_\nu$ is an 'even' operator and is not significantly modified when a nucleon is immersed in a nuclear medium. For an isovector current, $F_1^{(1)} = 1$ and $F_2^{(1)} = 3.7$, so the anomalous contribution dominates and the medium modification is rather minor. On the other hand for an isoscalar current $F_1^{(0)} = 1$ and $F_2^{(0)} = -0.12$ the anomalous contribution is small and the medium modification is significant. As mentioned, the difficulty for these relativistic models is that

8.3 RELATIVISTIC THEORIES OF NUCLEAR MOMENTS

there is no experimental evidence to support these strong enhancements to the isoscalar moments.

In a finite nucleus, the potentials are functions of the nucleon's radial coordinate, r, and the medium-modified mass, M^*, becomes a function of r. We restrict the discussion to the case of spin-saturated closed-shell nuclei then the mean fields deduced in the Hartree approximation are spherically symmetric. The nucleon wavefunction can then be expressed as

$$\psi_\alpha = \frac{1}{r}\begin{pmatrix} iG_\alpha(r) \\ F_\alpha(r)\boldsymbol{\sigma}\cdot\boldsymbol{p}/r \end{pmatrix} Y_{j_\alpha l_\alpha m_\alpha} \tag{8.26}$$

where Y_{jlm} is the spinor spherical harmonic and $G_\alpha(r)$ and $F_\alpha(r)$ are radial wavefunctions for the upper and lower components normalized to

$$\int_0^\infty dr[|G_\alpha|^2 + |F_\alpha|^2] = 1 \tag{8.27}$$

and obtained from the coupled equations

$$\left(\frac{\partial}{\partial r} + \frac{\kappa_\alpha}{r}\right)G_\alpha = (M + E_\alpha + V_\sigma - V_\omega)F_\alpha, $$
$$\left(\frac{\partial}{\partial r} - \frac{\kappa_\alpha}{r}\right)F_\alpha = (M - E_\alpha + V_\sigma + V_\omega)G_\alpha, \tag{8.28}$$

where $\kappa_\alpha = \mp(j_\alpha + \tfrac{1}{2})$ for $j_\alpha = l_\alpha \pm \tfrac{1}{2}$.

The meson field equations are simply radial Laplace equations

$$(-\nabla^2 + m_\sigma^2)V_\sigma = -g_\sigma^2\rho_S \equiv \frac{-g_\sigma^2}{4\pi r^2}\sum_\alpha (2j_\alpha + 1)(G_\alpha^2 - F_\alpha^2),$$
$$(-\nabla^2 + m_\omega^2)V_\omega = g_\omega^2\rho_B \equiv \frac{g_\omega^2}{4\pi r^2}\sum_\alpha (2j_\alpha + 1)(G_\alpha^2 + F_\alpha^2), \tag{8.29}$$

where the sum, α, runs over the occupied orbitals and the factor $(2j_\alpha + 1)$ represents the occupancy of the orbital. Here ρ_S is known as the scalar density, $\rho_S = \sum_\alpha \bar\psi_\alpha \psi_\alpha$, and ρ_B the baryon density, $\rho_B = \sum_\alpha \psi_\alpha^\dagger \psi_\alpha$. The relativistic expression for the single-particle magnetic moment is the sum of the Dirac and anomalous pieces

$$\mu = \mu_D + \mu_A,$$
$$\mu_D = \mp\frac{2j+1}{2j+2}\frac{Mc}{\hbar}\int rF(r)G(r)\,dr; \qquad j = l \pm \tfrac{1}{2}, \tag{8.30}$$
$$\mu_A = \pm\tfrac{1}{2}F_2\left(\frac{j}{j+1}\right)^{(1\mp 1)/2}\left[\int G^2\,dr + \left(\frac{j}{j+1}\right)^{\pm 1}\int F^2\,dr\right]; \qquad j = l \pm \tfrac{1}{2}.$$

Note the Dirac piece depends linearly on the lower component $F(r)$ in the spinor and is strongly medium modified. In Table 8.1, we give some

Table 8.1 Relativistic impulse approximation (RIA) and RPA vertex corrections in the σ–ω model for the isoscalar magnetic moments of ground states in closed-shell-plus (or minus)-one nuclei. From Ichii et al.[34]

		$A=15$	$A=17$	$A=39$	$A=41$
RIA	(Dirac)	0.331	1.643	0.994	2.338
	(anomalous)	0.021	−0.060	0.037	−0.060
	(sum)	0.352	1.583	1.031	2.278
RPA vertex correction	(Dirac)	−0.146	−0.147	−0.389	−0.346
	(anomalous)	0.000	−0.000	0.000	0.000
	(sum)	−0.146	−0.147	−0.389	−0.346
RIA + RPA vertex correction	(Dirac)	0.185	1.496	0.605	1.992
	(anomalous)	0.021	−0.060	0.037	−0.060
	(sum)	0.206	1.437	0.642	1.932
Schmidt values	(Dirac)	0.167	1.500	0.600	2.000
	(anomalous)	0.020	−0.060	0.036	−0.060
	(sum)	0.187	1.440	0.636	1.940
Experiment	(sum)	0.218	1.414	0.707	1.921

sample calculations for closed-LS-shells-plus (or minus)-one nuclei where the results are labelled RIA standing for relativistic impulse approximation. The results are taken from Ichii et al.[34] Similar values have been obtained by Shepard et al.,[35] Bouyssy et al.,[36] and others.[37] Compared with the non-relativistic impulse approximation, labelled 'Schmidt' in the Table 8.1, it is seen there is a strong enhancement to the Dirac component, μ_D, but relatively little change to the anomalous component, μ_A, of the single-particle magnetic moment. In the case of the Dirac moment μ_D, which is directly proportional to the product of the large and small components, $G(r)$ and $F(r)$, the use of the Dirac equation to eliminate $F(r)$ shows that μ_D is approximately proportional to:[36]

$$\int \frac{M}{M^*(r)} G^2(r)\, dr$$

neglecting terms in $dM^*(r)/dr$. The value of $M^*(r)/M$ is roughly a constant in the nuclear interior and around 0.6, but increases to 1.0 at the nuclear surface. Because the valence particle wavefunction, $G^2(r)$, in a closed-shell-plus-one nucleus is peaked at the nuclear surface where $M^*(r)$ is closer to M the enhancement in μ_D is less than in a closed-shell-minus-one nucleus where the hole wavefunction samples much more the nuclear interior. Thus Table 8.1 shows the enhancements to be more significant for $p_{1/2}$ and $d_{3/2}$ hole states in $A=15$ and $A=39$ respectively than for the $d_{5/2}$ and $f_{7/2}$ particle states in $A=17$ and $A=41$.

8.3 RELATIVISTIC THEORIES OF NUCLEAR MOMENTS

Note that the sum RIA results are seriously in disagreement with experiment compared to the corresponding non-relativistic Schmidt values.

The resolution of this problem, as we have mentioned, concerns performing relativistic Hartree calculations for $A \pm 1$ nucleons, which is much harder than doing it for a closed-shells A-nucleons case. One approximate way of correcting an A-nucleons calculation is to allow for excitations from the core creating particle–antiparticle vibrations. Kurasawa and Suzuki[38] show that coupling a single-particle state in a nuclear medium to such a vibrational state through meson exchange (mainly the isoscalar ω meson) produces a cancelling mechanism that restores the single-particle electromagnetic current to its free-nucleon value. The results of Ichii et al.[34] and Shepard et al.[35] follow this method. The calculation involves the relativistic random phase approximation (RPA). Bentz et al.[39] come to the same conclusion from more general discussions starting with a Ward identity in which the coupling to a vibrational state represents a vertex correction. Indeed this approach emphasizes that the Lorentz invariance of the energy and momentum of a nucleon in a nuclear medium must be maintained in relativistic theories.[40] The problem is really one of maintaining self-consistency. A mean field calculation, which is self-consistent only for the core nucleons, will lead to an enhancement in the properties of an extra-core nucleon, whereas a fully self-consistent calculation in the core-plus-one system will produce the correct result. To add RPA corrections to a non-self-consistent result is an approximate way to achieve self-consistency. McNeil et al.,[35] Brown et al.,[41] and Nakayama et al.[42] all discuss the problem in terms of the Landau–Migdal quasiparticle approach to relativistic nuclear matter in which not only the response of the single nucleon to the external electromagnetic field is evaluated but also the response of the background 'spectator' nucleons. This again can be thought of as a response from a particle–antiparticle excitation. In the language of quantum liquids, this is sometimes called 'backflow'.[43] Finally, one other approach to this problem is to begin with a non-relativistic calculation to which relativistic corrections are added perturbatively. This amounts to a consideration of meson-exchange-current (MEC) corrections (see Chapter 4) in which nucleon–antinucleon excitations are introduced through pair graphs, e.g. Fig. 4.2(b). The calculation is mainly concerned with σ- and ω-exchange pair graphs. This approach is followed by Ichii et al.,[39] Blunden,[44] and Delorme and Towner.[45] The σ-pair graph produces the enhancement characteristic of the RIA results, while the ω-pair graph gives in first order the leading term in the relativistic RPA correction.

In Table 8.1, the RPA results of Ichii et al.[34] are shown. The enhancements of magnetic moments in the relativistic impulse approximation is reduced by these RPA-type vertex corrections to reasonable values. The vertex correction is purely isoscalar, all the while the calculation is restricted to σ and ω mesons. In order to treat isovector quantities one is naturally led to include isovector mesons such as π and ρ and to treat the many-body problem not just in the Hartree approximation but in Hartree–Fock. But then the corresponding vertex corrections are no longer as simple as the RPA-type vertex and it may be very difficult to sum up the higher-order corrections.[34]

Finally, relativistic corrections to quadrupole moments are very small even in the Walecka $\sigma-\omega$ model.[33] This is because the quadrupole moment derives principally from the time component of the Dirac current, γ_4, and this is an 'even' operator. Thus the strongly medium-modified lower components, $F(r)$, in the nucleon spinor only enter quadratically as a correction in $F^2(r)/G^2(r)$ in the expression for the quadrupole moment.

8.4 References

1. Bertsch, G. F. (1968). *Phys. Lett.*, **26B**, 130.
2. Castel, B. and Parikh, J. C. (1969). *Phys. Lett.*, **29B**, 341.
3. Brown, B. A., Chung, W., and Wildenthal, B. H. (1980). *Phys. Rev.*, **C21**, 2600.
4. Brown, B. A., Radhi, R., and Wildenthal, B. H. (1984). *Phys. Rep.*, **C101**, 313.
5. Jaqaman, H. R. and Zamick, L. (1984). *Phys. Rev.*, **C30**, 1719.
6. Casten, R. F. and Cizewski, J. A. (1978). *Nucl. Phys.*, **A309**, 477.
7. Baker, F. T. (1985). *Phys. Rev.*, **C32**, 1430.
8. Pelte, D. and Smilansky, U. (1973). *Minerva Conference*, Rehovot.
9. Goldring, G. (1974). Electromagnetic moments. In *Proc. int. conf. on Nuclear Physics, Munich, 1973*, (ed. J. de Boer and H. J. Mang) North Holland, Amsterdam.
10. Warner, D. D. and Casten, R. F. (1983). *Phys. Rev.*, **C28**, 1798.
11. Baker, F. T. (1985). *Phys. Rev.*, **C32**, 1430.
12. Morrison, I. (1986). *J. Phys. G: Nucl. Phys.*, **12**, L201.
13. Bohr, A. and Mottelson, B. (1969). *Nuclear structure*, Vol. 1, p. 383. Benjamin, Reading, MA.
14. Preston, M. A. (1972). *Physics of the nucleus*. Addison-Wesley, Reading, MA.
15. Kantele, J. (1983). *Phys. Rev. Lett.*, **51**, 92; (1981). *Heavy ions and nuclear research*, p. 391. Harwood Academic Press, New York.
16. Henry, E. A. and Stoeffl, W. (1984). *Nucl. Instrum. Meth.*, **227**, 77.
17. Fenyes, T. (1986). In *Nuclear structure, reactions and symmetries* (ed. R. A. Meyer and V. Paar), p. 567. World Scientific, Singapore.
18. Heyde, K. and Meyer, R. A. (1988). *Phys. Rev.*, **C37**, 2170.
19. Estep, R. J., et al. (1987). *Phys. Rev.*, **C35**, 1485.
20. Blaizot, J. P. (1980). *Phys. Rep.*, **64**, 171; Blaizot, J. P., Gogny, D., and Grammaticos, B. (1976). *Nucl. Phys.*, **A265**, 315.

21. Bethe, H. A. (1971). *Ann. Rev. Nucl. Sci.*, **21**, 93.
22. Beiner, M., Flocard, H., Giai, N. V., and Quentin, P. (1975). *Nucl. Phys.*, **A238**, 29.
23. Myers, W. and Swiatecki, W. (1969). *Ann. Phys. (N.Y.)*, **55**, 626.
24. Treiner, J., Krivine, H., Bohigas, O., and Martorell, J. (1981). *Nucl. Phys.*, **A371**, 253; Co, G. and Speth, J. (1986). *Phys. Rev. Lett.*, **57**, 547.
25. Cavedon, J. M., *et al.* (1987). *Phys. Rev. Lett.*, **58**, 195.
26. Sharma, M. M., *et al.* (1988). *Phys. Rev.*, **C38**, 2562.
27a. Molitoris, J. J. and Stocker, H. (1985). *Phys. Rev.*, **C32**, 346; Sano, M., Gyulassy, M., Wakai, M., and Kitazoe, Y. (1985). *Phys. Lett.*, **156B**, 27.
27b. Baron, E., Cooperstein, J., and Kahana, S. (1985). *Nucl. Phys.*, **A440**, 744.
28. Zamick, L. (1973). *Phys. Lett.*, **45B**, 313; (1975). *Nucl. Phys.*, **A249**, 63; (1987). *Z. Phys.*, **A237**, 409.
29. Sagawa, H. (1989). In *Symmetry violations in subatomic physics*, (ed. B. Castel and P. J. O'Donnell), pp. 93–112. World Scientific, Singapore.
30. Lui, W., *et al.* (1980). *Phys. Lett.*, **93B**, 31; Bertrand, F., *et al.* (1979). *Phys. Lett.*, **30B**, 198; Buenerd, M., *et al.* (1980). *Phys. Rev. Lett.*, **45**, 1667; Buenerd, M. (1984). In *Proc. of int. symposium on highly excited states and nuclear structure, Orsay, 1983*, (ed. M. Marty and N. van Giai), *Jour. de Physique*, **C4**, 115.
31. Towner, I. S. (1977). *A shell model description of light nuclei* Clarendon Press, Oxford.
32. Serot, B. D. and Walecka, J. D. (1986). *Adv. Nucl. Phys.*, **16**, 1.
33. Walecka, J. D. (1974). *Ann. Phys. (N.Y.)*, **83**, 491.
34. Ichii, S., Bentz, W., Arima, A., and Suzuki, T., (1987). *Phys. Lett.*, **B192**, 11.
35. Shepard, J. R., Rost, E., Cheung, C. Y., and McNeil, J. A. (1988). *Phys. Rev.*, **C37**, 1130; McNeil, J. A., *et al.* (1986). *Phys. Rev.*, **C34**, 746.
36. Bouyssy, A., Marcos, S., and Mathiot, J. F. (1984). *Nucl. Phys.*, **A415**, 497.
37. Noble, J. V. (1979). *Phys. Rev.*, **C20**, 1188; Ohtsubo, H., Sano, M., and Morita, M. (1973). *Prog. Theor. Phys.* **49**, 877; Miller, L. D. (1975). *Ann. Phys. (N.Y.)*, **91**, 40; Bawin, M., Hugues, C., and Strobel, G. L. S. (1983). *Phys. Rev.*, **C28**, 456.
38. Kurasawa, H. and Suzuki, T. (1985). *Phys. Lett.*, **165B**, 234; Kurasawa, H. and Suzuki, T. (1986). *Nucl. Phys.*, **A454**, 527.
39. Bentz, W., Arima, A., Hyuga, H., Shimizu, K., and Yazaki, K. (1985). *Nucl. Phys.*, **A436**, 593; Ichii, S., Bentz, W., and Arima, A. (1987). *Nucl. Phys.*, **A464**, 575.
40. Baym, G. and Chin, S. A. (1976). *Nucl. Phys.*, **A262**, 527.
41. Brown, G. E., Weise, W., Baym, G., and Speth, J. (1987). *Comm. Nucl. Part. Phys.*, **17**, 39.
42. Nakayama, K., Drozdz, S., Krewald, S., and Speth, J. (1987). *Nucl. Phys.*, **A470**, 573.
43. Pines, D. and Nozieres, P. (1966). *Theory of quantum liquids*. Benjamin, New York.
44. Blunden, P. G. (1987). *Nucl. Phys.*, **A464**, 525.
45. Delorme, J. and Towner, I. S. (1987). *Nucl. Phys.*, **A475**, 720.

CONCLUSION

> We shall not cease from exploration
> And the end of all our exploring
> Will be to arrive where we started
> And know the place for the first time.
> T. S. Eliot, *Four Quartets*

We have now arrived, after a meandering journey through the nuclear physics of the 1980s, at our final station. Let us pause here and reflect on how nuclear moments have contributed to the elucidation of some major problems of nuclear structure and discuss also some remaining difficulties and future prospects for their solution.

In recent years it has become increasingly clear that certain nuclear properties, especially magnetic properties, cannot be fully accounted for by models invoking only nucleon degrees of freedom. Extensions beyond the simple-impulse approximation are clearly needed with the most obvious being the inclusion of meson degrees of freedom. Although meson-exchange currents (MEC) were first introduced more than 30 years ago, it has only been in the past decade that they have become firmly established. It has also become popular in recent years to talk about quark degrees of freedom. This raises a problem of overlap between quark, meson, and nucleon variables and a possibility of double counting that is difficult to sort out. A number of hybrid models have appeared in which quark degrees of freedom are relevant at short distances and nucleon and meson degrees of freedom at larger distances. In this book we decided to stay with the more classic treatment of nucleons-only nuclear-structure calculations augmented with MEC corrections. The price paid is that some short-range phenomenology has to be introduced in the MEC calculation.

Direct confirmation of MEC contributions can be found in the very light systems, such as the deuteron, ^{3}He, and ^{3}H where data on moments, electron scattering, and electro-disintegration all demonstrate the need for a theory beyond the simple impulse approximation. In heavier nuclei, however, the indications are less clear because of the complexities of nuclear structure. But in simple cases, such as closed-shell-plus-one nuclei, wavefunction corrections evaluated to second order in perturbation theory and MEC corrections are of comparable magnitude and often of opposite sign. In some cases, such as the isovector magnetic moment in mass 17 and 41 these corrections almost exactly cancel each other leaving only a very small displacement from the single-particle Schmidt value.

Direct confirmation of MEC effects in heavy nuclei comes from the beautiful idea of Yamazaki to measure the magnetic moment of the two-proton 11^- isomer state in ^{210}Po. This measurement determines the effective proton $g_L^{(\pi)}$ value in a nucleus and found not the free proton value of 1.0 but an enhanced value of 1.16 ± 0.01 anticipated by the theories. Recently a similar experiment in a two-neutron 10^- isomer in ^{190}Os deduced a result $-0.12 \leq g_L^{(\nu)} \leq -0.07$ again in agreement with theoretical expectation that the shift in the neutron g_L value is less than the shift in the proton value.

In medium and heavy nuclei away from closed shells, microscopic calculations are impractical and collective models are a necessity. Here one of the most notable successes is the interacting boson model, in its second version, in which proton and neutron degrees of freedom are differentiated. This model explains rather well the systematics of the g-factors of 2^+ states in even–even nuclei, especially the departure from the Z/A prediction of the simple geometric models, that are seen in series of isotopes. The model also predicts a 1^+ state in deformed even–even nuclei at around 3 MeV excitation with considerable M1 strength. Dubbed the 'scissors mode' resonance, this excitation has been found in electron and photon scattering in the rare earths and actinides.

Turning now to quadrupole moments, we have seen that the shell model has been particularly successful in reproducing the magnitudes and signs of quadrupole moments of 2^+ states in the sd shell. Otherwise, the prolate–oblate transition at around $A = 28$ has been a particularly difficult test for self-consistent calculations like the projected Hartree–Fock and band-mixing models. In ^{208}Pb it was the abnormally large $Q(3^-)$ value which elicited a great deal of attention, with various core–particle and RPA calculations all proving unable to come within calling distance of experiment until the problem was finally resolved in 1977 with a definitive measurement and re-evaluation of the Coulomb scattering experiment.

One of the major remaining problems involving quadrupole moments has been the calculation of the E2 effective charge from first principles. There is some very elegant formalism based on perturbation theory that yields a formal solution to the problem. But when applied to specific cases, such as the neutron effective charge in ^{17}O, the calculation in lowest orders is not very successful, and going beyond second order proves to be impractical. This failure has been one of the major disappointments of recent many-body physics research.

Another long-standing problem in heavy nuclei is connected with the search for the M1 strength in ^{208}Pb. The one-particle one-hole (1p–1h) sum rule estimates a summed $B(M1; 0^+ \to 1^+) \simeq 50\mu_N^2$ in this M1 resonance. RPA correlations reduces this by about 20%, mixing with background 2p–2h states, a further 20% and MEC corrections yet another 20%. But it is hard to get the theoretical estimate any lower than $20\mu_N^2$. For decades this estimate has been a factor of two higher than had been experimentally observed. But during the writing of this book, this 'missing strength' problem has been resolved by the University of Illinois group, who with highly polarized tagged photons identified additional M1 strength in ^{208}Pb.

We started this book invoking Blin-Stoyle's *Theories of nuclear moments* written more than thirty years ago when meson-exchange theories were barely in existence and relativistic theories certainly not yet dreamed of. In retrospect it seems on the whole as if the degree of sophistication of nuclear models has kept pace with the improvements in our ability to determine nuclear moments. This situation is of course no guarantee for the future, for as Steven Weinberg reminded us a couple of years ago in his Dirac Memorial Lectures 'Our existing theories work well, which is certainly a reason to be happy; but we should also be sad because the fact that they work so well is now revealed as very little assurance that any future theory will look at all like them'.

APPENDIX A

A.1 Notation and definitions

If A_μ and B_μ are two Lorentz four-vectors we adopt a metric such that

$$A \cdot B = \mathbf{A} \cdot \mathbf{B} - A_0 B_0 = \mathbf{A} \cdot \mathbf{B} + A_4 B_4$$

with $A_4 = iA_0$. Thus the rest mass of a particle, with $\hbar = c = 1$, is given by the equation

$$p^2 = \mathbf{p}^2 - E^2 = -m^2.$$

A.1.1 Spin-0

For a spin-0 particle (scalar or pseudoscalar) the equation of motion is the Klein–Gordon equation:

$$(p^2 + m^2)\phi(\mathbf{r}, t) = \left(\nabla^2 - \frac{\partial^2}{\partial t^2} - m^2\right)\phi(\mathbf{r}, t) = 0.$$

The solutions of this equation may be expanded in plane waves and creation and annihilation operators $a(\mathbf{p})$ and $a^\dagger(\mathbf{p})$:

$$\phi(x) \equiv \phi(\mathbf{r}, t) = \frac{1}{(2\pi)^{3/2}} \int \frac{d\mathbf{p}}{(2p_0)^{1/2}} (a(\mathbf{p})e^{ip \cdot x} + a^\dagger(\mathbf{p})e^{-ip \cdot x}),$$

with $p \cdot x = \mathbf{p} \cdot \mathbf{r} - p_0 t$. The creation and annihilation operators satisfy commutation relations

$$[a(\mathbf{p}), a(\mathbf{p}')]_- = [a^\dagger(\mathbf{p}), a^\dagger(\mathbf{p}')]_- = 0,$$
$$[a(\mathbf{p}), a^\dagger(\mathbf{p}')]_- = \delta^3(\mathbf{p} - \mathbf{p}').$$

We call $|0\rangle$ the vacuum state defined so that

$$a(\mathbf{p})|0\rangle = 0$$

and the one-particle state is

$$|\mathbf{p}\rangle = a^\dagger(\mathbf{p})|0\rangle$$

normalized such that

$$\langle \mathbf{p}'|\mathbf{p}\rangle = \langle 0| a(\mathbf{p}')a^\dagger(\mathbf{p})|0\rangle = \delta^3(\mathbf{p} - \mathbf{p}').$$

Similarly we can take matrix elements of the field operators, $\phi(x)$, between one-particle states. For example

$$\langle 0| \phi(x) |p\rangle = \langle 0| \phi(0) |p\rangle e^{ip\cdot x} = (2p_0)^{-1/2}(2\pi)^{-3/2} e^{ip\cdot x}$$

A.1.2 Spin-$\frac{1}{2}$

For spin-$\frac{1}{2}$ the fundamental equation is that of Dirac. If $\psi(x)$ is the free Dirac field (ψ is a four-dimensional column vector), we shall write its equation in the form

$$(\gamma_\mu \partial_\mu + m)\psi(x) = 0.$$

The γ are 4×4 traceless matrices obeying anticommutation relations:

$$\gamma_\mu \gamma_\nu + \gamma_\nu \gamma_\mu = 2\delta_{\mu\nu}.$$

We shall always use Hermitian γs and when necessary use the explicit representation

$$\gamma_4 = \begin{pmatrix} I & 0 \\ 0 & -I \end{pmatrix} \quad \gamma = \begin{pmatrix} 0 & -i\sigma \\ i\sigma & 0 \end{pmatrix}$$

where I is a 2×2 unit matrix and σ the 2×2 Pauli spin matrices. We also define γ_5 as

$$\gamma_5 = \gamma_1 \gamma_2 \gamma_3 \gamma_4 = \begin{pmatrix} 0 & -I \\ -I & 0 \end{pmatrix}$$

and $\sigma_{\mu\nu} = -i(\gamma_\mu \gamma_\nu - \gamma_\nu \gamma_\mu)/2$. The free Dirac field, $\psi(x)$, can be expanded in terms of creation and annihilation operators and the Dirac wavefunctions $u_s(p)$ and $v_s(p)$; $u_s(p)$ is a solution of the momentum space equation

$$(i\gamma \cdot p + m)u_s(p) = 0$$

and represents a free spin-$\frac{1}{2}$ particle of spin s moving in the direction p, while $v_s(p)$, which satisfies the Dirac equation for negative energy and momentum

$$(i\gamma \cdot p - m)v_s(p) = 0,$$

is interpreted as representing the antiparticle of mass m moving in the direction $-p$ with s. Thus

$$\psi(x) = \frac{1}{(2\pi)^{3/2}} \int dp \sum_s (a_s(p)u_s(p)e^{ip\cdot x} + b_s^\dagger(p)v_s(p)e^{-ip\cdot x}).$$

The as and bs satisfy anticommutation relations

$$[a_s(p), a_{s'}^\dagger(p')]_+ = [b_s(p), b_{s'}^\dagger(p')]_+ = (m/E)\delta_{ss'}\delta^3(p-p'),$$

where the factor m/E follows from the chosen normalization below. To $\psi(x)$ we may associate $\psi^\dagger(x)$, its Hermitian conjugate field, and

$$\bar\psi(x) = \psi^\dagger(x)\gamma_4,$$

A.1 NOTATION AND DEFINITIONS

where $\bar{\psi}(x)$ satisfies a Dirac equation

$$\bar{\psi}(x)(-\partial_\mu\gamma_\mu + m) = 0.$$

It, too, can be expanded in plane-wave fields

$$\bar{\psi}(x) = \frac{1}{(2\pi)^{3/2}} \int dp \sum_s (a_s^\dagger(p)\bar{u}_s(p)e^{-ip\cdot x} + b_s(p)\bar{v}_s(p)e^{ip\cdot x}).$$

The chosen normalization for the Dirac spinors is

$$\bar{u}_s(p)u_{s'}(p) = -\bar{v}_s(p)v_{s'}(p) = \delta_{ss'},$$

and hence

$$u_s^\dagger(p)u_{s'}(p) = v_s^\dagger(p)v_{s'}(p) = (E/m)\delta_{ss'}.$$

The plane-wave solutions for the free-particle Dirac equation are

$$u_s(p) = \left(\frac{E+m}{2m}\right)^{1/2} \begin{pmatrix} \chi_s \\ \frac{\boldsymbol{\sigma}\cdot\boldsymbol{p}}{E+m} \chi_s \end{pmatrix},$$

$$v_s(p) = \left(\frac{E+m}{2m}\right)^{1/2} \begin{pmatrix} \frac{\boldsymbol{\sigma}\cdot\boldsymbol{p}}{E+m} \chi_s \\ \chi_s \end{pmatrix},$$

where E is the positive square root of $E^2 = \boldsymbol{p}^2 + m^2$. The spinor χ_s is the two-element Pauli spinor: $\chi_{1/2}^\dagger = (1\ 0)$ and $\chi_{-1/2}^\dagger = (0\ 1)$. Note that $u_s(p)$ and $v_s(p)$ are not orthogonal solutions to the Dirac equation $u_s^\dagger(p)v_s(p) \neq 0$. This is because $i\not{p}$ is not a Hermitian matrix. However, $u_s(p)$ and $v_s(-p)$ are orthogonal being eigenfunctions of a Hermitian matrix $(i\gamma_4\boldsymbol{\gamma}\cdot\boldsymbol{p} + m\gamma_4)$ with eigenvalues $\pm E$. Therefore $u_s(p)$ and $v_s(-p)$ form a complete orthogonal set of eigenvectors implying a completeness relation

$$\sum_s \{u_s(p)u_s^\dagger(p) + v_s(-p)v_s^\dagger(-p)\} = \frac{E}{m}\mathbf{1}.$$

Similarly we have two further projection identities

$$\sum_s u_s(p)\bar{u}_s(p) = (-i\not{p} + m)/2m,$$

$$\sum_s v_s(p)\bar{v}_s(p) = (-i\not{p} - m)/2m,$$

where $\not{p}$ represents $\gamma_\mu p_\mu$.

Again let $|0\rangle$ be the vacuum state, then the one-particle state of momentum $\boldsymbol{p}$ and spin s is

$$|p,s\rangle_+ = a_s^\dagger(p)|0\rangle$$

and the one-antiparticle state of momentum $-\mathbf{p}$ and spin s

$$|\mathbf{p}, s\rangle_- = b_s^\dagger(\mathbf{p}) |0\rangle.$$

Then matrix elements of the field operators, $\psi(x)$ and $\bar{\psi}(x)$ in one-particle states are

$$\langle 0| \psi(x) |\mathbf{p}, s\rangle_+ = \langle 0| \psi(0) |\mathbf{p}, s\rangle_+ e^{ip \cdot x} = (2\pi)^{-3/2} u_s(\mathbf{p}) e^{ip \cdot x},$$
$$\langle 0| \bar{\psi}(x) |\mathbf{p}, s\rangle_- = \langle 0| \bar{\psi}(0) |\mathbf{p}, s\rangle_- e^{ip \cdot x} = (2\pi)^{-3/2} \bar{v}_s(\mathbf{p}) e^{ip \cdot x}.$$

A.1.3 Spin-1

The standard relativistic treatment of spin-1 particles is in terms of a vector field $V_\mu(x)$ whose free Lagrangian is

$$\mathcal{L} = -\tfrac{1}{4}(\partial_\mu V_\nu - \partial_\nu V_\mu)^2 - \tfrac{1}{2}m^2 V_\mu^2$$
$$= -\tfrac{1}{2}(\partial_\mu V_\nu)^2 + \tfrac{1}{2}\partial_\mu V_\nu \partial_\nu V_\mu - \tfrac{1}{2}m^2 V_\mu^2.$$

The first and third term form a generalization of the Klein–Gordon Lagrangian for scalar fields; the reason for the second term will become clear in a moment. The field equation derivable from this Lagrangian is

$$\left(\nabla^2 - \frac{\partial^2}{\partial t^2} - m^2\right) V_\mu(x) - \partial_\mu \partial_\nu V_\nu(x) = 0.$$

For $m^2 \neq 0$ contraction with one more derivative gives $\partial_\mu V_\mu = 0$, so that the field equation is equivalent to two equations

$$\left(\nabla^2 - \frac{\partial^2}{\partial t^2} - m^2\right) V_\mu(x) = 0, \qquad \partial_\mu V_\mu(x) = 0.$$

The reason for the second term in the Lagrangian is now clear; its coefficient is chosen such that one finds a Klein–Gordon equation for each of the four components of V_μ separately, together with a subsidiary restriction. The latter restricts the number of independent plane-wave solutions to three, i.e.

$$V_\mu(x) = \varepsilon_\mu(k) e^{ik \cdot x}, \qquad k^2 = -m^2, \qquad k \cdot \varepsilon(k) = 0,$$

which is precisely the correct number to describe the three degrees of freedom of a spin-1 particle. Here $\varepsilon_\mu(k)$ is the polarization four-vector and k the particle's momentum. Again we can expand $V_\mu(x)$ in terms of creation and annihilation operators

$$V_\mu(x) = \frac{1}{(2\pi)^{3/2}} \int \frac{d\mathbf{k}}{(2k_0)^{1/2}} \sum_\lambda \varepsilon_\mu(k, \lambda)(a_\lambda(k) e^{-ik \cdot x} + a_\lambda^\dagger(k) e^{ik \cdot x}),$$

where λ distinguishes the three directions of polarization.

A massless spin-1 particle, the photon, is described by the Lagrangian

$$\mathcal{L} = -\tfrac{1}{4}(\partial_\mu A_\nu - \partial_\nu A_\mu)^2$$

which is invariant under a local gauge transformation

$$A_\mu(x) \to A_\mu(x) + \partial_\mu \xi(x).$$

(Note, we use $A_\mu(x)$ to denote a massless vector field.) The main consequence of this invariance is that the theory depends on a smaller number of fields. Correspondingly the number of plane-wave solutions is also reduced in comparison to the massive case. The photon vector field can be expanded in terms of creation and annihilation operators as before

$$A_\mu(x) = \frac{1}{(2\pi)^{3/2}} \int \frac{d\mathbf{k}}{(2k_0)^{1/2}} \sum_\lambda \varepsilon_\mu(\mathbf{k},\lambda)(a_\lambda(\mathbf{k})e^{-ik\cdot x} + a_\lambda^\dagger(\mathbf{k})e^{ik\cdot x}),$$

but λ takes only two values distinguishing the two directions of polarization (both transverse to $\mathbf{k}$). The polarization four-vector satisfies $k \cdot \varepsilon(\mathbf{k}) = 0$ as before, and in addition $\varepsilon_0(\mathbf{k},\lambda) = 0$. The as and $a^\dagger$s satisfy commutation relations, as given in the scalar field case.

A.2 Feynman rules

Write the scattering matrix

$$S_{fi} = \delta_{fi} + R_{fi}$$

where f and i are asymptotic free states. Thus δ_{fi} means 'no scattering' and R_{fi} is the reaction matrix that contains a delta function expressing conservation of total energy and momentum. It is convenient to define a T-matrix

$$R_{fi} = -i(2\pi)^4 \delta(\Sigma p_j - \Sigma p'_j) T_{fi}$$

where T_{fi} can be thought of as the matrix element of a potential between the initial and final state, $T_{fi} = (\psi_f, V\psi_i)$. Note that when an electromagnetic current is involved, T_{fi} represents the matrix element of $-j_\mu \varepsilon_\mu$ where j_μ is the electromagnetic current and ε_μ the polarization four-vector of the photon. The differential cross-section for a reaction involving the scattering of two particles with momenta p_1 and p_2 into a final state of n particles of momenta p'_j ($j = 1, \ldots n$) is given by the general expression

$$d\sigma = \frac{(2\pi)^4 \delta(p_1 + p_2 - \Sigma_j p'_j)}{4[(p_1 \cdot p_2)^2 - m_1^2 m_2^2]^{1/2}} |T_{fi}|^2 \prod_{j=1}^n \frac{d^3 p'_j}{(2\pi)^3 2E'_j},$$

where $d^3 p'_j / [(2\pi)^3 2E'_j]$ is the Lorentz-invariant volume element in the phase-space factor of the particle number j. If the frame of reference is chosen such that the two incoming particles are collinear then the factor $[(p \cdot p_2)^2 - m_1^2 m_2^2]^{1/2}$ is simplified to $E_1 E_2 |v_{12}|$, where v_{12} is the relative velocity of the incoming particles. The observable cross-sections are obtained by integration over those momentum variables in the final state which are not observed. Similarly, depending on whether or not the incoming particles have non-vanishing spin and whether or not they are polarized, the appropriate average over initial spin projections must be taken. If the spin orientations in the final state are not distinguished, the expression is summed over them.

Similarly the decay rate of a particle of mass m_1, momentum p_1 for decay into a final state of n particles is

$$d\Gamma = \frac{(2\pi)^4}{2E_1} \delta^4\left(p_1 - \sum_j p_j'\right) |T_{fi}|^2 \sum_{j=1}^{n} \frac{d^3p_j'}{(2\pi)^3 2E_j'}.$$

Again, depending on what shall be observed, there is an integration over final momenta of all unobserved final particles and possibly a sum over final spin projections and an average over initial spin orientations.

The rules for the R-matrix are as follows:

1. *Diagrams.* Draw all connected diagrams of the process under consideration to the order in the coupling constant that one wishes to calculate. External and internal fermion lines are provided with arrows which point in the direction of the flow of negative charge. The momenta of internal lines are chosen such as to follow the arrow. All factors prescribed by the following rules must be written down in the order from 'final' to 'initial' (i.e. top to bottom).

2. *External lines.* For each external, incoming fermion, f^-, write a spinor $u_s(p)$ in momentum space, for each incoming antifermion, f^+, write $\bar{v}_s(p)$. Similarly for an outgoing f^- write $\bar{u}_{s'}(p')$, for each outgoing f^+ write $v_{s'}(p')$. Because of our chosen normalization of spinors of $\bar{u}u = 1$ rather than $u^\dagger u = 1$ there is an additional factor of $(m/E)^{1/2}$ for each external fermion line. For an incoming or outgoing spin-1 boson write a polarization vector $\varepsilon_\mu(k, \lambda)$ with the index μ to be contracted with the Lorentz vector index of the vertex to which the boson is attached.

3. *Vertices.* For each vertex write a factor

$$i(2\pi)^4 \delta(\Sigma_i k_i) \times \text{(coefficient of field operators in Lagrangian)}$$

where the convention is such that k_i denotes the incoming momenta of each field into the vertex.

4. *Internal lines.* An internal line is represented by a propagator:

$$\Delta(k^2) = \frac{-i}{(2\pi)^4} \frac{1}{k^2 + m^2}, \quad \text{for spin-0};$$

$$\Delta(k^2) = \frac{-i}{(2\pi)^4} \frac{1}{i\slashed{k} + m}, \quad \text{for spin-}\tfrac{1}{2};$$

$$\Delta_{\mu\nu}(k^2) = \frac{-i}{(2\pi)^4} \frac{\delta_{\mu\nu} + k_\mu k_\nu / m^2}{k^2 + m^2}, \quad \text{for spin-1, massive};$$

$$= \frac{-i}{(2\pi)^4} \frac{\delta_{\mu\nu}}{k^2 + m^2}, \quad \text{for spin-1, massless};$$

where the direction of k is chosen according to (1).

A.2 FEYNMAN RULES

5. *Integrations.* All internal momenta must be integrated over. In all cases this leads to a delta function, $\delta(\Sigma p_j - \Sigma p'_j)$, representing the conservation of energy and momentum.

6. *Factors.* Include a factor $(-)^P$ where P is the permutation of fermions in the final state, as well as a factor $(-)^L$ if L is the number of closed fermion loops.

These rules lead to the reaction matrix R_{fi}. The required T-matrix is obtained after pulling out the factor $-i(2\pi)^4\delta(\Sigma p_j - \Sigma p'_j)$.

INDEX

addition theorem 5–8, 23
Alaga's model 215
algebraic structure
   IBM-1 170
   IBM-2 177
alignment 39
anharmonicity in vibrations 214
   particle-vibration model 215
      octupole mode 216
   time-dependent Hartree–Fock 220
Arima–Horie effect 65, 192

backbending 159, 163
band mixing 155, 207
BCS theory 157, 207
   gap equations 158
beta decay, *see also* Gamow–Teller
   relation to M1 transitions 27
   relation to magnetic moments 29
beta vibration 169, 173
Bohr Hamiltonian 169, 170
breathing modes 241, 245
Brown–Bolsterli model 196
Buck–Perez plot 28–32

coexistence 239
collective coordinates
   in rotational model 146
   in vibrational model 212
compressibility, *see* incompressibility
core polarization 55–86
   blocking 73
   first order 56–79
   in MEC calculations 132–6
   intermediate-state summations 58, 81, 133
   RPA calculations 67–70
   second order 79–84
Coriolis coupling 154, 159, 228

Coulomb excitation 46
cranking model 227
current conservation 91, 98, 113, 119

decoupling parameter
   energy spectra, a 148, 156
   M1 moments, b 152, 156
deformation parameter
   $\alpha_{\lambda\mu}$ 167, 212, 230
   β 156, 166, 167
   δ 155, 165, 184
deuteron
   D-state probability 9
   magnetic moment 8
   quadrupole moment 10
diamagnetic correction 44, 49
Dirac equation 260
   deduced from Lagrangian 98
   in nuclear matter 250
Dirac form-factor, F_1 91–2, 95, 111, 119
dynamical deformation 240

E0 transitions, *see* monopole transitions
E2 transitions
   ^{40}Ca region 191
   core polarization 193
   definition 16
   effective charge 191
   in axially symmetric nuclei 151
   in harmonic vibration model 213
   in IBM-1 173
   in IBM-2 179
   in sd shell 198
   in superdeformed band 167
   schematic model 196
   shell model 200
   single particle 188

268 INDEX

E2/M1 mixing ratio
  in odd-mass deformed nuclei 164
E4 transitions 236–8
effective charge
  core polarization calculation 192
  definition 18, 190
  for bosons in IBM-2 179
  in E4 transitions 236
  in particle-vibration coupling 216
  mass dependence 194
  relation to polarizability 194
effective mass
  effect on nuclear incompressibility 243
  in relativistic theories 250
  in RPA calculations 73
effective operators 18
  boson g-factor 174, 181
  Gamow–Teller operator 27, 31, 82
  M1 operator 19–22, 24, 64, 74, 82, 96, 130
electric form factor, G_E 111, 119
electromagnetic transition probabilities
  definition 16
  hermiticity 17
  Weisskopf unit 17–18
electronic field gradient 37, 44
Euler–Lagrange equations 98

Feynman diagram rules 263
F-spin 176, 178

gamma vibration 169, 173
Gamow–Teller
  operator 26, 72
  strength, quenching of 31, 73, 84, 136
  transitions in closed-shell-plus-one nuclei 82
  transitions in ^{208}Pb 72–3, 136–8
g-factors, see also gyromagnetic ratio
  definition 1, 36
  of Dirac particle 4, 88–90
  of j^n configuration 8
  of two-particle systems 23
  of 2^+ states 181
G-parity 116

gyromagnetic ratio
  g_B 174
  g_K 152, 156
  g_L 1, 4, 19–22, 24, 25, 64–7, 69, 82, 130, 135
  g_P 19–22, 64–7, 69, 82, 130, 135
  g_R 152, 160
  g_S 1, 4, 19–22, 25, 64–7, 69, 82, 130, 135

Harmonic oscillator
  deformed 155
  Hamiltonian in vibration model 212
  radial wavefunctions 127–8
Hartree–Fock theory
  in E4 transitions 237
  insertions, in core polarization 57
  isotope shifts 245
  nuclear incompressibility 242
  quadrupole moments 207
  rearrangement energy 246
  relativistic theories 248
  time-dependent 220
hexadecapole transitions 236–8
hyperfine interaction
  electric quadrupole 37
  magnetic dipole 36

Ikeda sum rule 72
impulse approximation 127
  relativistic IA 252
incompressibility 240
interacting boson model 170–85
  IBM-1 170
  IBM-2 175
  in E4 transitions 238
intrinsic
  Hamiltonian 153
  quadrupole moment 148, 155, 229
  in superdeformed band 167
isobar currents 114, 136
isospin 3, 90
  in π-N interaction Lagrangian 104
  in transition probabilities 24
  isoscalar operators 4, 25
  isovector operators 4, 25
isotope shift 245

Klein–Gordon equations 259
Knight-shift correction 45
Kumar–Baranger theory 169, 230

Lagrangian 98
  for Dirac particle 87, 98
  for electromagnetic current 87, 91
  for nucleons 90
  for pions 103
  for $\pi N\Delta$-interaction 114
  for πNN-interaction
    pseudoscalar coupling 103
    pseudovector coupling 103
  for πNNγ-interaction 106
  for $\pi\pi\gamma$-interaction 112
  for ρNN-interaction 117
  for $\rho\pi\gamma$-interaction 117
  for ωNN-interaction 118
  for $\omega\pi\gamma$-interaction 118
Landau–Migdal parameters 60
Landé formula 7
Larmor frequency 38, 42
laser spectroscopy 50
lifetime, definition 17
Lindhard–Winther theory 50

M1 transitions
  definition 16
  in axially symmetric nuclei 153
  in closed-shell-plus-one-nuclei 66, 69
  in collective models 212, 215
  in IBM-1 173
  in odd-mass deformed nuclei 157, 164
    pairing factors 158
  in ^{208}Pb 70–3, 136–8
  quenching 136
magic numbers 12, 165
  in deformed nuclei 165
magnetic electron scattering 142
magnetic field at nucleus
  hydrogen-like atom 36
  transient fields 49
magnetic form factor, G_M 111
magnetic moment
  definition 1
    construction of MEC operator 120–6

  derivation from field theory 92–5
  relativistic corrections 96–7
  deuteron 8
  ^{3}He, ^{3}H 10
  in axially symmetric nuclei 152, 153
  in backbending nuclei 161
  in closed-shell-plus-one nuclei 20, 66, 69, 82, 126–32, 134
    relativistic theories 249
  in IBM-1 174
  in IBM-2 181
  in j^n configurations 8, 75–9
  in odd-mass deformed nuclei 157, 163
  in odd-mass nuclei 12
  isospin notation 4
  of two-particle configuration 22–4
  single-particle value 13
meson-exchange currents 87–143
  core polarization 132
  graphs of pion range 103–20
    current graph 112
    isobar current 114
    pair graph 107
    ρ-π graph 116
    seagull graph 106
    ω-π graph 118
  in deuteron 9, 140
  in few nucleons 11, 139
  in neutron capture 139
minimal coupling 87, 91, 100
  in πNN Lagrangian 106
  in $\pi\pi$ Lagrangian 112
  nonuniqueness 89, 100
minimal substitution, see minimal coupling
moment of inertia 148
  dynamical 166
  in cranking model 228
  kinematical 165, 229
  of uniform rigid sphere 166
monopole transitions 238
  isoscalar giant resonance 241, 245
  isotope shifts 245
  nuclear incompressibility 240
  polarization charge 245
Morpurgo's rule 25

Nilsson model 154

Noether's theorem 99
Nordheim's rule 13
nuclear magnetic resonance 38, 47
  on beta-emitters 47
nuclear magneton 1
nucleon–nucleon interaction, *see* residual interaction

one-pion exchange potential 105
  monopole form factors 120
ω–π graphs 118
optical pumping 39, 51
orbital g-factor, *see* gyromagnetic ratio, g_L
orientation, methods of production
  low temperatures 39
  optical pumping 39, 51
  nuclear reactions 40, 41, 47
orientation coefficients 41

pairing factors 158
paramagnetic correction 45
particle-rotor model 153
  vibration model 215
Pauli form factor, F_2 91–2, 95, 111
Pauli spin operator 27
perturbed angular correlation 44
perturbed angular distribution 41
  time-integrated 43
phonon excitations 168
polarization 39
propagators
  for bosons 264
  for fermions 108, 264
  for spin-3/2 isobars 115

quadrupole moment
  definition 3
    derivation from field theory 95
    relativistic corrections 97
  in anharmonic vibration model 215, 216
  in axially symmetric nuclei 151
  in harmonic vibration model 213
  in IBM-1 174
  in IBM-2 179
  in j^n configurations 189

  in p shell 204
  in sd shell 198
  in time-dependent Hartree–Fock theory 226
  isospin notation 4
  of deuteron 9
  of 3^- states 216
  of two-particle systems 190
  single-particle value 188
quark model
  for isobar couplings 114
  for nucleon magnetic moment 4
quenching
  of Gamow–Teller strength 31, 73, 84, 136
  of M1 strength 70–3, 136–8
  of spin g-factor 21, 67, 84, 136

random phase approximation, RPA
  effective charge 197
  in closed-shell nuclei 70–3, 136–8
  in core polarization 67–70
  in time-dependent Hartree–Fock 223
  isoscalar monopole resonance 244
  relativistic RPA 253
Rarita–Schwinger equation 114
reduced transition probabilities, *see* electromagnetic transition probabilities
relativistic corrections to moments 96–7
relativistic theory of moments 248
reorientation in Coulomb excitation 46
residual interaction
  definition 56
  finite range 65, 67, 71, 81, 205
  zero range 60, 71, 196
    density dependence 242
    velocity dependence 243
ρ–π graphs 116
rotational alignment, *see* backbending
rotational model
  deformed γ-unstable, O(6) limit 173
  energies and spectra 148, 158
    signature-dependent terms 148
  intensity relations 151
  SU(3) limit 173

rotation matrices 146, 149
rotation–vibration coupling 169

Sachs moment 102, 126
Schmidt diagrams 13–15
scissor mode 183
shape mixing 239
shell model 12
  large basis 205
  p-shell calculations 204
  sd-shell calculations 22, 198–204, 236
side bands (or s-bands) 159
σ–ω model 249
signature 147
spin g-factor, *see* gyromagnetic ratio, g_s
sum-rule method 225
superdeformation 164

Talmi Hamiltonian 177
Tamm–Dancoff approximation, TDA
  effective charge 197
  in closed-shell nuclei 70–3
time-dependent Hartree–Fock 220
transient field 49
transition isospin operator 114

vibrational model 212
  anharmonic 214
    U(5) limit 173
  β vibration 169, 173
  γ vibration 169, 173
  E2 transitions 213, 215
  mass parameter, D_λ 168, 212
  particle-vibration coupling 215
  quadrupole moment 213, 215, 216
  restoring force parameters, C_λ 168, 212
  spectra 167

weak-coupling model 215
Weisskopf unit 17–18
width, gamma
  definition 17
Wigner–Eckart theorem 2

Young tableaux 177
yrast states 159

zero-range interaction, *see* residual interaction

The manufacturer's authorised representative in the EU for product safety is Oxford University Press España S.A. of el Parque Empresarial San Fernando de Henares, Avenida de Castilla, 2 – 28830 Madrid (www.oup.es/en or product. safety@oup.com). OUP España S.A. also acts as importer into Spain of products made by the manufacturer.

www.ingramcontent.com/pod-product-compliance
Ingram Content Group UK Ltd.
Pitfield, Milton Keynes, MK11 3LW, UK
UKHW022153230426
12049UKWH00003BA/69